Advanced Structured Materials

Volume 118

Common engineering materials reach in many applications their limits and new developments are required to fulfil increasing demands on engineering materials. The performance of materials can be increased by combining different materials to achieve better properties than a single constituent or by shaping the material or constituents in a specific structure. The interaction between material and structure may arise on different length scales, such as micro-, meso- or macroscale, and offers possible applications in quite diverse fields.

This book series addresses the fundamental relationship between materials and their structure on the overall properties (e.g. mechanical, thermal, chemical or magnetic etc) and applications.

The topics of *Advanced Structured Materials* include but are not limited to

- classical fibre-reinforced composites (e.g. glass, carbon or Aramid reinforced plastics)
- metal matrix composites (MMCs)
- micro porous composites
- micro channel materials
- multilayered materials
- cellular materials (e.g., metallic or polymer foams, sponges, hollow sphere structures)
- porous materials
- truss structures
- nanocomposite materials
- biomaterials
- nanoporous metals
- concrete
- coated materials
- smart materials

Advanced Structured Materials is indexed in Google Scholar and Scopus.

More information about this series at http://www.springer.com/series/8611

Aamir Hussain Bhat · Imran Khan ·
Mohammad Jawaid · Fakhreldin O. Suliman ·
Haider Al-Lawati · Salma Muhamed Al-Kindy
Editors

Nanomaterials for Healthcare, Energy and Environment

Springer

Editors
Aamir Hussain Bhat
Department of Applied Sciences
Higher College of Technology
Muscat, Oman

Mohammad Jawaid
Laboratory of Biocomposite Technology
Universiti Putra Malaysia
Serdang, Selangor, Malaysia

Haider Al-Lawati
Department of Chemistry
College of Science
Sultan Qaboos University
Muscat, Oman

Imran Khan
Department of Chemistry
College of Science
Sultan Qaboos University
Muscat, Oman

Fakhreldin O. Suliman
Department of Chemistry
College of Science
Sultan Qaboos University
Muscat, Oman

Salma Muhamed Al-Kindy
Department of Chemistry
Sultan Qaboos University
Muscat, Oman

ISSN 1869-8433 ISSN 1869-8441 (electronic)
Advanced Structured Materials
ISBN 978-981-13-9832-2 ISBN 978-981-13-9833-9 (eBook)
https://doi.org/10.1007/978-981-13-9833-9

This Springer imprint is published by the registered company Springer Nature Singapore Pte Ltd.
The registered company address is: 152 Beach Road, #21-01/04 Gateway East, Singapore 189721, Singapore

 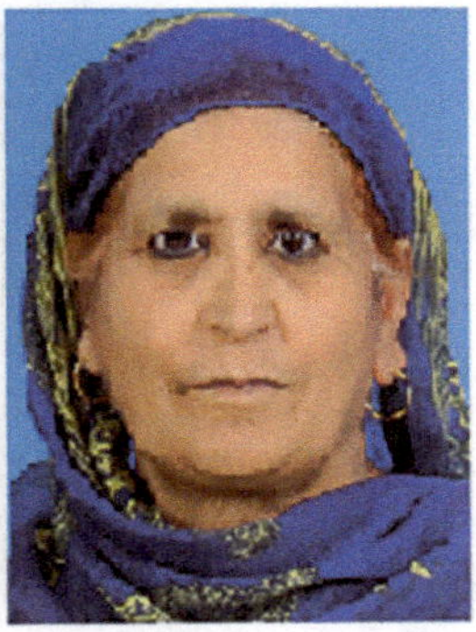

Dedicated to my Grandmother and Mother,
both of whom have been my great Inspiration

Preface

Nanotechnology has emerged as an important field of modern scientific society due to its diverse range of applications in the area of electronics, material sciences, biomedical sciences, energy science, food, environmental detection and monitoring and other applications. The advantages of the use of nanomaterials, which are related to their properties that are completely different from the bulk materials, make them extremely attractive and give them enormous potential. Various nanomaterials like titanium dioxide, nanoclays, nanotubes, nanodendrimers, graphene, ferritin, porphyrinogens, noble metals, etc., find their applications in various fields.

Although there have been extensive interdisciplinary activities, major collaborative efforts are needed to jointly address some of the most challenging issues in life and medical sciences. Nanomaterials also find great deal of applications in environmental control and remediation. The biosynthetic route of synthesis of nanomaterials is taking a central stage nowadays. Nanomaterials lead to new-generation devices for more efficient, cost-effective and reliable solar energy conversion.

Therefore, realizing the importance of nanomaterials with their application in these three major sectors, this edited book is hoped to fill the gap of knowledge in the field of nanotechnology. Generally, conventional materials find their use in the fields of energy, health and environment, but the nanomaterials often have properties that are significantly different and efficient from the properties of the same matter at the bulk scale and also have an enormous potential economic impact. Nanomaterials are used in a variety of, manufacturing processes, products and healthcare including paints, filters, insulation and lubricant additives. High-quality filters may be produced using nanostructures; these filters are capable of removing particulate as small as a virus. Nanomaterials are being used in modern and human-safe insulation technologies; in the past, they were found in asbestos-based insulation. As a lubricant additive, nanomaterials have the ability to reduce friction in moving parts. This book is basically intended to address the holistic approach in terms of nanomaterial applications by taking into consideration various stakeholders using nanomaterials. It is hoped that publication of this book will provide the readers new knowledge and understanding on the broad range of nanomaterials and their applications.

Last but not least, we are highly thankful to all the authors who contributed chapters and provide their valuable ideas and knowledge in this edited book. We attempt to gather all the scattered information of authors from diverse fields around the world (Malaysia, Portugal, Japan, India, Saudi Arabia and Oman) in the areas of nanomaterials and finally complete this venture in a fruitful way. We greatly appreciate contributor's commitment for their support to compile our ideas in reality.

We are highly thankful to Springer Nature Team, Heidelberg Platz, Berlin, Germany, for their generous cooperation at every stage of the book production.

Muscat, Oman — Aamir Hussain Bhat
Muscat, Oman — Imran Khan
Serdang, Malaysia — Mohammad Jawaid
Muscat, Oman — Fakhreldin O. Suliman
Muscat, Oman — Haider Al-Lawati
Muscat, Oman — Salma Muhamed Al-Kindy

Contents

Editors and Contributors

About the Editors

Dr. Aamir Hussain Bhat is currently working as assistant professor at Department of Applied Sciences, Higher College of Technology, Muscat, Oman. He was born on 4 June 1980 in Baramulla, India. He received his highest degree of doctorate from Indian Institute of Technology Kharagpur, which ranks among the prestigious institutes of India. He has four years of postdoctoral experience at Universiti Sains Malaysia and around five years of teaching experience in the capacity of assistant professor in chemistry at Universiti Teknologi Petronas, Malaysia. He was awarded Prime Ministerial Postdoctoral Fellowship by the Ministry of Higher Education, Malaysia, for his excellence in the field of research. His research interests include polymer bio-nanocomposites and techniques to characterize them, nanofluids for oil well-drilling applications, isolation and application of nanocellulosic materials, nanomaterial synthesis using top-down and bottom-up approach, nanocoating using geopolymers and wastewater treatment using bio-sorbent-based die removal and micro-algae for heavy metal adsorption. He has been principal investigator of many government-funded projects. The prominent among them are Fundamental Research Grant Scheme (FRGS) entitled "New Malaysian Green Nano fluid for drilling at high temperature and pressure", YUTP Grant on "Biopolymer blend of Poly (lactic acid) and Poly (hydroxybutyrate co-valerate) based nano bio-composites reinforced with nanocrystalline cellulose with potential application in packaging" and Graphene Oxide Additives in Water Based Drilling Fluid for Enhanced Performance of Fluid Loss Control again funded by YUTP. He has also worked on joint industrial project with PRSB based on "Synthesis and Characterization of Nanoparticles for Enhanced Oil Recovery". His research group includes four Ph.D. and two MS students and has mentored 13 final-year undergraduate students. He has served as an international/national examiner for many research dissertations. He is also a member of various scientific chemical societies, prominent among them is American Nano Society. He has published 39 full-length research papers in highly reputed international scientific journals with a citation of more than 1350 and 23 chapters with highly reputed publishers. He is serving as a reviewer for several high-impact ISI journals of Elsevier, Springer, Wiley, Taylor and Francis, Sage, etc.

Dr. Imran Khan is currently working an assistant professor in Department of Chemistry, Sultan Qaboos University, Muscat, Oman, and earlier worked as post-doctoral fellow in the group named as Process and Product Applied Thermodynamics (PATh), in the Associated Laboratory Center for Research in Ceramics and Composite Materials (CICECO), Department of Chemistry, University of Aveiro, Portugal. Also, he was the principal investigator of Exploratory Research and Development Projects, funded by Fundação para a Ciência e a Tecnologia (FCT), Portugal, on development of a sustainable technology for the extraction and purification of chlorophylls from biomass in year 2014. His area of research interests includes solution chemistry, study thermophysical behaviour of pure liquids and liquid mixtures with ionic liquids, surfactant and polymer, as well as extraction and separation using ionic liquid. He worked as a visiting scientist for three months in the Department of Chemistry, University of Delhi, India, to study the effect of polymer on the ionic liquid solution funded by FCT, Portugal, to expand the collaboration between India and Portugal. Previously, he worked on the effect of filler on pressure-sensitive adhesive as postdoctoral fellow at Universiti Sains Malaysia, Penang, Malaysia, in the year 2011–2013, and published many research articles. He also worked in the Department of Chemistry, Durban University of Technology, Durban, South Africa, in the year 2010 in the area of solution chemistry. He has published more than 51 scientific papers in international peer-reviewed journals and 16 chapters, and has an h-index of 16. He also presented many scientific research papers in various international conferences and is also a member of various scientific chemical societies.

Dr. Mohammad Jawaid is currently working as a fellow researcher (associate professor), at Biocomposite Technology Laboratory, Institute of Tropical Forestry and Forest Products (INTROP), Universiti Putra Malaysia, Serdang, Selangor, Malaysia, and also a visiting professor at Department of Chemical Engineering, College of Engineering, King Saud University, Riyadh, Saudi Arabia, since June 2013. He is also a visiting scientist to TEMAG Laboratory, Faculty of Textile Technologies and Design at Istanbul Technical University, Turkey. Previously, he worked as a visiting lecturer, Faculty of Chemical Engineering, Universiti Teknologi Malaysia (UTM), and also worked as a expatriate lecturer under UNDP project with Ministry of Education of Ethiopia at Adama University, Ethiopia. He received his Ph.D. from Universiti Sains Malaysia, Malaysia. He has more than 10 years of experience in teaching, research and industries. His area of research interests includes hybrid-reinforced/filled polymer composites, advanced materials: graphene/nanoclay/fire retardant, lignocellulosic-reinforced/filled polymer composites, modification and treatment of lignocellulosic fibres and solid wood, nanocomposites and nanocellulose fibres, and polymer blends. So far he has published 8 books, 17 chapters, and more than 140 international journal papers and 5 published review papers under top 25 hot articles in ScienceDirect during 2013–2015. He is also the deputy editor-in-chief of Malaysian Polymer Journal, guest editor of special issue—Current Organic Synthesis and Current Analytical Chemistry, Bentham Publishers, UK, and editorial board member—Journal of Asian Science Technology and Innovation. Beside that, he is also a reviewer of several high-impact ISI journals of Elsevier, Springer, Wiley, Saga, etc. Presently, he is supervising 15 Ph.D. students and 5 master students in the fields of hybrid composites, green composites, nanocomposites, natural fibre-reinforced composites, etc. Four Ph.D. and three master students graduated under his supervision in 2014–2016. He has several research grants at university and national levels on polymer composites of around RM 700,000 (USD 175,000). He also delivered plenary and invited talks in international conferences related to composites in India, Turkey, Malaysia, Thailand, and China. Beside that, he is also a member of technical committee of several national and international conferences on composites and material science.

Prof. Fakhreldin O. Suliman obtained an M.Sc. (1992) and a Ph.D. (1996) in analytical chemistry from King Fahd University of Petroleum and Minerals, Dhahran, Saudi Arabia. He was awarded a Science and Technology Agency (STA) Fellowship at the National Institute for Environmental Studies, Japan, in 1998. He joined the Department of Chemistry at Sultan Qaboos University in Sultanate of Oman, in 1999, where he is now professor and head of department. His research interests include developing fluorescent probes for sensing, miniaturization and automation of analytical techniques, supramolecular chemistry and molecular modelling. He has a teaching experience of more than twenty years in tertiary education. He has supervised and co-supervised more than thirty postgraduate students. His publication record in international peer-refereed journals exceeds hundred papers with an h-index of 21. Recently, he became a fellow of the Royal Society of Chemistry (RSC). He has been involved as a principal and co-principal investigator in many research projects and served as a reviewer to many high-impact journals of well-known publishers such as RSC, Elsevier and Wiley.

Dr. Haider Al-Lawati is an associate professor in the Department of Chemistry, Sultan Qaboos University, Sultanate of Oman. He completed his degree of doctorate from University of Hull, UK, in 2007. After obtaining his Ph.D., he started planning to establish a new research group in microfluidics area.

In May 2009, he successfully obtained His Majesty (HM) Grant for a project entitled "Developing Microfluidic Systems for Routine Analysis of Pharmaceutical Samples" with a budget of 207,254 $ RO. He was able to establish the first research laboratory in the field of microfluidics at SQU and in the Gulf region. Additionally, the grant helped a great deal in creating an excellent research environment and a strong research group. The group participated in Oman innovation fair in 2011 and received award for the best innovation in the exhibition. These heavy research activities requested further investment, and an extension for a year was granted with a budget of 26,316 $.

In June 2011, a new research proposal was submitted to The Research Council (TRC), Oman. The research utilizes microfluidics as an efficient mixing device as a chemiluminescence detection system for a capillary HPLC. The project was highly appreciated by the referees, and based on their comments, the TRC accepted to fund the project with a budget of 372,020 $ for three-year duration (2012–2015). Since 2015, he was able to attract additional two research funds: one from TRC and the other was funded by His Majesty Grant Funds with total budget of 628,109 $, in addition to some other funds like internal grants and few external funds. He published around 50 articles in refereed journals and presented many papers in various international conferences. He also supervised a number of Ph.D. and master students. He received several awards among these: first, GCC award of excellence in chemistry, 7 December 2015. This award is the most prestigious award in GCC. The awards cover areas such as science, medicine, industry, literature, politics and diplomacy, economics, youth and sports, security and philanthropy which are of crucial importance to the progress and welfare of the GCC states, the Arab world and humanity at large. A candidate is eligible to vie only once for these awards. The awards are given once every five years at a ceremony held under auspices of the leader of the host country of the GCC Supreme Council session.

A GCC state can nominate seven scholars for a prize through a formal letter to the GCC Secretariat. The GCC Secretariat and the secretariat of each competition assess the achievements of the nominees before selecting the winners and then table a name list of the winners to the GCC Ministerial and the GCC Supreme Council for final endorsement. He also received the National Research Award in the research area Culture, Basic and Social Sciences, 5 October 2016, TRC, Oman. In May 2014, his student received Marlene DeLuca Award in the 18th International Symposium on Biochemiluminescence and Chemiluminescence. Finally, he was heading the Department of Chemistry for three years from September 2014.

Prof. Salma Muhamed Al-Kindy is currently the dean of the College of Science, and a professor of analytical chemistry. She obtained her B.Sc. (Honours, 1982) in chemistry from American University of Cairo, Egypt, and Ph.D. in chemistry from Loughborough University, UK, in 1987. She started her academic career in 1989 at SQU where she became the first female professor in the university's history, and the first Omani national with a doctorate to join the Department of Chemistry. She was awarded a Matsumae International Fellowship by the Matsumae International Foundation in 1996, where she spent time at the Department of Bioanalytical Chemistry in Tokyo University, Japan, working on developing methods for the analysis of enantiomeric drugs. Her research interest has been in developing analytical protocols for the monitoring of analytes in complex matrices. She focuses her research on the development of analytical methodology and instrumentation for drug analysis in pharmaceutical and biological matrices, monitoring of organic pollutants and toxic metal ions in water using luminescence techniques in combination with HPLC, and flow systems such as, FIA and SIA and in developing sensitive and selective method for the essay of pharmaceutical components using microfluidic systems. Currently, she is developing methods to remove hazardous chemical by-products from wastewater using green chemistry approach. In April 2010, she became the first Omani national elected as a member of the prestigious World Academy of Science for Sustainable Development (TWAS). She was Oman's recipient of the United States Department of State's Award for outstanding female scientist in 2013 and has been inducted into the State Department's Middle East and North Africa (Mena) Women in Science Hall of Fame. She was recently awarded a Fellowship of Royal Society of Chemistry (FRSC). In 2014, she was awarded a medal by The World Academy of Science for Sustainable Development (TWAS) for her contribution to science, and she delivered a medal lecture during TWAS general meeting in Muscat last October. She recently received a "Lifetime Achievement Award in Chemistry" by the Venus International Foundation (VIF), in recognition for her contribution, research excellence and accomplishments in the field of chemistry. Furthermore, she has published more than 88 scientific papers in reputable scientific journals and has contributed to many international scientific conferences and seminars worldwide. She was recently invited to attend the General Assembly and Conference of Organization for Women in Science for the Developing World (OSWD) where she gave a presentation on research which was well received. Her recent paper was chosen as a cover page for Analytical Methods Journal published by Royal Society of Chemistry.

Contributors

Sadiya Anjum Bioengineering Laboratory, Department of Textile Technology, Indian Institute of Technology, New Delhi, India

Leonard Sean Anthony Centre of Innovative Nanostructures and Nanodevices (COINN), Universiti Teknologi PETRONAS, Seri Iskandar, Perak Darul Ridzuan, Malaysia

Sayyed Jaheera Anwar School of Fundamental Science, Universiti Malaysia Terengganu, Kuala Terengganu, Malaysia

M. K. Md Arshad School of Microelectronic Engineering, Universiti Malaysia Perlis, Arau, Perlis, Malaysia

Khalid Mujasam Batoo King Abdullah Institute for Nanotechnology, King Saud University, Riyadh, Saudi Arabia

Sandra C. Bernardo Chemistry Department, CICECO—Aveiro Institute of Materials, University of Aveiro, Aveiro, Portugal

Pankaj Bharmoria Chemistry Department, CICECO—Aveiro Institute of Materials, University of Aveiro, Aveiro, Portugal

Aamir Hussain Bhat Department of Applied Sciences, Higher College of Technology, Muscat, Oman

Irshad Ul Haq Bhat School of Fundamental Science, Universiti Malaysia Terengganu, Kuala Terengganu, Malaysia

Showkat Ahmad Bhawani Department of Chemistry, Faculty of Resource Science and Technology, UNIMAS, Kota Samarahan, Sarawak, Malaysia

Meena Bisht CICECO—Aveiro Institute of Materials, Chemistry Department, University of Aveiro, Aveiro, Portugal

Mara G. Freire Chemistry Department, CICECO—Aveiro Institute of Materials, University of Aveiro, Aveiro, Portugal

Subash C. B. Gopinath Institute of Nano Electronic Engineering, Universiti Malaysia Perlis, Kangar, Perlis, Malaysia;
School of Bioprocess Engineering, Universiti Malaysia Perlis, Arau, Perlis, Malaysia

Rashid Ilmi Department of Chemistry, Sultan Qaboos University, Muscat, Oman

Farah Khan Independent Researcher, Muscat, Oman

Imran Khan Department of Chemistry, College of Science, Sultan Qaboos University, Muscat, Oman

Norani Muti Mohamed Centre of Innovative Nanostructures and Nanodevices (COINN), Universiti Teknologi PETRONAS, Seri Iskandar, Perak Darul Ridzuan, Malaysia;
Department of Fundamental and Applied Sciences, Universiti Teknologi PETRONAS, Seri Iskandar, Perak Darul Ridzuan, Malaysia

Abdul Moheman Department of Chemistry, Gandhi Faiz-e-Aam College (Affiliated to M. J. P. Rohilkhand University), Shahjahanpur, India

Márcia C. Neves Chemistry Department, CICECO—Aveiro Institute of Materials, University of Aveiro, Aveiro, Portugal

Veeradasan Perumal Centre of Innovative Nanostructures and Nanodevices (COINN), Universiti Teknologi PETRONAS, Seri Iskandar, Perak Darul Ridzuan, Malaysia;
Department of Mechanical Engineering, Universiti Teknologi PETRONAS, Seri Iskandar, Perak Darul Ridzuan, Malaysia

Prabakaran Poopalan School of Microelectronic Engineering, Universiti Malaysia Perlis, Arau, Perlis, Malaysia

Santheraleka Ramanathan Institute of Nano Electronic Engineering, Universiti Malaysia Perlis, Kangar, Perlis, Malaysia

Mohamed Shuaib Mohamed Saheed Department of Fundamental and Applied Sciences, Centre of Innovative Nanostructures and Nanodevices (COINN), Universiti Teknologi PETRONAS, Seri Iskandar, Perak Darul Ridzuan, Malaysia

Aabid Hussain Shalla Department of Chemistry, Islamic University of Science & Technology, Awantipora, Jammu and Kashmir, India

Dhananjay K. Sharma Department of Mechanical Engineering, TEMA, University of Aveiro, Aveiro, Portugal

Ana C. A. Sousa Chemistry Department, CICECO—Aveiro Institute of Materials, University of Aveiro, Aveiro, Portugal

Emamalar Subramaniam Faculty of Earth Science, Universiti Malaysia Kelantan, Campus Jeli, Jeli, Kelantan, Malaysia

Abu Tariq School of Distance Education, Universiti Sains Malaysia, Penang, Malaysia

Mohd Amil Usmani Department of Chemistry, Gandhi Faiz-E-Aam College, Shahjahanpur, India

Sónia P. M. Ventura Chemistry Department, CICECO—Aveiro Institute of Materials, University of Aveiro, Aveiro, Portugal

Optical Applications of Nanomaterials

Pankaj Bharmoria and Sónia P. M. Ventura

Abstract Riding on their size tunable properties, "*Nanomaterials*" have emerged as darling materials of 21st century for plethora of practical applications including optical. The nonlinear optical properties and optical emission of nanomaterial's, enhances with the decrease in particle size due to the "*quantum confinement effect.*" Therefore, the quantum mechanical effects emerge at the nanoscale which ultimately dictates the optical properties of nanomaterials. This book chapter will delineate the conceptual basis of optical applications of nanomaterials, subject to their size and material specific optical properties, including examples for conceptual demonstration. Considering the broad width of applications this book chapter is particularly focussed on biosensing and photovoltaic applications of nanomaterials.

Keywords Nanomaterials · Optical applications · Quantum confinement effect · Biosensing · Photovoltaics

1 Introduction

Nanomaterial is any particle, aggregate or agglomerate (natural or manufactured) with one or more than one dimensions in the size range of 1–100 nm (European Commission 2011). Historically, nanomaterial's, are being used by humans since 600 BC, in the form of carbon nanotubes and cementite nanowires in Hindvi steel developed in Southern India (Sanderson 2006). However, the modern world witnessed their scientific revolution for practical utility after the invention of scanning tunnelling microscope (STM) in 1981 by Gerd Binnig and Heinrich Rohrer (awarded Nobel prize in 1986) (Binnig and Rohrer 1986; Nobelprize.org. 1996 and discovery of fullerenes in 1985 by Richard Smalley, Robert Curl, and Harold Kroto (awarded Nobel Prize in 1996) (Kroto et al. 1985). Thereafter, in early 2000, the research on nanotechnology picked up impetus and growing unabated till date with a plethora of practical applications (Bhagyaraj et al. 2018; Guo and Tan 2009). Thanks to their

P. Bharmoria (✉) · S. P. M. Ventura
Chemistry Department, CICECO—Aveiro Institute of Materials,
University of Aveiro, 3810-193 Aveiro, Portugal
e-mail: pankajbharmoria@gmail.com; pankaj@ua.pt

A. H. Bhat et al. (eds.), *Nanomaterials for Healthcare, Energy and Environment*,
Advanced Structured Materials 118, https://doi.org/10.1007/978-981-13-9833-9_1

unique physical, chemical, mechanical and optoelectronic properties which offer tunability as *per* size and dimensions (Guo and Tan 2009). Among these, the optical properties have gained a lot of interest, namely because of the advanced optical applications of nanomaterials. This is due to the fact that the optical properties are a function of their internal electronic structures, which can be tuned by altering their size and dimension as *per* the required optical application. However, before going deep into details we must understand the basics of material optics and *quantum confinement effect* of nanomaterials, responsible for their exceptional optical properties required for optical applications.

Material optics generally deals with the interaction of light with matter which results in manipulation of the flow of light involving reflection, refraction, absorbance, fluorescence, dispersion, frequency alterations and focusing or splitting of an optical beam (McGraw-Hill 1993). The characteristics of the optical material are a strict function of the light wavelength used. Since the dimensions of nanomaterials are defined in the nanoscale, which is sometimes lower than the wavelength of light, the understanding of the light matter interaction becomes a key step under the scope of optical applications of nanomaterials. Thus, the optical properties of materials may be categorized in two main sets, the linear and non-linear properties.

Linear optics and linear optical properties: The linear optics deals with "weak light", which upon interaction with medium is deflected or delayed but does not undergoes to a frequency change, thus following the superposition principle (Zhang and Wang 2017). According to the superposition principle when two waves undergo overlap in a space time then the optical property of resulting wave would be an algebraic sum of individual wave. In simple algebraic equation if the wave A gives signal X and the wave B gives signal Y, then upon superposition of A and B (A + B) the resulting wave would give the signal X + Y (i.e. A + B = X + Y). In linear optics, the light wave induces vibration in the molecules followed by the emission of the light, having the same frequency of the incident light, then interfering with the original light (Fig. 1). The optical response of materials scales linearly with the amplitude of the electric field of light when electric field associated with the radiation is small as shown in Eq. (1) (Suresh and Arivuoli 2012).

$$\vec{P} = \varepsilon_o \chi \vec{E} \tag{1}$$

Here, P is induced by dipole moment *per* unit volume, which defines polarization, ε_o is the permittivity of free space, χ is polarization susceptibility and E is electric field amplitude. The arrow above P and E is indicating their vector nature. The general linear optical properties of materials are reflection, refraction and diffraction, which are utilized for practical optical applications such as in phase shifters, beam splitters and recently, in quantum computing (Turner et al. 2013; Wu et al. 2017; Pittman et al. 2004). Lenses, mirrors, wave-plates and diffraction gratings are some examples describing the well-known linear optical materials.

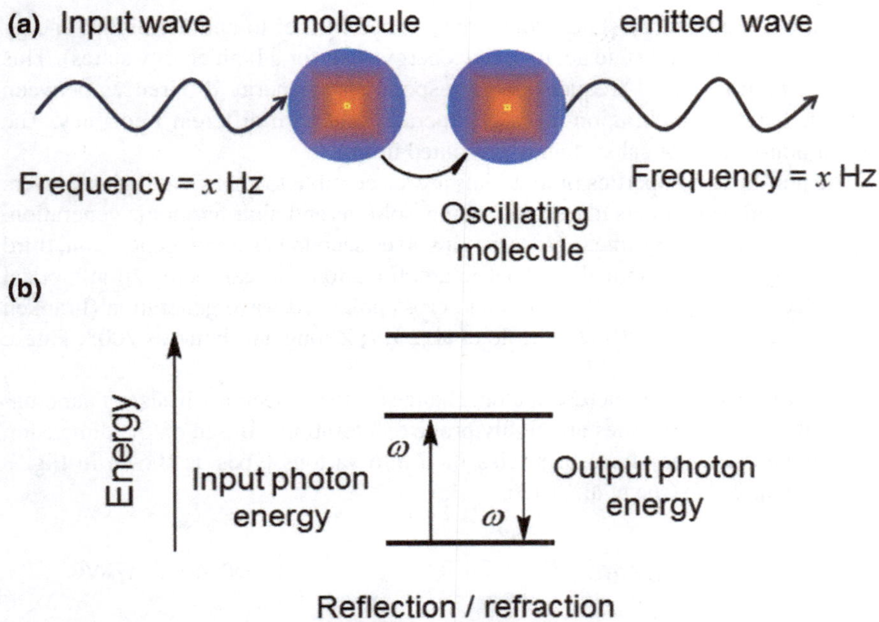

Fig. 1 Presentation of linear optical system; **a** showing no change in frequency of input light wave after being emitted by vibrating the molecule with whom it interacted and **b** no change in the photon energy after the same event

Non-linear optics and non-linear optical properties: The non-linear optics deals with the "intense light", which upon interaction with material changes its optical properties (Luca 2010; Lewis et al. 1941). Unlike linear optical systems the non-linear optical systems does not follow superposition principle (i.e. $A + B \neq X + Y$). As a result of non-linear effect the incident light undergoes a change in optical properties like polarization, frequency, phase or path of incident light because the polarization density of medium responds non-linearly to the electric field of light (Zhang and Wang 2017). This behaviour is observed when the optical electric field strength of light is very high and comparable to that of intra-atomic electric field. In this case, the induced polarization is given by Eq. (2) (Agrawal 2013).

$$\vec{P} = \chi^{(1)}\vec{E} + \chi^{(2)}\vec{E}^2\chi^{(3)}\vec{E}^3 + \cdots \tag{2}$$

$\chi^{(1)}$ is the linear polarization susceptibility of materials which is applicable for lenses. $\chi^{(2)}$ and $\chi^{(3)}$ are non-linear polarization susceptibilities of the materials, which defines second order effects like a second harmonic generation and third order effects such as third harmonic generation, stimulated Raman scattering, four wave mixing and intensity dependence of the index of refraction. In non-linear optics, the light does not follow superposition principle. At high irradiance many molecules are excited to a high energy state, which are excited further to another higher energy states (the first excited state act on a low energy state for a high energy states). This causes vibrations at all frequencies corresponding to energy differences between populated states which, upon mixing, generate light with different frequency. The typical non-linear optical system is presented in Fig. 2.

The non-linear properties of materials are susceptible to change at higher powers inducing nonlinear effects like self-focusing, solitons and high-harmonic generation. The most common non-linear processes involves second harmonic generation, third harmonic generation, optical parametric amplification, optical rectification, optical Kerr effect, multi photon absorption and cross polarised wave generation (Franken et al. 1961; Heinz et al. 1982; Ciriolo et al. 2017; Zhong and Fourkas 2008; Palese et al. 1994).

Since the electronic structure of atom changes with the decrease in size of nanomaterials, its optical properties are highly prone to alterations. Based on the dimension of nanomaterials, they have been classified into various types as shown in Fig. 3 (Tiwari et al. 2012; Cha et al. 2013).

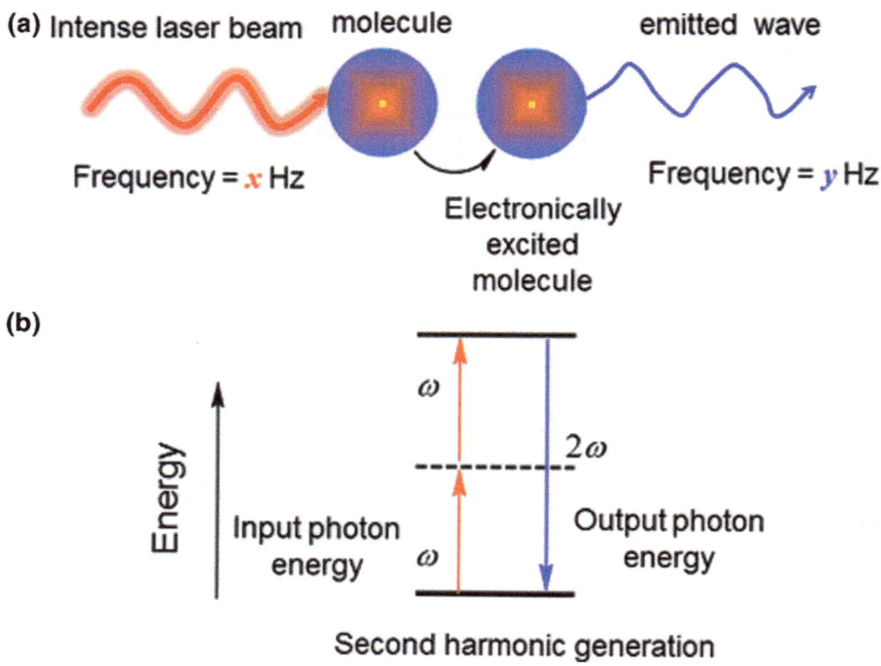

Fig. 2 Presentation of the non-linear optical system; **a** showing change in frequency of input light wave after being emitted by vibrating molecule with whom it interacted and **b** change in the photon energy after same event, exemplified by a second harmonic generation

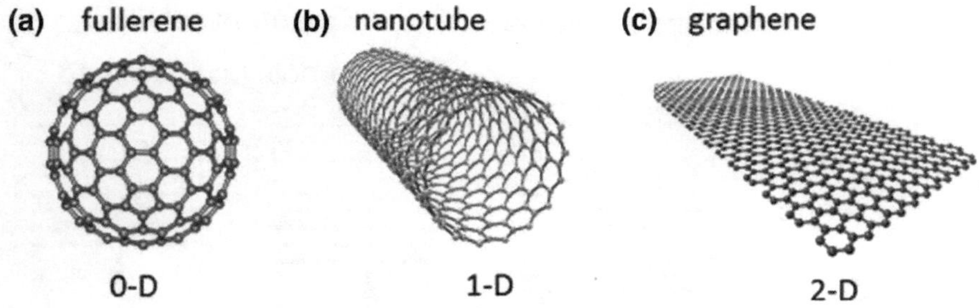

Fig. 3 Nanomaterials of various dimensions **a** 0-dimension (fullerene); **b** 1-dimension (nanotube) and **c** 2-dimension (graphene)

(i) **Zero dimension nanomaterials** (confined into three dimensions), e.g. *nanoparticles, nanoshells, nanocapsules, nanorings, fullerenes and quasi crystals.*
(ii) **One dimension nanomaterials** (confined into two dimensions), e.g. *nanorods, nanofilaments, nanotubes, quantum wires, and nanowires.*
(iii) **Two dimension nanomaterial** (confined into two dimensions), e.g. *discs, platelets, ultrathin films, super lattices, graphene and quantum wells.*

The alterations in optical properties of nanomaterial with reduced dimensionality are usually defined by "*quantum confinement effect.*"

Quantum confinement effect: The increase in energy difference between band gap and energy states of a material due to the origin of discrete energy spectrum, when one of its dimension approaches the size below 5 nm, is called as "*quantum confinement effect*" (Zorman et al. 1995; Takei et al. 2011) (Fig. 4). Consequently, both optical and electronic properties of nanomaterials deviates compared to the bulk material, where the energy levels remain continuous. The '*quantum confinement effect*' arises due to the spatial confinement of electrons in the conduction band, and holes in the valence band when diameter of the particle approaches de Broglie wavelength of electron ($\lambda_{electron} = 1.23$ nm). This will cause the quantization of their energy and momentum with restricted motion since in such a situation they follow the principle of quantum mechanical motion rather than classical mechanics (Anas et al. 2014). The situation becomes similar to the particle in one dimension box. In this situation it becomes very intriguing to confine the probing light for measuring the optical property of a single nanoparticle or nanowire, since the surrounding substrate is always going to interfere with the optical measurement. If the substrate is not photo-luminescent and its absorption range doesn't lie on the frequency range of the probing light, then its effects on the measurement of non-linear optical properties are automatically discarded. However, it affects the linear optical properties like transmission and reflection, just to mention a few. Therefore, the correction of linear optical measurements is inevitable, which can be performed by subtracting the linear optical properties of the substrate alone from the nanomaterial laden substrate.

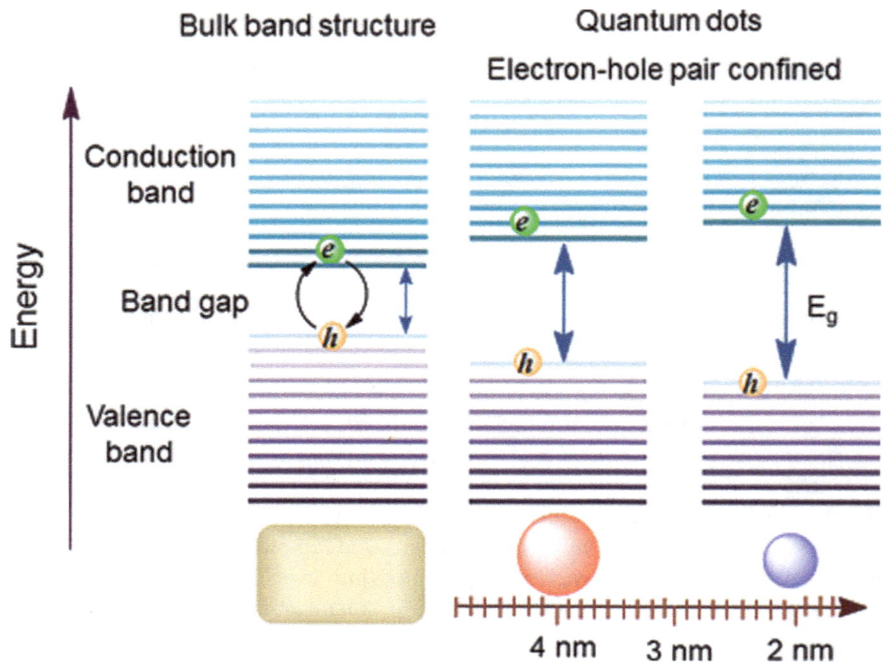

Fig. 4 Presentation of *quantum confinement effect*, wherein the band gap between valence and conduction band increases while decreasing the material size to quantum level due to electron-hole confinement. The holes remain confined in the discrete valence band and electrons remain confined in the discrete conduction band

The "*quantum confinement effect*" has been exploited immensely by tailoring the optical properties of the nanomaterials by tuning their crystal dimensions and the chemistry of their surfaces (Clancy et al. 2018). However, developing technologies for the utilization of these nanomaterials becomes the key factor to achieve their practical applications. The optical applications of nanomaterials include bio-sensing, solar cells, photovoltaics, imaging, non-linearity, photonics and optoelectronics (Keshea and Khakpoor 2017; Jang et al. 2016; Carey et al. 2015; Joarder et al. 2018; Zheng and Zhang 2016). Therefore, in the following section, the optical applications of nanomaterials specifically in Biosensing and Photovoltaics shall be discussed.

2 Optical Applications of Nanomaterials

2.1 Nanomaterials in Optical Biosensing

The high optical sensitivity leading to lower detection limits of analytes by nanomaterials is pointed out as an advantage favouring their application for biosensing applications. Their highly specific surface along with their large surface area could enable the increase in immobilization rate of different types of bio-receptors. In fact, a typical biosensor is composed of a bio-receptor, a transducer, a signal processor and an interface (Li and Liu 2017; Wen et al. 2015; Borisov and Wolfbeis 2008), as depicted in Fig. 5. By its turn, an optical biosensor is comprised of an optical transducer and a bio-receptor. The bio-receptors at the surface senses physical or chemical change when 'in natura', which are transported/transduced to the transducer resulting in changes of the properties of light, namely absorption, fluorescence, transmission, reflection, refraction, phase, amplitude, frequency and polarization (Lara and Perez-Potti 2018).

As observed, nanomaterials perform the function of a transducer in a biosensor, providing a high electrical conductivity and optical sensitivity to a very small detection limit. The biomolecules can be immobilized on the nanomaterial surface either via non-covalent interactions such as electrostatic, H-bonding and π-π stacking, or via covalent cross-linking, e.g. amide coupling reactions. The covalent binding is particularly useful in terms of stability and reproducibility of the surface functionalization. However it has a major drawback of uncontrolled functionalization, consequently changing the principal recognition site, whereas the non-covalent immobilization possesses advantages regarding the maintenance of the properties

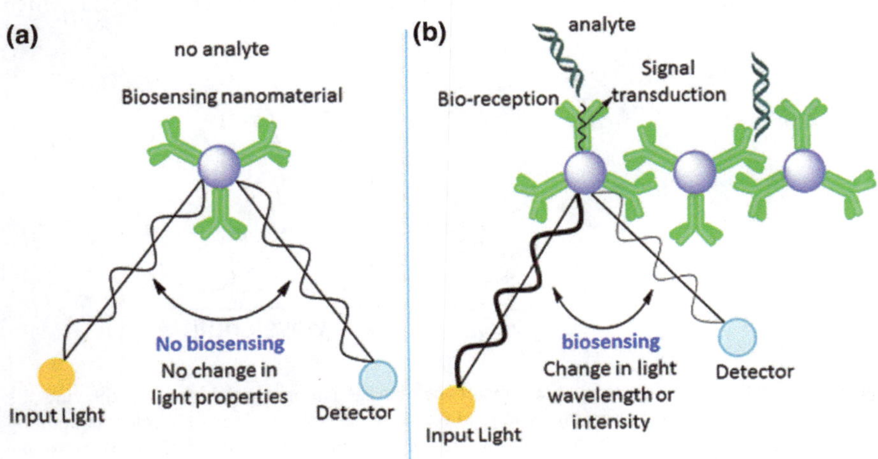

Fig. 5 Schematic representation of optical bio-sensing by using a nanomaterial **a** without and **b** with an analyte, thus demonstrating characteristic changes in the properties of light

of both the biomolecule and nanomaterial. The supramolecular or coordinate binding is another promising technique offering advantages in terms of reversibility and allowing regeneration of the transducer element and bio-receptor (Holzinger et al. 2014; Kwiatkowski et al. 2015; Zeng et al. 2015; Nehra and Singh 2015). Optical biosensors are usually categorized based on the type of optical properties or optical phenomena they measure, namely the absorption-based biosensors (Lobnik and Baldini 2006; Frederix et al. 2003), the fluorescence-based biosensors (e.g. FRET (förster resonance energy transfer) (Zadran et al. 2012; Shi et al. 2015; Zhang et al. 2005), the surface plasmon resonance (SPR) biosensors (Algar and Krull 2010) and the fibre optic-based biosensors (Shi et al. 2015).

Absorption-based biosensors involve alterations in absorbance or absorption coefficients of the nanomaterials upon biosensing (Fig. 6) (Frederix et al. 2003). The absorption spectrum of noble metal nanoparticles like gold, silver and platinum, by their turn, shows a characteristic bulk inter-band absorption due to the promotion of electrons from d-levels to an empty fermi energy level. This inter-band absorption usually occurs at a short wavelength due to the influence of the electron density of d and s states in the conduction band. In such situation, when the biomolecules approach the nanoparticle under observation, the latest absorbs the probing light in the entire bulk area of particles, thus resulting in an increase in the dielectric constant at the nanoparticle surface. The biomolecule adsorbed at the nanoparticle surface causes an increase in the electromagnetic field at the particle position, thus leading to the increase in transition probability of light absorption for bulk transition. This effect is mainly observed when the size of the nanoparticle is lower than the wavelength of the probing light. Because of the differences in size, the internal optical modes of the nanoparticle are not pure as its lateral extent is too small. Therefore, the electric field of internal operator to such nanoparticles is determined by the extended modes, which are a function of the dielectric constant of the surrounding environment, which

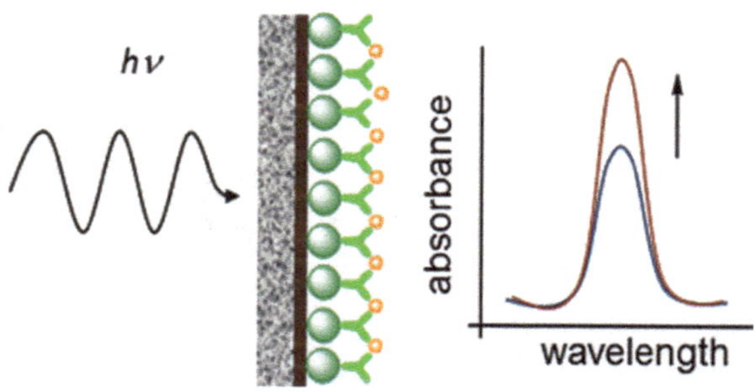

Fig. 6 Schematic representation of an absorbance-based nanomaterial biosensor showing the absorbance increase upon biosensing. The gold or silver nanoparticles (green circles) containing the bioreceptor (Y-shape) are adsorbed on a quartz substrate with the assistance of a mercaptosilane adhesion layer, which upon exposure to light in the presence of the biomolecule (orange circles) leads to an increase in absorbance. Reproduced with permission from Frederix et al. (2003). Copy right 2003 American Chemical Society

affect the absorption characteristics upon biosensing (Frederix et al. 2003). Meanwhile, the absorption of the nanoparticle increases due to the increase in dielectric constant upon biosensing as shown in Fig. 6.

Fuorescence-based biosensors are the most common in biosensing applications because of their high sensitivity, simplicity, and diversity. The advent of nanotechnology has replaced the organic dyes used in biosensing with nanomaterial's possessing enhanced fluorescence properties. Fluorescence biosensors operate via different photo-physical phenomena like **FRET**, fluorescence quenching, fluorescence lifetimes and multi-photon microscopy. For biomolecular assays **FRET** is the most useful phenomena regarding its sensitivity for nano-scale distances between the molecules under study. **FRET** involves non-radioative transfer of energy from donor chromophore to the acceptor chromophore via dipole-dipole coupling. The efficiency of this energy transfer is inversely proportional to the sixth power of the distance between donor and acceptor, making **FRET** extremely sensitive to small changes in distance (Fig. 7a). It can be defined by a simple energy transfer Eq. (3) (Shi et al. 2015).

$$E = \frac{1}{1 + (r/R_o)^6} \tag{3}$$

where E is the donor to acceptor energy transfer efficiency, r denotes the donor-acceptor distance and R_o denotes the donor-acceptor distance for 50% of fluorescence energy transfer. For FRET the R_o is generally 10–100 Å. The quantum nanomaterials have recently gained momentum for FRET assay based biosensing in the form of semiconductor quantum dots (Zhang et al. 2005; Algar and Krull 2010), graphene quantum dots (Shi et al. 2015; Zhao et al. 2013), upconversion nanoparticles (Ye et al. 2014), graphene oxide (Ye et al. 2014) and gold nanoparticles (Wang et al. 2012), due to their exceptional optical properties and penetration depth in vivo as

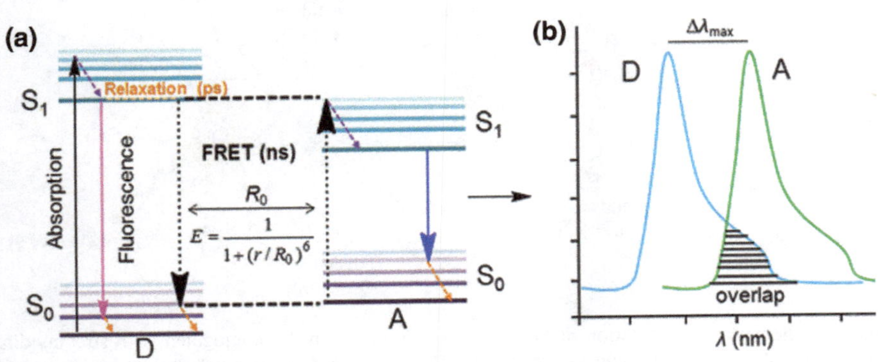

Fig. 7 Schematic representation of **a** Jablonski diagram of **FRET** and **b** Spectral overlap of donor's emission and acceptors absorption spectrum required for **FRET**. In figure, D = donor and A = acceptor

previously discussed. Nanomaterials can be used both as donor and acceptor for FRET based biosensing (Zhang et al. 2005; Algar and Krull 2010; Shi et al. 2015; Zhao et al. 2013; Ye et al. 2014; Wang et al. 2012). In this direction, Zhang et al. (Zhang et al. 2005) utilized semiconductor quantum dots of CdSe-ZnS coated with streptavidin as donor for the detection of DNA using Cy5 as an reporter dye. The co-hybridized structure consisting of DNA sandwiched with biotin as capture probe and Cy5 as an reporter probe when in the vicinity of streptavidin coated CdSe-ZnS quantum dots (nanosensor), FRET effect was observed by the appearance of very distinct signals due to energy transfer from CdSe-ZnS to Cy5 (Fig. 8). The CdSe-ZnS came closer to the co-hybrid structure due to the strong affinity of streptavidin with biotin. More importantly, the nanosensor displayed remarkable sensitivity for DNA with a detection limit up to 4.8 femto molar even with $R_0 = 69.4$ Å, which is an exceptional result (Zhang et al. 2005).

Similarly, a number of other FRET based nano biosensors like graphene quantum dots (donor) and gold nanoparticles (acceptor) for the detection of food-borne pathogen *Staphylococcus aureus* specific gene sequence (Shi et al. 2015), graphene quantum dots (donor) and graphene (acceptor) for the detection of immunoglobulin (Zhao et al. 2013), upconversion nanoparticles (donor) and gold nanoparticles (acceptor) for the detection of the H7 hemagglutinin gene (Ye et al. 2014) and gold nanoparticles (acceptor) and graphene oxide (donor) for the detection of integrin avb3 protein or integrin over-expressing cancer cells (Wang et al. 2012) were also reported, where a significantly low detection limit was evidenced.

Surface plasmon resonance (SPR) biosensor is another class of optical biosensors operating on *the principle of linear optics* (Zeng et al. 2014). Basically, the SPR is a physical phenomenon dealing with resonance of oscillating conduction electrons at the interface of materials with opposite permittivity, under the application of probing light generating a non-radiative surface plasmon electromagnetic wave at the interface. This electric field vector of the plasmon wave is driven in the direction of

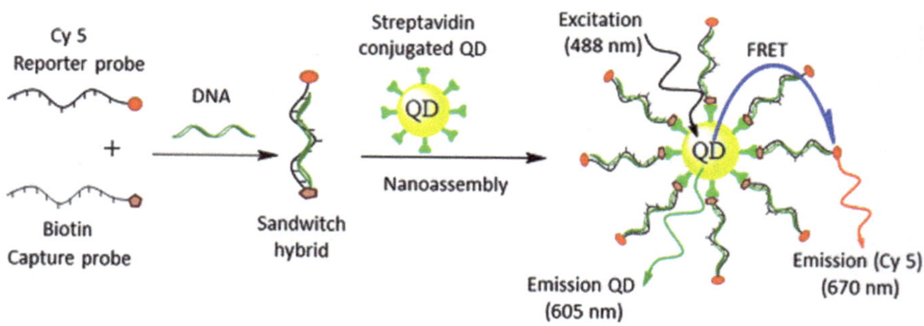

Fig. 8 Schematic representation of DNA sensing by quantum dot conjugated with streptavidin nano-assembled with biotin and Cy5 dye via FRET. The DNA is sandwiched with biotin which is a capture probe and Cy5, which is a reporter probe via nanoassembly. The FRET between QD and Cy5 upon intercalation of DNA upon nanoassembly leads to DNA sensing. Reproduced with permission from Zhang et al. (2005). Copyright 2005 Nature Publishing Group

negative permittivity possesing conducting materials interface and hence, it travels parallel to the direction of interface (Fig. 9). Due to conducting nature of the material, the travelling SPR oscillations are very sensitive to any physicochemical alteration in the vicinity of the interface (e.g. the adsorption of biomolecules), which is exploited for biosensing applications (Fig. 9) (Sabban 2011).

SPR based sensors measure the change in refractive index near the sensing surface and detector. When a monochromatic light is applied at the nanoparticle sensing surface, it is reflected at an angle called as *SPR angle*. At SPR angle, the reflected light is reduced to its minimum level. However, during this process, the monochromatic light passes its energy to the nanoparticle surface which, upon excitation, generates oscillating electrons called plasmons. The electrical field vector of this plasmon wave is sensitive to changes in environment at the sensing surface due to biomolecule adsorption. Therefore, when the biomolecule bind to the sensing surface it results in the change of refractive index at the surface, thus changing the SPR angle due to the change in electric field vector of the plasmonic wave as per ***Fresnel*** equations and ***Snell's*** law (Fig. 9). The photodiode detects the change in SPR angle and expresses the signals as response units (RU), which are directly proportional to total mass of bound ligand and 1 picogram *per* square millimetre. The RU versus time plot is called as sensorgram (Vachali et al. 2015).

SPR sensors are classified into (i) propagating surface plasmon resonance (PSPR) and (ii) localized surface plasmon resonance (LSPR). In PSPR (Fig. 9), a thin film of a metal is exposed to a monochromatic light source generating a resonating surface

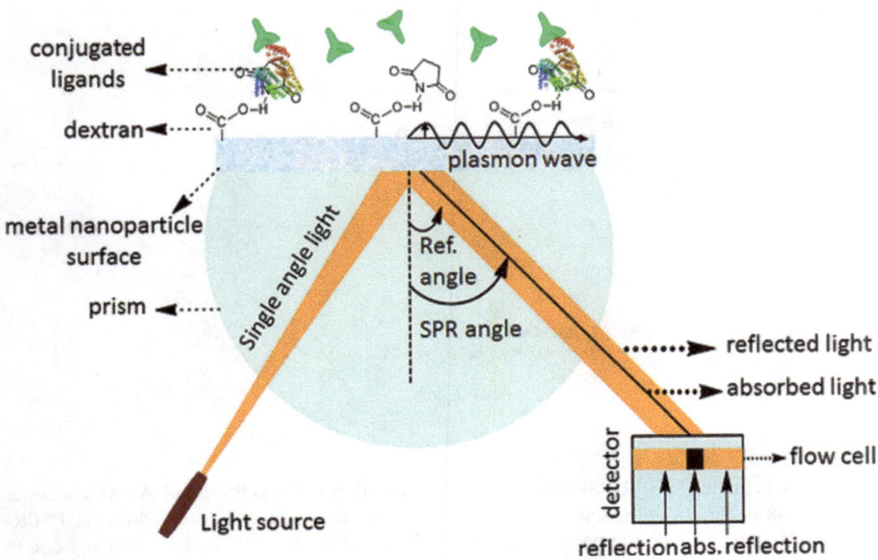

Fig. 9 Schematic representation of biomolecule sensing by SPR spectroscopy. Reproduced with permission from Sabban (2011). Copyright 2011, University of Sheffield

plasmon electromagnetic wave, which migrates up to hundreds of micrometres in the parallel direction to the conducting metal surface (Sabban 2011; Zeng et al. 2013). Contrary to PSPR, in LSPR the generated SPR oscillations do not propagate, instead they remain localized allowing consequent alterations on the resonance of plasmon in size and dimension of the nanomaterial (Zeng et al. 2011). However, and despite having high spectral tunabilities, the sensitivity of LSPR is one order of magnitude lower than the one found for PSPR. Moreover, both LSPR and PSPR have detection limits of analytes up to 1 pM and 8 kDa (Kabashin et al. 2009; Law et al. 2011). Nanomaterials have played a significant role in overcoming these limitations to raise the sensitivity of SPR, by appropriately coupling the LSPR signals of nanoparticles with the PSPR wave generated on the thin film metal surface upon light excitation as sensor. For example, Wang et al. (2009) have recently shown the sensing of adenosine by a gold nanoparticles-aptamer conjugate as schematically represented in Fig. 10. In Fig. 10a, the conjugated anti-adenosine-aptamers-Au-NPs were stabilized with citrate showing complementary hybridization with ssDNA molecules on SPR Au-thin film resulting in high SPR signals with time. However, in the presence of adenosine the adenosine hybridized with antiadenosine-aptamers on Au-NPs which did not recognize and hence did not bind to ssDNA on SPR Au-thin film resulting in change of the electric field of plasmon wave and ultimately caused decrease in SPR signals (Fig. 10b). Interestingly, the Au-NPs-antiadenosine-aptamers showed selectivity towards adenosine in the presence of other nucleotides, namely uridine, cytidine, and guanosine (Matharu et al. 2009). Other prominent

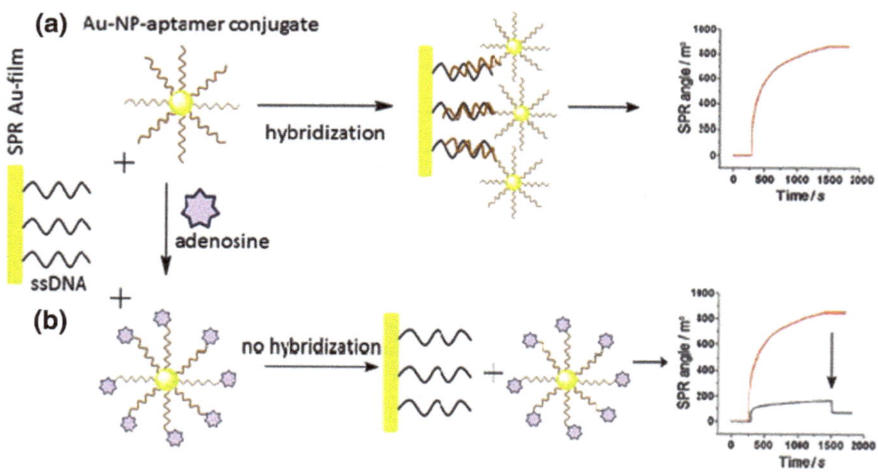

Fig. 10 A schematic of LSPR-PSPR coupled biosensor with Au-film as PSPR and Au NPs-aptamer conjugates as LSPR. In the absence of adenosine a large SPR angle is observed due to PSPR-LSPR coupling whereas in the presence of adenosine the SPR angle decreased drastically due to non-coupling caused by non-complexation of Au NPs-aptamer conjugates with ss-DNA coated metal thin film due to adenosine bindng to Au NPs-aptamer conjugates surface thus indicating its biosensing. Reproduced with permission from Wang et al. (2009). Copyright 2009 Elsevier

example of a nanomaterial based SPR biosensors includes the quantitative detection of the activity of RsaI endonuclease up to 5×10^{-8} M, indicated by a dip in SPR signals upon cleavage of probe DNA1 on Au film hybridized with cDNA labelled Au-NPs (Luan et al. 2011). The sensing of antiglutamic acid decarboxylase antibody (anti-GAD Ab), a diabetic mellitus indicator in blood serum up to femto molar level by using gold nanoparticle-anti immunoglobulin G-Horseradish peroxidase conjugate was also reported. The detection is demonstrated by a double enhancement of SPR signals (Cao and Sim 2007), and poly(amidoamine) functionalized Au-NPs as amplifier immobilized on modified gold surface as SPR sensor for the detection of insulin in human serum up to 5 pico molar (Frasconi et al. 2010). Besides Au-NPs, other nanomaterials like liposome nanoparticles, magnetic nanoparticles, graphene and carbon nanotubes are being also explored to enhance the SPR signals (Wink et al. 1998; Hoshino et al. 2011; Cui et al. 2012). They function by coupling with a metallic film like Au-NPs, but the SPR signal enhancement is induced by an increase in mass loading at the surface and the transfer of charge from the nanomaterial to the thin metal, augmenting the induced field and thus enhancing the SPR signals. Following this idea, Lee et al. have studied and reported the detection of human erythropoietin (EPO) and human granulocyte macrophage colony-stimulating factor (GM-CSF) molecules in the range of 0.1–1000 ng mL^{-1} using a sandwich model comprised of carbon nanotubes (CNTs) conjugated with polyclonal antibody and Au sensing film with the target analyte in between (Fig. 11) (Lee et al. 2010). The process resulted in SPR signals 30-fold higher for both the analytes when CNTs were used as amplification tag for detection. For detailed literature on nanomaterials based SPR sensing please go through Ref. (Zeng et al. 2014).

Fiber optic biosensors, also called as "**Optrodes**" are another class of biosensors whose sensitivity, dynamic range, robustness and lifetime can be enhanced upon nano-coating with nanomaterial (Nanomaterials in biosensors 2018). Optical fibers are flexible silica or plastic wires with a diameter range within a micrometer and possesses sustaining power against harsh environment for remote sensing. They are composed of inner core, coated with cladding having different refractive indexes (RI). Basically, the signal transmission in the optical fibers occurs by total internal reflection (TIR), which can be transmitted to long distances. The RI of core (n_1) should be higher than that of cladding (n_2) for signal transmission via TIR (Fig. 12) (Kodaira et al. 2018; Urrutia et al. 2015).

According to the principle of light propagation in optic fiber, when a light strikes the interface of wire (having dissimilar RI in the core and cladding) with the angle of incidence greater than the critical angle as *per* Snell's law ($\theta_c = \sin^{-1}[n_1/n_2]$), the light is reflected via TIR consequently propagating along the fiber (Fig. 13a) (Marazuela and Moreno-Bondi 2002). The biosensing applications of fiber optics can be understood considering that when a light of certain intensity strikes the interface of a material with dissimilar refractive indexes, it experiences an exponential decrease with the penetration depth going from the interface towards the low refractive index part of the material. The penetration depth (dp) is generally defined as the distance corresponding to the decrease in the amplitude of electric field (E) to 1/e

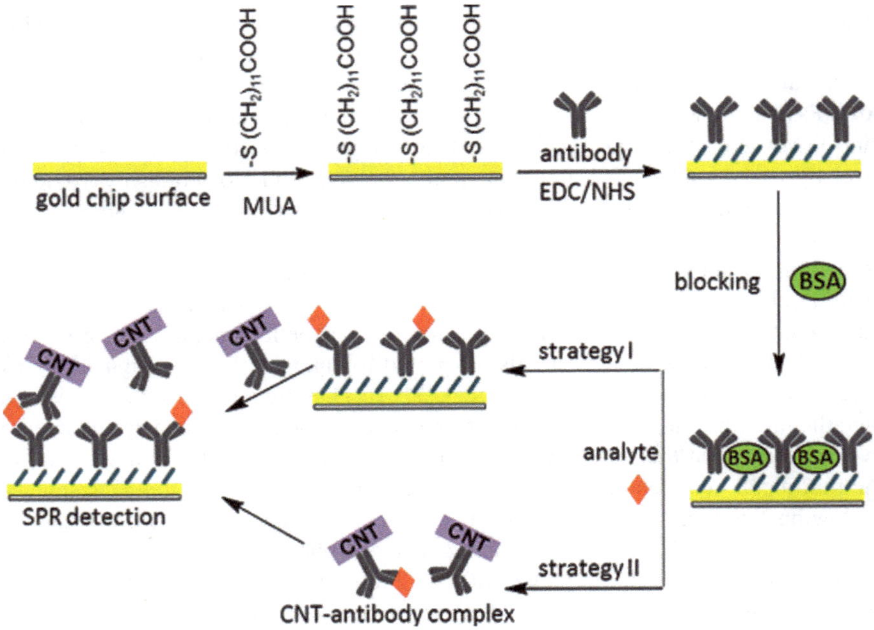

Fig. 11 Pictorial presentation of CNT-antibody complex induced enhanced SPR signal for biosensing of human erythropoietin and human granulocyte macrophage colony stimulating factor. Reproduced with permission from Lee et al. (2010). Copyright Elsevier

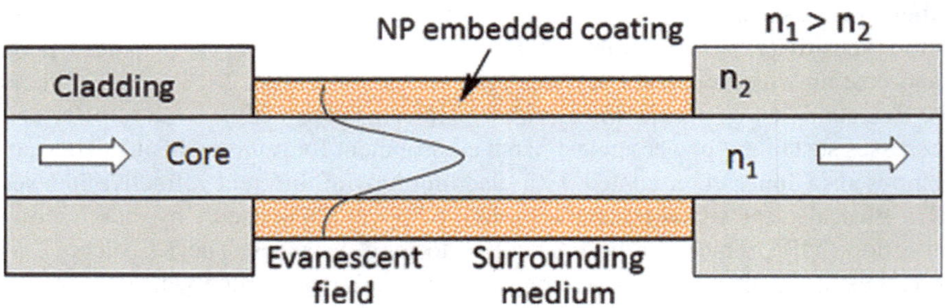

Fig. 12 Pictorial presentation of typical nanoparticle coated by an absorbance fiber optic sensor designed by removal of cladding

(0.37) at the interface. The dp depends on the wavelength and angle of incidence of the light striking at the interface (Fig. 13b) (Marazuela and Moreno-Bondi 2002). The evanescent wave produced interacts with the molecules within the penetration depth of light and produces net flow of energy across the reflecting surface called as attenuation in reflectance, utilized for the development of attenuated reflectance biosensors. If the molecules interacting with the evanescent wave are fluorescent they undergo emission which can be directed back on to the optic fiber and transported

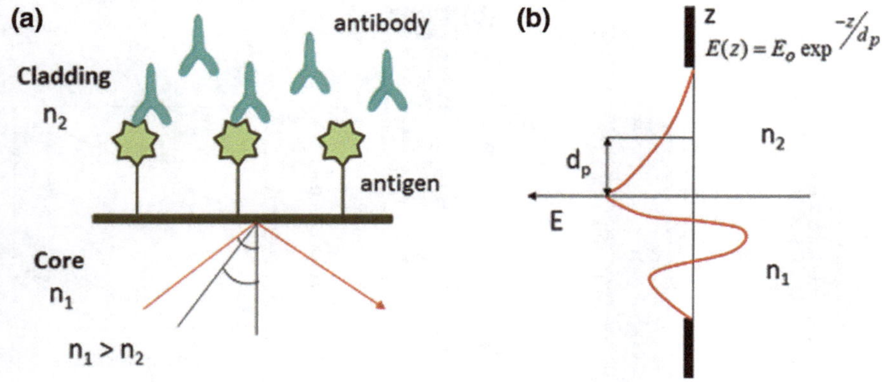

Fig. 13 Schematic demonstration of **a** the phenomenon of total internal reflection in a material with dissimilar refractive index between the core and cladding when the light strikes its interface at an angle larger than the critical angle, n_1 and n_2 denote the refractive indexes; and **b** represents the variation of amplitude of electric field (E) at the interface of both core and cladding. The amplitude of the electric field decreases exponentially with an increase in penetration depth (dp) in the medium with lower refractive index. Reproduced with permission from Marazuela and Moreno-Bondi (2002). Copyright © 2002, Springer-Verlag

to the detector to sense the biomolecule via TIR fluorescence. Besides fluorescence, other optical phenomenon like, absorption, interferometer and SPR can be used as well for biosensing via **Fiber optic** (Urrutia et al. 2015). Further, the sensitivity of detection can be enhanced by replacing the cladding with nanoparticles coating (Fig. 12). The high surface area of nanoparticles in the sensitive region coupled with diverse optical properties allows reaching lower limits of detection. For example, an interferometer based fiber optic biosensor for the detection of DNA was reported by Yin et al. (2013) (Fig. 14).

The DNA sensor is composed of polyelectrolyte multilayers of polyethylenimine (PEI) and poly-acrylic acid (PAA) and ssDNA fabricated on optic fiber. The sensor works on the principle of the interference of DNA base pairs for perfect matching leading to sensing. The sensitivity of the biosensor is up to 0.27 nm/matched base pairs at a concentration of 1 mM. Moreover, the sensor can be utilized for quantification of a number of matched bases single standard DNA chains (Yin et al. 2013). Nowadays, there are a number of nanomaterials coupled optic fiber biosensors operating via different optical phenomena (Zibaii et al. 2014; Tseng et al. 2008; Carrasquilla et al. 2011; Bakalova et al. 2005; Makrides et al. 2005; Huang et al. 2015).

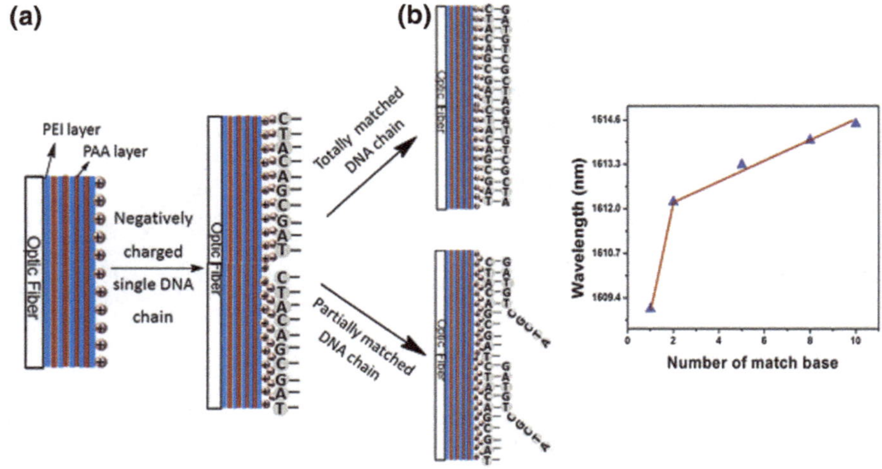

Fig. 14 **a** Pictorial presentation of the DNA sensing by a Fiber optic biosensor and **b** plot showing the shift in wavelength during single standard DNA sensing in solution by thin-core fiber modal interferometer based DNA sensor. Reproduced with permission from Yin et al. (2013) copyright 2013 Royal Society of Chemistry

2.2 *Nanomaterials in Photovoltaics*

Solar energy generated via *photovoltaics* is considered amongst the cleanest source of energy on earth, since it doesn't emit any hazardous gases into the atmosphere responsible for the greenhouse effect. **Photovoltaics** deal with the conversion of light into electricity using semiconducting materials. *Photovoltaic Effect* is a physiochemical phenomenon where voltage and electrical currents are created across the p-n junction in a semiconducting material due to absorption of radiant energy (Fig. 15) (Williams 1960). It was discovered by a French experimental physicist, Edmund Becquerel in 1839, with an electrolytic cell made up of two metal electrodes (Green 1990).

The photovoltaic system is composed of solar panels where each one comprises a certain number of solar cells responsible for the generation of electrical power (Fig. 15). Since the invention of the first practical silicon solar cells by Bells lab in 1954 (Chapin et al. 1954), the silicon based solar energy has grown unabated, as justified by the fact that 90% of the working solar panels around the world use crystalline silicon (c-Si) as a semiconducting material. The large abundance of silicon on earth also makes solar energy as an easily scalable energy resource. However, despite having so many advantages, the contribution of solar energy to world's electricity production until 2016 was just 1.8%, with projections of 2.8% being made for 2018 (Fig. 16) (International Energy Agency 2018).

The key culprit behind this issue is "Shockley Queisser limit" or SQ limit of single junction solar cells. *SQ* limit is generally associated to single p-n junction solar cells, and corresponds to a maximum theoretical efficiency of the solar cell to convert

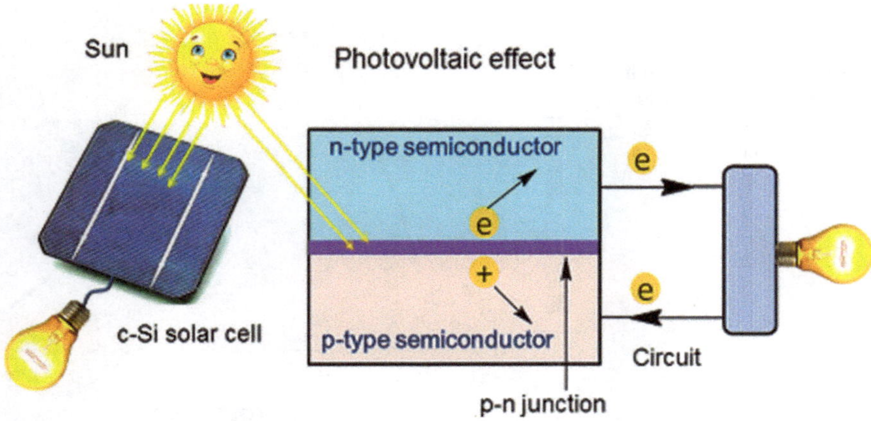

Fig. 15 Pictorial presentation of the photovoltaic effect leading to generation of electricity in a single p-n junction silicon solar cell

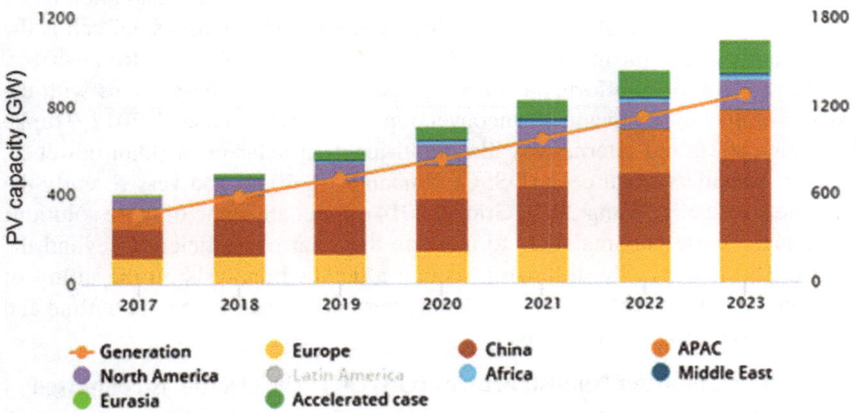

Fig. 16 Solar photovoltaics generation and cumulative capacity by region, 2017–2023 (International Energy Agency 2018)

sunlight into electricity at solar irradiance (actual power of sunlight on a sunny day). Numerically, it is defined for a solar cell with a band gap of 1.34 eV and corresponds to 33.7% (337 W m^{-2}) at 1000 W m^{-2} of sun light (Rühle 2016). *SQ* limit (Fig. 17) was originally calculated by William Shockley and Hans-Joachim Queisser at Shockley Semiconductor in 1961, with a value of 30% at a band gap of 1.1 eV (Rühle 2016; Shockley and Queisser 1961). For a monocrystalline silicon solar cells (c-Si), which accounts for 90% of the commercially available solar cells, the *SQ* limit is 24% and their practical efficiency is 15–20%. Major causes of *SQ* limit are the various losses in channels which include, (i) thermal losses due to conversion of thermal energy to entropy work, (ii) losses due to the mismatch between absorption and emission

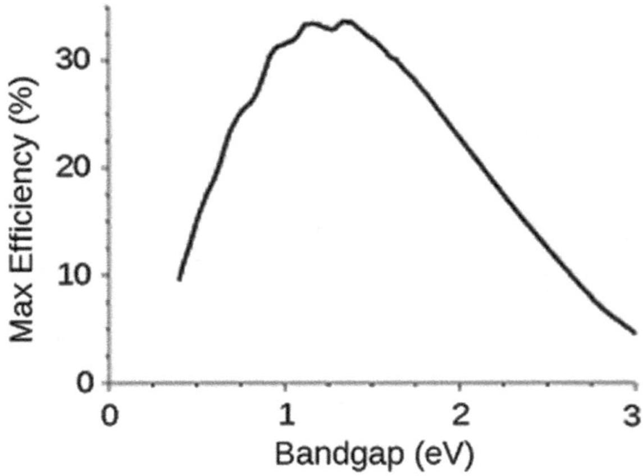

Fig. 17 Plot showing variation of maximum efficiency of a solar cell with a bandgap corresponding to *SQ* limit of a single p-n junction solar cell

angle, (iii) losses due to dissipation of energy due to re-emission from the solar device and (iv) transmission losses due to the non-absorption of sub-band gap photons by solar cells. However, the biggest of these losses in a single junction solar cell is the non-absorption due to the transmission of sub-band gap level photons (Rühle 2016). Over the years, a lot of efforts have been made to overcome these issues with the assistance of upconversion and downconversion materials (Asahi et al. 2017; Trupke and Green 2002), and alternatives like multi-junction solar cells (Dimroth et al. 2014), dye sensitized solar cells (DSSC) (Ghann et al. 2017) and very recently the Perovskite solar cells (Wang 2019; Grätzel 2014). If not all, some of these solutions and alternatives use nanomaterials to increase the solar cell efficiency beyond the *SQ* limit. Therefore, in the following section a detailed analysis on the utility of nanomaterials in photovoltaics development considering the conceptual outline and recent examples will be performed.

Upconverting Quantum Nanostructures (QN-UC): The QN-UC is comprised of a semiconductor nanocrystal containing quantum dots of two different band gaps connected by a tunnelling barrier, which is in fact a semiconductor rod (Shang et al. 2015). Deutsch et al. (2013) demonstrated that the UC in semiconductor quantum nanostructures can proceed via subsequent absorption of photons by two step mechanisms (Fig. 18) (Deutsch et al. 2013). The absorption of the first photon occurs in quantum dots with a lower energy band gap, which results in the generation of an inter-band electron-hole pair with a confined hole and delocalized electron in a compound semiconductor crystal. The subsequent absorption of second photon results in intra-band excitation of hole to cross the energy barrier to high energy state either via direct or indirect absorption process of photon by the hole. In the direct process, a hole confined in the lower energy core absorbs photons to undergo intra-band excitation (Fig. 18a). In the indirect process Auger mediation occurs wherein the absorption of second photon leads to the generation of another electron-hole pair (Fig. 18b). The generated electron-hole pair further undergoes recombination and simultaneous transfer of non-radioactive energy to the hole confined in quantum

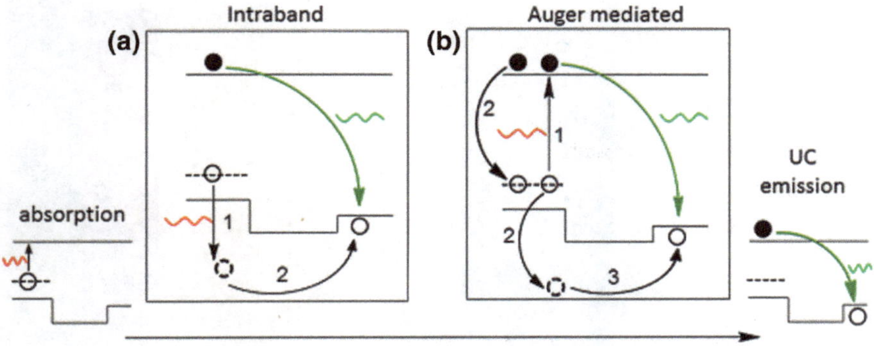

Fig. 18 Schematic representation of the mechanisms of photon upconversion in quantum nanostructures reported by Deutsch et al. **a** Intraband hole absorption mechanism and **b** upconversion mediated by Auger-mechanism. Reproduced with permission from Deutsch et al. (2013). Copyright © 2013, Springer Nature

dots with small band gaps to help it cross the energy barrier to quantum dots with large band gap. This process is followed by radiative recombination with delocalized electron leading to upconversion luminescence. Such a stable inorganic crystalline structure with tunable spectral properties via size tuning of the nanostructures holds high promise as photovoltaic material subject to enhancement in upconversion range and efficiency (Shang et al. 2015; Deutsch et al. 2013).

Besides UN-QC, the **Rare earth doped UC nanomaterials (RED-UC)** have also been used to increase the efficiency of solar cells beyond *SQ* limit (Shalav et al. 2005, Shao et al. 2015). For example, the RED upconversion materials doped with single erbium (Er^{3+}) are desirable to increase the efficiency of crystalline silicon (c-Si) based solar cells regarding the overlap of their absorption range with the sub-band gap energy of silicon from 145 to 1580 nm and their emission wavelength in the range of 540–980 nm, which is suitable for producing excitons of silicon (Fig. 19). These rare earth metals are generally doped into a variety of host materials, namely $NaYF_4$, YF_3, CaF_2, Y_2O_3, $BaCl_2$ and glass ceramics, just to mention a few. Among these, the $NaYF_4$ has been extensively used as nanohost for RED-UC nanoparticles. The rare earth metal ions act as suitable sensitizer-emitter on account of their distinguished absorption-emission characteristics due to d-d, f-f and d-f electronic transitions.

Different strategies and geometric configurations have been designed to exploit the maximum efficiency of RED upconversion nanoparticles by avoiding the limiting factors or photo-physical processes. Shao et al., have overcome the losses due to cross relaxation processes between different rare earth metal ions by the introduction of inert $NaYF_4$ layers in between the upconverting relams doped with Er^{3+}, Ho^{3+} and Tm^{3+} ions, therefore developing the so called multi-layer core/shell nanoparticles with a typical configuration, of $NaYF_4$:10% Er^{3+}@$NaYF_4$@$NaYF_4$:10% Ho^{3+}@$NaYF_4$@$NaYF_4$: 1% Tm^{3+}@$NaYF_4$. The developed strategy led to two fold enhancement in upconversion efficiency, as compared to nanoparticles without inert

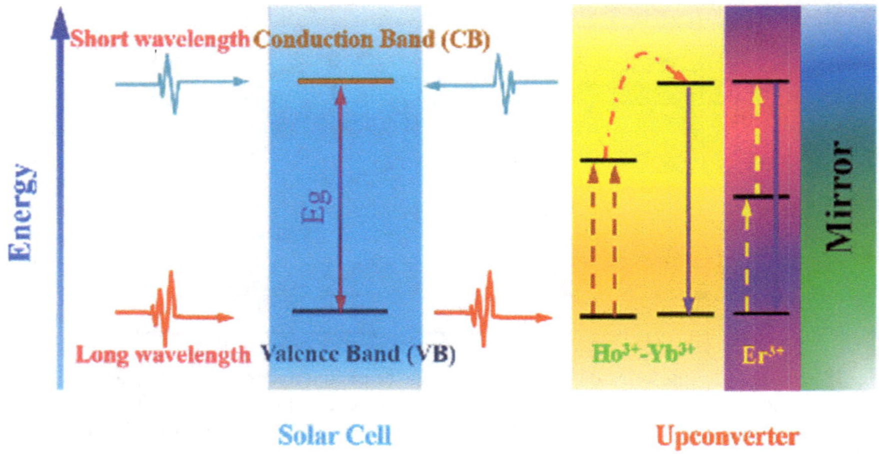

Fig. 19 Schematic representation of the mechanism behind the c-Si solar cells upon fabrication with two RED upconversion material (Ho^{3+}-Yb^{3+}-doped and Er^{3+} doped). Adopted from Shang et al. (2015). Copyright © 2015, MDPI

$NaYF_4$ layers in between. Additionally, due to the presence of multiple rare earths metal ions (Er^{3+}, Ho^{3+} and Tm^{3+}) with wide NIR spectral range of around 270 nm in a single nanoparticle there is an advantage in terms of the photosensitization of c-Si, thus increasing their efficiency (Shao et al. 2015). In fact, these ions could be excited between 1120 and 1190 nm for Ho^{3+}, between 1190 and 1260 nm for Tm^{3+} and between 1450 and 1580 nm for Er^{3+}. However, such systems have limitations in terms of high excitation intensities (~10^4 W cm^{-2}) which are much higher than the solar irradiance at A. M. 1.5 solar spectrum i.e. 1 mW cm^{-2}.

Recent research in photon upconversion materials operating via triplet-triplet annihilation phenomenon (TTA-UC), which uses quantum dots as sensitizer and organic dyes as acceptors have produced better results at lower excitation intensities near to sub solar irradiance, proving their efficiency in near future regarding the development of PV technologies (Wu et al. 2016; Okumura et al. 2016). A typical TTA-UC (Fig. 20) system is comprised of donor and acceptor chromophores either in the form of same or different molecules. Upon excitation of sensitizer at low energy, its triplet excitons transfer their energy to low lying triplet excitons of the acceptors via Dexter energy transfer (DET). Then, the sensitized triplets of acceptors relay their triplet energy to other acceptor in a chain reaction operating either via triplet diffusion or triplet energy migration. During this triplet energy relay process when two sensitized acceptors collide they undergo triplet-triplet annihilation. The annihilation results in the formation of high energy singlet state (S_1) with an energy of two triplets and relaxes to ground state via delayed anti-stokes photoluminescence. Since the DET occurs via electron exchange mechanism, the molecules or quantum dots in TTA-UC matrix must be present within the distance of 1–10 Å (Baluschev et al. 2006).

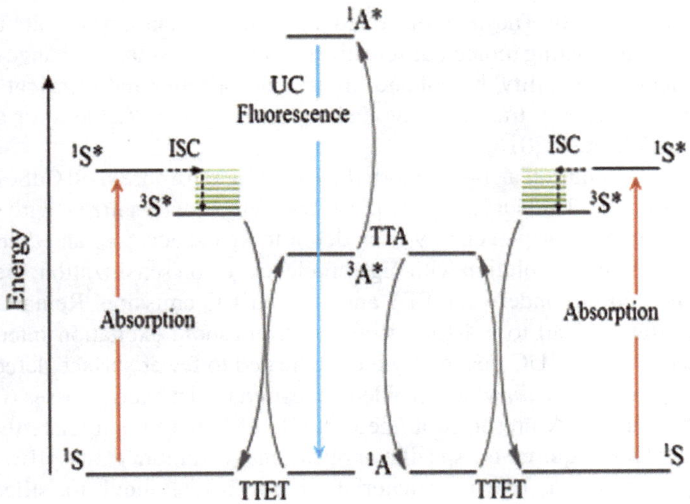

Fig. 20 Jablonski diagram describing TTA-UC process. S and A stands for sensitizer and acceptor, respectively

In this case, Wu et al. have reported solid-state infrared-to-visible upconversion sensitized by lead sulphide (PbS) colloidal nanocrystals (Wu et al. 2016). The solid film is composed of lead sulphide nanocrystal as sensitizer, rubrene as acceptor and dibenzotetraphenylperiflanthene DBP) as collector-emitter (Fig. 21). Upon excitation of nano-crystals at 808 nm in the film displayed UC emission at 612 nm with quantum yield of $1.2 \pm 0.2\%$ at excitation intensity of 12 mW cm^{-2}, which is close to

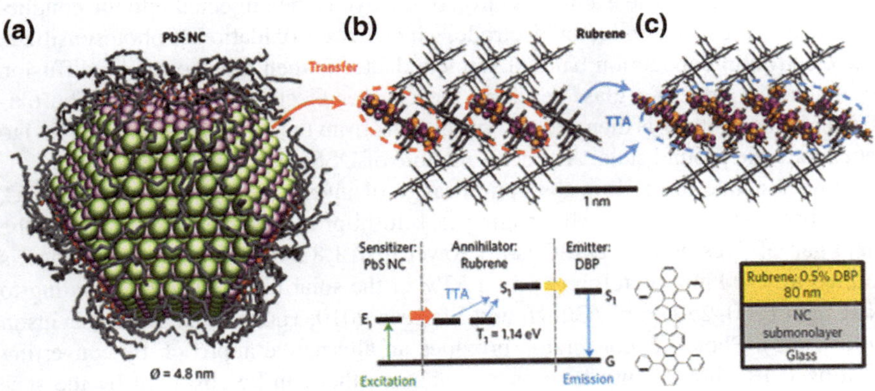

Fig. 21 Image representing the NTR to Vis TTA-UC in solid state operating through sensitization via lead sulphide (PbS) nanocrystal with rubrene as acceptor and dibenzotetraphenylperiflanthene (DBP) as collector-emitter. Adopted with permission from Wu et al. (2016). Copyright © 2015, Springer Nature

sun excitation intensity. They demonstrated that colloidal nanocrystals are an attractive alternative to existing molecular sensitizers, given their small exchange splitting, wide wavelength tunability, broadband infrared absorption, and transient observations of efficient energy transfer, thus making such systems suitable for solar cell applications (Wu et al. 2016).

Moreover, Okumura et al. have reported TTA-UC using a core shell CdSe quantum dots as sensitizers. The surface of CdSe QDs were functionalized with acceptor molecules to relay the triplet energy from donor to ligand acceptor and then to other acceptor moving in the solution with high efficiency. Upon sensitization, the moving acceptors in solution underwent TTA and showed UC emission. Remarkably, the developed strategy lead to a 4-fold decrease in threshold excitation intensity and 50-fold increase in the UC quantum yield, dedicated to fewer surface defects in the core shell QD as sensitizer, which avoided undesired triplet energy losses (Okumura et al. 2016). This work presented a fine example of how to manipulate the optical properties of QDs required for specific applications (Okumura et al. 2016).

The application of UC nanomaterial is not just limited to silicon solar cells, and they have also been applied for **Dye sensitized solar cells (DSSC)**. First reported by Grätzel in 1991, DSSC have emerged as potential alternative to conventional c-Si solar cells as third generation photovoltaic cells (O'Regan and Grätzel 1991). A recent report on the emerging future market of DSSC is discussed (https://www.millioninsights.com/industry-reports/dye-sensitized-solar-cells-dssc-market). DSSC are described as being alternatives with a high potential, since these have better physical and optical properties. Examples are given regarding the DSSC simplicity in fabrication, transparency and colour and workability in low-light conditions (non-direct sunlight and gray skies). In a typical mechanism, the incident photon is absorbed by the photosensitizer dye adsorbed on the semiconductor surface, due to which the photosensitizer is excited to higher energy singlet state. The excited electron of the dye is then injected into the conduction band of the semiconductor electrode, which causes oxidation of photosensitizer. The electron in conduction band of semiconductor is then transported by diffusion to attached conducting glass reaching the counter electrode via circuit. The oxidized photosensitizer is then accepting electrons from the redox mediator (I^-/I^{3-}) to regenerate the ground state. A general module of DSSC is shown in Fig. 22a.

However, due to the low absorption range of investigated dyes used in DSSC, it has been proved to be a challenging task to improve their efficiency. Since the band gap of dyes used in DSSC is not lower than 1.8 eV, their absorption range is limited to <700 nm, therefore wasting 52% of the solar radiation corresponding to NIR light (700–2500 nm) (Bünzli and Eliseeva 2010; Hafez et al. 2011; Levinson et al. 2005). Photon upconversion provides an alternative approach by converting unutilized IR photons into high-energy photons that can be absorbed by the sensitizing dye (Fig. 22b). For example, the rare earth doped upconverting materials nanoparticles of colloidal β-$NaYF_4$:2% Er^{3+}/20% Yb^{3+} were used by Yuan et al. in a DSSC (Yuan et al. 2012). The nanoparticles, due to their small size (20 nm) and mesoporous nature of TiO_2 layer, diffused smoothly through it for enhanced interaction with the sensitizer dyes leading to efficient energy transfer without much

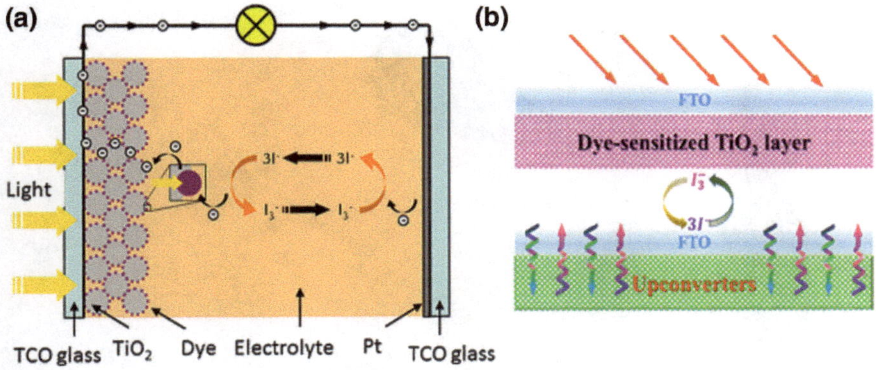

Fig. 22 **a** A general module of dye sensitised solar cell (DSSQ). **b** UC nanomaterials fabricated DSSC. Adopted from Shang et al. (2015). Copyright 2015, MDPI

modification of device fabrication process. Further improvement in this process was brought in by Yang's research group by attachment of YF3:Yb^{3+}/Er^{3+} nanoparticles on the surface of TiO_2 thin film which lead to an increase in power conversion efficiency of 18–6.76% (Li et al. 2014), therefore signifying the utility of nanomaterials in PV cell. Recent developments in this direction include the application of TTA-UC solid matrices to increase the efficiency of DSSC (Hill and Hanson 2017; Amemori et al. 2016).

Besides the conventional silicon based solar cells (c-Si) and DSSC, in the past decade two new types of solar cells have been under development; the (i) Colloidal quantum dots solar cells (C-QD) discovered in 2005 (McDonald et al. 2005; Carey et al. 2015; Ye et al. 2017; Pan et al. 2018) and (ii) Perovskite solar cells (PSC) discovered in 2009 (Kojima et al. 2009; Best Research-Cell Efficiencies (NREL, accessed 02 January 2019; Saule Technologies 2018; Nature Energy 2019). At the optimistic front, these solar cells have shown a great promise to replace conventional silicon solar cells for practical applications both in terms of easiness of fabrication and cost, however at the realistic front, silicon solar cells with improved efficiency are still going to dominate the solar energy world at least for the next decades from now. A schematic presentation of various optical applications of nanomaterials is depicted in Fig. 23.

3 Conclusion

In this book chapter we presented the conceptual basis of optical application of nanomaterials in biosensing and photovoltaics. The exceptionally tunable, linear and non-linear optical properties of nanomaterials, like absorbance, reflectance, interference, fluorescence and resonance have been exploited to develop biosensors with a detection limit reaching up to femto molar (fM) range. Moreover the absorption and

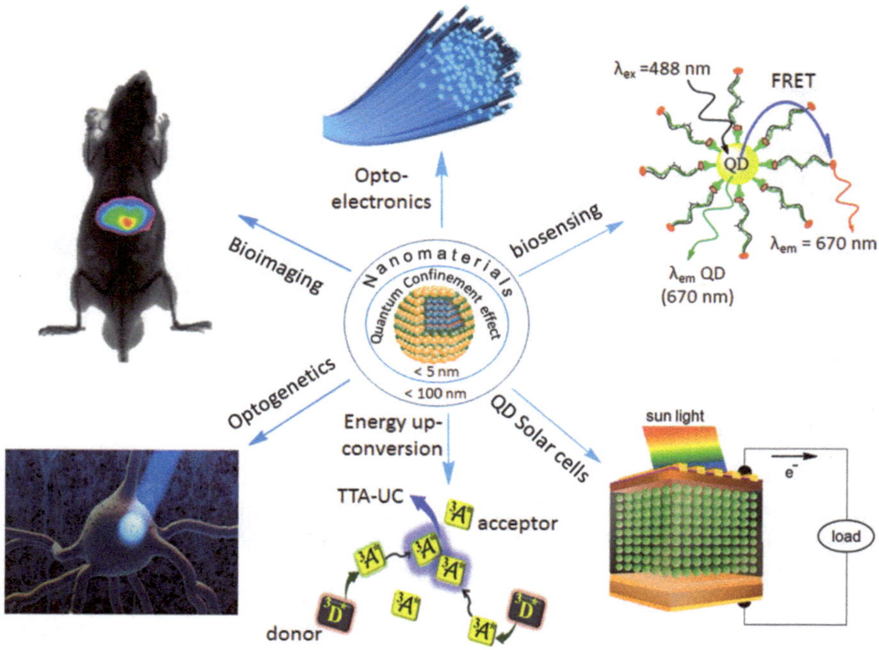

Fig. 23 A pictorial presentation of various optical applications of nanomaterials

emission spectral range of nanomaterials has been exploited to develop upconverting materials in accordance with the band gap of solar cells to increase their efficiency. More recently, the nanomaterials have been used to develop new solar cells in the form of colloidal quantum dot solar cells (CQD). We conclude that nanomaterials as a recent development have inspired great changes in the daily life science and technology, and based on their promising characteristics, they are expected to shape the materialistic future in health, security, warfare, food, energy and ultimately the economy sector.

References

Agrawal G (2013) Nonlinear fiber optics, fifth edn. Chapter 1. Elsevier Inc., pp 1–25

Algar R, Krull UJ (2010) Developing mixed films of immobilized oligonucleotides and quantum dots for the multiplexed detection of nucleic acid hybridization using a combination of fluorescence resonance energy transfer and direct excitation of fluorescence. Langmuir 26:6041–6047

Amemori S, Sasaki Y, Yanai N, Kimizuka N (2016) Near infrared-to-blue photon upconversion by exploiting direct S-T absorption of a molecular sensitizer. J Am Chem Soc 138:872–875

Anas MM, Othman AP, Gopir G (2014) First-principle study of quantum confinement effect on small sized silicon quantum dots using density-functional theory. AIP Conf Proc 1614:104–109

Asahi S, Teranishi H, Kusaki K, Kaizu T, Kita (2017) Two-step photon up-conversion solar cells. Nat Commun 8(14962):1–9

Bakalova R, Zhelev Z, Ohba H, Baba Y (2005) Quantum dotbased western blot technology for ultrasensitive detection of tracer proteins. J Am Chem Soc 127:9328–9329

Baluschev S, Miteva T, Yakutkin V, Nelles G, Yasuda A, Wegner G (2006) Up-conversion fluorescence: noncoherent excitation by sunlight. Phys Rev Lett 97:143903

Best Research-Cell Efficiencies (NREL, accessed 02 January 2019); https://www.nrel.gov/pv/assets/pdfs/pv-efficiencychart.20181221.pdf

Bhagyaraj SM, Oluwafemi OS, Kalarikkal N, Thomas S (2018) Applications of nanomaterials: advances and key technologies, 1st edn. Woodhead Publishing

Binnig G, Rohrer H (1986) Scanning tunneling microscopy. IBM J Res Dev 30:355–369

Borisov SM, Wolfbeis OS (2008) Optical biosensors. Chem Rev 108:423–461

Bünzli JCG, Eliseeva SV (2010) Lanthanide NIR luminescence for telecommunications, bioanalyses and solar energy conversion. J Rare Earths 28:824–842

Cao C, Sim SJ (2007) Signal enhancement of surface plasmon resonance immunoassay using enzyme precipitation-functionalized gold nanoparticles: a femto molar level measurement of anti-glutamic acid decarboxylase antibody. Biosens Bioelectron 22:1874–1880

Carey GH, Abdelhady AL, Ning Z, Thon SM, Bakr OM, Sargent EH (2015) Colloidal quantum dot solar cells. Chem Rev 115:12732–12763

Carrasquilla C, Xiao Y, Xu CQ, Li Y, Brennan JD (2011) Enhancing sensitivity and selectivity of long-period grating sensors using structure-switching aptamers bound to golddoped macroporous silica coatings. Anal Chem 83:7984–7991

Cha C, Shin SR, Annabi N, Dokmeci MR, Khademhosseini A (2013) Carbon-based nanomaterials: multi-functional materials for biomedical engineering. ACS Nano 7:2891–2897

Chapin DM, Fuller CS, Pearson GL (1954) A new silicon p-n junction photocell for converting solar radiation into electrical power. J Appl Phys 25:676–677

Chapter 1. Nanomaterials in biosensors: fundamentals and applications. Copyright © 2018 Elsevier Inc.

Ciriolo AG, Negro M, Devetta M, Cinquanta E, Faccialà D, Pusala A, Silvestri SD, Stagira S, Vozzi C (2017) Optical parametric amplification techniques for the generation of high-energy few-optical-cycles IR pulses for strong field applications. Appl Sci 7:265-1-28

Clancy AJ, Bayazit MK, Hodge SA, Skipper NT, Howard CA, Shaffer MSP (2018) Charged carbon nanomaterials: redox chemistries of fullerenes, carbon nanotubes, and graphenes. Chem Rev 118:7363–7408

Cui SM, Pu HH, Lu GH, Wen ZH, Mattson EC, Hirschmugl C, Gajdardziska-Josifovska M, Weinert M, Chen JH (2012) Fast and selective room-temperature ammonia sensors using silver nanocrystal-functionalized carbon nanotubes. ACS Appl Mater Interfaces 4:4898–4904

Deutsch Z, Neeman L, Oron D (2013) Luminescence upconversion in colloidal double quantum dots. Nat Nanotechnol 8:649–653

Dimroth F, Grave M, Beutel P, Fiedeler U, Karcher C, Tibbits TND, Oliva E, Siefer G, Schachtner M, Wekkeli A (2014) Wafer bonded four-junction GaInP/GaAs//GaInAsP/GaInAs concentrator solar cells with 44.7% efficiency. Prog Photovolt Res Appl 22:277–282

European Commission (2011) Nanomaterials, 18 Oct 2011. http://ec.europa.eu/environment/chemicals/nanotech/

Franken P, Hill A, Peters C, Weinreich G (1961) Generation of optical harmonics. Phys Rev Lett 7:118–119

Frasconi M, Tortolini C, Botre F, Mazzei F (2010) Multifunctional au nanoparticle dendrimer-based surface plasmon resonance biosensor and its application for improved insulin detection. Anal Chem 82:7335–7342

Frederix F, Friedt JM, Choi KH, Laureyn W, Campitelli A, Mondelaers D, Maes G, Borghs G (2003) Biosensing based on light absorption of nanoscaled gold and silver particles. Anal Chem 75:6894–6900

Ghann W, Kang H, Sheikh T, Yadav S, Chavez-Gil T, Uddin FJ (2017) Fabrication, optimization and characterization of natural dye sensitized solar cell. Sci Rep 7(41470):1–12

Grätzel M (2014) The light and shade of perovskite solar cells. Nat Mater 13:838–842

Green MA (1990) Photovoltaics: coming of age. In: 21st IEEE photovoltaic specialists conference. Orlando, USA, pp. 1-8

Guo Z, Tan L (2009) Fundamental and applications of nanomaterials. Artech House, ISBN-13: 978-1-59693-262-3

Hafez H, Saif M, Abdel-Mottaleb MSA (2011) Down-converting lanthanide doped TiO_2 photoelectrodes for efficiency enhancement of dye-sensitized solar cells. J Power Sources 196:5792–5796

Heinz TF, Chen CK, Ricard D, Shen YR (1982) Spectroscopy of molecular monolayers by resonant 2nd-harmonic generation. Phys Rev Lett 48:478–481

Hill SP, Hanson K (2017) Harnessing molecular photon upconversion in a solar cell at sub-solar irradiance: role of the redox mediator. J Am Chem Soc 139:10988–10991

Holzinger M, Goff AL, Cosnier S (2014) Nanomaterials for biosensing applications: a review. Front Chem 2:63-1-10

Hoshino K, Huang YY, Lane N, Huebschman M, Uhr JW, Frenkel EP, Zhang XJ (2011) Microchip-based immunomagnetic detection of circulating tumor cells. Lab Chip 11:3449–3457

Huang Y, Tang F, Liang X, Chen G, Xiao H, Azarmi F (2015) Steel bar corrosion monitoring with long-period fiber grating sensors coated with nano iron/silica particles and polyurethane. Struct Heal Monitor 14:178–189

Jang YH, Jang YJ, Kim S, Quan LN, Chung K, Kim DH (2016) Plasmonic solar cells: from rational design to mechanism overview. Chem Rev 116:14982–15034

Joarder B, Yanai N, Kimizuka N (2018) Solid-state photon upconversion materials: structural integrity and triplet-singlet dual energy migration. J Phys Chem Lett 9:4613–4624

Kabashin AV, Evans P, Pastkovsky S, Hendren W, Wurtz GA, Atkinson R, Pollard R, Podolskiy VA, Zayats AV (2009) Plasmonic nanorod metamaterials for biosensing. Nat Mater 8:867–871

Keshea BA, Khakpoor AA (2017) Opto-electronic properties of nano-electronic materials. Mater Res 20:1248–1253

Kodaira S, Korposh S, Lee SW, Batty WJ, James SW, Tatam RP (2018) Fabrication of highly efficient fibre-optic gas sensors using SiO_2/polymer nanoporous thin films. In: Proceedings of the 3rd international conference on sensing technology (ICST'08), pp 481–485

Kojima A, Teshima K, Shirai Y, Miyasaka T (2009) Organometal halide perovskites as visible-light sensitizers for photovoltaic cells. J Am Chem Soc 131:6050–6051

Kroto HW, Heath JR, O'Brien SC, Curl RF, Smalley RE (1985) C60: Buckminsterfullerene. Nature 318:162–163

Kwiatkowski M, Bezverkhyy I, Skompska M (2015) ZnO nanorods covered with a TiO_2 layer: simple solegel preparation, and optical, photocatalytic and photoelectrochemical properties. J Mater Chem A 3:12748–12760

Lara S, Perez-Potti A (2018) Applications of nanomaterials for immunosensing. Biosensors 8:104-1-21

Law WC, Yong KT, Baev A, Prasad PN (2011) Sensitivity improved surface plasmon resonance biosensor for cancer biomarker detection based on plasmonic enhancement. ACS Nano 5:4858–4864

Lee EG, Park KM, Jeong JY, Lee SH, Baek JE, Lee HW, Jung JK, Chung BH (2010) Carbon nanotube-assisted enhancement of surface plasmon resonance signal. Anal Bio Chem 408:206–211

Levinson R, Berdahl P, Akbari H (2005) Solar spectral optical properties of pigments-Part I: Model for deriving scattering and absorption coefficients from transmittance and reflectance measurements. Sol Energy Mater Sol C 89:319–349

Lewis GN, Lipkin D, Magel TT (1941) Reversible photochemical processes in rigid media. A study of the phosphorescent state. J Am Chem Soc 63:3005–3018

Li X, Liu X (2017) Group III nitride nanomaterials for biosensing. Nanoscale 9:7320–7341

Li L, Yang Y, Fan R, Jiang Y, Wei L, Shi Y, Yu J, Chen S, Wang P, Yang B (2014) A simple modification of near-infrared photon-to-electron response with fluorescence resonance energy transfer for dye-sensitized solar cells. J Power Sources 264:254–261
Lobnik A, Baldini F (ed) (2006) Absorption-based sensors. Opt Chem Sens 5, 224:77–98
Luan QF, Zhou KB, Tan HN, Yang D, Yao X (2011) Au-NPs enhanced SPR biosensor based on hairpin DNA without the effect of nonspecific adsorption. Biosens Bioelectron 26:2473–2477
Luca R (2010) Schiff base metal complexes for second order nonlinear optics. Chim Ind 3:118–122
Makrides SC, Gasbarro C, Bello JM (2005) Bioconjugation of quantum dot luminescent probes for western blot analysis. BioTechn 39:501–505
Marazuela MD, Moreno-Bondi MC (2002) Fiber-optic biosensors-an overview. Anal Bioanal Chem 372:664–682
Matharu Z, Bandodkar AJ, Sumana G, Solanki PR, Ekanayake EM, Kaneto K (2009) Low density lipoprotein detection based on antibody immobilized selfassembled monolayer: investigations of kinetic and thermodynamic properties. J Phys Chem B 113:14405–14412
McDonald SA, Konstantatos G, Zhang S, Cyr PW, Klem EJD, Levina L, Sargent EH (2005) Solution-processed PbS quantum dot infrared photodetectors and photovoltaics. Nat Mater 4:138–142
McGraw-Hill Encyclopedia of Science and Technology, 5th edn. McGraw-Hill (1993)
Nature Energy (2019) A decade of perovskite photovoltaics. Nat Energy 4:1
Nehra A, Singh KP (2015) Current trends in nanomaterial embedded field effect transistor-based biosensor. Biosens Bioelectron 74:731–743
O'Regan B, Grätzel MA (1991) Low-cost, high-efficiency solar cell based on dye-sensitized. Nature 353:737–740
Okumura K, Mase K, Yanai N, Kimizuka N (2016) Employing core-shell quantum dots as triplet sensitizers for photon upconversion. Chem Eur J 22:7721–7726
Palese S, Schilling L, Miller RJD, Staver PR, Lotshaw WT (1994) Femtosecond optical Kerr effect studies of water. J Phys Chem 98:6308–6316
Pan Z, Rao H, Mora-Seró I, Bisquert J, Zhong X (2018) Quantum dot-sensitized solar cells. Chem Soc Rev 47:7659–7702
Pittman TB, Jacobs BC, Franson JD (2004) Quantum computing using linear optics. Appl Tech Dig 25:84–90 Johns Hopkins
Press Release: the 1996 Nobel Prize in Physics. Nobelprize.org. 15 Oct 1986
Renewables 2018. Solar energy, International Energy Agency
Rühle S (2016) Tabulated values of the Shockley-Queisser limit for single junction solar cells. Sol Energ 130:139–147
Sabban S (2011) Development of an in vitro model system for studying the interaction of *Equus caballus* IgE with its high affinity FcεRI receptor. University of Sheffield, Ph.D thesis, pp 91. http://etheses.whiterose.ac.uk/2040/2/Sabban,_Sari.pdf
Sanderson K (2006-11-15) Sharpest cut from nanotube sword. Nat News
Saule Technologies (2018) Saule Technologies and Skanska change construction industry. https://sauletech.com/2018/12/13/sauletechnologies-and-skanska-change-construction-industry
Shalav A, Richards BS, Trupke T (2005) Application of $NaYF_4:Er^{3+}$ up-converting phosphors for enhanced near-infrared silicon solar cell response. Appl Phys Lett 86:013505
Shang Y, Hao S, Yang C, Chen G (2015) Enhancing solar cell efficiency using photon upconversion materials. Nanomaterials 5:1782–1809
Shao W, Chen G, Ohulchanskyy TY, Kuzmin A, Damasco J, Qiu H, Yang C, Hans Å, Prasad PN (2015) Lanthanide-doped fluoride core/multishell nanoparticles for broad-band upconversion of infrared light. Adv Opt Mater 3:575–582
Shi J, Tian F, Lyu J, Yang M (2015a) Nanoparticle based fluorescence resonance energy transfer (FRET) for biosensing applications. J Mater Chem B 3:6989–7005
Shi J, Chan C, Pang Y, Ye WW, Tian F, Lyu J, Zhang Y, Yang M (2015b) A fluorescence resonance energy transfer (FRET) biosensor based on graphene quantum dots (GQDs) and gold nanoparticles (AuNPs) for the detection of mecA gene sequence of Staphylococcus aureus. Biosens Bioelectron 67:595–600

Shockley W, Queisser HJ (1961) Detailed balance limit of efficiency of p-n junction solar cells. J Appl Phys 32:510–519

Suresh S, Arivuoli D (2012) Nanomaterials for nonlinear optical (NLO) applications: a review. Rev Adv Mater Sci 30:243–253

Takei K, Fang H, Kumar SB, Kapadia R, Gao Q, Madsen M, Kim HS, Liu CH, Chueh YL, Plis e, Krishna S, Bechtel HA, Guo J, Javey A (2011) Quantum confinement effects in nanoscale-thickness InAs membranes. Nano Lett 11:5008–5012

Tiwari JN, Tiwari RN, Kim KS (2012) Zero-dimensional, one-dimensional, two-dimensional and three-dimensional nanostructured materials for advanced electrochemical energy devices. Prog Mater Sci 57:724–803

Trupke T, Green MA (2002) Improving solar cell efficiencies by down-conversion of high-energy photons. J Appl Phys 92:1668

Tseng YT, Chuang YJ, Wu YC, Yang CS, Wang MC, Tseng FG (2008) A gold- nanoparticle-enhanced immune sensor based on fiber optic interferometry. Nanotech 19:345501

Turner MD, Saba M, Zhang Q, Cumming BP, Schröder-Turk GE, Gu M (2013) Miniature chiral beamsplitter based on gyroid photonic crystals. Nat Photonics 7:801–805

Urrutia A, Goicoechea J, Arregui FJ (2015) Optical fiber sensors based on nanoparticle-embedded coatings. J Sens, 1–18

Vachali PP, Li B, Bartschi A, Bernstein PS (2015) Surface plasmon resonance (spr)-based biosensor technology for the quantitative characterization of protein-carotenoid interactions. Arch Biochem Biophys 15:66–72

Wang B (2019) First commercial perovskite solar late in 2019 and the road to moving the energy needle. Next Big Future, Saule Tech

Wang JL, Munir A, Zhou HS (2009) Au NPs-aptamer conjugates as a powerful competitive reagent for ultrasensitive detection of small molecules by surface plasmon resonance spectroscopy. Talanta 79:72–76

Wang Z, Huang P, Bhirde A, Jin A, Ma Y, Niu G, Neamati N, Chen X (2012) A nanoscale graphene oxide–peptide biosensor for real-time specific biomarker detection on the cell surface. Chem Commun 48:9768–9770

Wen J, Xu Y, Li H, Lu A, Sun S (2015) Recent applications of carbon nanomaterials in fluorescence biosensing and bioimaging. Chem Commun 51:11346–11358

Williams R (1960) Becquerel photovoltaic effect in binary compounds. J Chem Phys 32:1505–1514

Wink T, van Zuilen SJ, Bult A, van Bennekom WP (1998) Liposome-mediated enhancement of the sensitivity in immunoassays of proteins and peptides in surface plasmon resonance spectrometry. Anal Chem 70:827–832

Wu M, Congreve DN, Wilson MWB, Jean J, Geva N, Welborn M, Voorhis TV, Bulović V, Bawendi MG, Baldo MA (2016) Solid-state infrared-to-visible upconversion sensitized by colloidal nanocrystals. Nat Photon 10:31–34

Wu K, Guo C, Wang H, Zhang X, Wang J, Chen J (2017) All-optical phase shifter and switch near 1550 nm using tungsten disulfide (WS_2) deposited tapered fiber. Opt Exp 25:17639–17649 OSA Publishing

Ye WW, Tsang MK, Liu X, Yang M, Hao J (2014) Upconversion luminescence resonance energy transfer (LRET)-based biosensor for rapid and ultrasensitive detection of avian influenza virus H7 subtype. Small 10:2390–2397

Ye M, Gao X, Hong X, Liu Q, He C, Lin XLC (2017) Recent advances in quantum dot-sensitized solar cells: insights into photoanodes, sensitizers, electrolytes and counter electrodes. Sust Energ Fuels 1:1217–1231

Yin MJ, Wu C, Shao LY, Chan WKE, Zhang AP, Lu C, Tam HY (2013) Label-free, disposable fiber-optic biosensors for DNA hybridization detection. Analyst 138:1988–1994

Yuan CZ, Chen GY, Prasad PN, Ohulchanskyy TY, Ning ZJ, Tian H, Sund LC, Agren H (2012) Use of colloidal upconversion nanocrystals for energy relay solar cell light harvesting in the near-infrared region. J Mater Chem 22:16709–16713

Zadran S, Standley S, Wong K, Otiniano E, Amighi A, Baudry M (2012) Fluorescence resonance energy transfer (FRET)-based biosensors: visualizing cellular dynamics and bioenergetics. Appl Microbiol Biotechnol 96:895–902

Zeng SW, Yong KT, Roy I, Dinh XQ, Yu X, Luan F (2011) A review on functionalized gold nanoparticlesfor biosensing applications. Plasmonics 6:491–506

Zeng SW, Yu XW, Law CY, Hu ZR, Dinh XQ, Ho HP, Yong KT (2013) Size dependence of Au NP-enhanced surface plasmon resonance based on differential phase measurement. Sens Actuators, B 176:1128–1133

Zeng S, Dominique B, Ho-Pui H, Ken-Tye Y (2014) Nanomaterials enhanced surface plasmon resonance for biological and chemical sensing applications. Chem Soc Rev 43:3426–3452

Zeng S, Hu S, Xia J, Anderson T, Dinh XQ, Meng XM, Coquet P, Yong KT (2015) Graphene-MoS_2 hybrid nanostructures enhanced surface plasmon resonance biosensors. Sens Actuat B Chem 207:801–810

Zhang Y, Wang Y (2017) Nonlinear optical properties of metal nanoparticles: a review. RSC Adv 7:45129–45144

Zhang CY, Yeh HC, Kuroki MT, Wang TH (2005) Single-quantum-dot-based DNA nanosensor. Nat Mater 4:826–831

Zhao H, Chang Y, Liu M, Gao S, Yu H, Quan X (2013) A universal immunosensing strategy based on regulation of the interaction between graphene and graphene quantum dots. Chem Commun 49:234–236

Zheng X, Zhang L (2016) Photonic nanostructures for solar energy conversion. Energ Environ Sci 9:2511–2532

Zhong Q, Fourkas JT (2008) Optical Kerr effect spectroscopy of simple liquids. J Phys Chem B 112:15529–15539

Zibaii MI, Latifi H, Saeedian Z, Chenari Z (2014) Nonadiabatic tapered optical fiber sensor for measurement of antimicrobial activity of silver nanoparticles against *Escherichia coli*. J Photochem Photobiol B: Bio 135:55–64

Zorman B, Ramakrishna MV, Friesner RA (1995) Quantum confinement effects in CdSe quantum dots. J Phys Chem 99:7649–7653

Functional Nanomaterials for Smart Healthcare Applications

Sadiya Anjum and Rashid Ilmi

Abstract Bacterial infections have become a major concern worldwide and is the center of genuine research efforts towards development of the functional system against microbial attachment. Sometimes infection associated issues become life-threatening for the patient due to bacterial growth within a biofilm matrix. These bacterial accumulation leads to inefficient use of conventional antibiotics and this fact emphasizes the development of new treatment strategies for the infection control. Various class of bioactive agents such as metal nanoparticles (NPs), antibiotics, herbal components and essential oils are available for this wide antimicrobial spectrum against all type of microbes. Despite the fact, each category of the bioactive agents has limitations such as lack of binding ability of metal NPs, bacterial resistant, low bioavailability and hydrophobicity of the agents. However, the most crucial challenge in front of researchers is the modification of bioactive agents and reduces their toxicity while using as biomaterials. Designing bioactive agents within nanomaterial confinement provides broad prospects to develop functional NPs which may be bonded to any biomaterial surface through nanogel functionality, either coated or blended with polymeric substrates/surfaces and make them antimicrobial. The aim of this book chapter is to present the current perspective, approaches and challenges on the design of functional and bioactive nanomaterials with diverse functionality, their bioactivity against a wide spectrum of microbes so that it leads to the development of bioactive materials for human healthcare.

Keywords Nanoparticles · Nanogels · Healthcare

S. Anjum (✉)
Bioengineering Laboratory, Department of Textile Technology,
Indian Institute of Technology, New Delhi 110016, India
e-mail: sadiya2203@gmail.com

R. Ilmi (✉)
Department of Chemistry, Sultan Qaboos University,
P.O. Box 36, Al Khod 123, Muscat, Oman
e-mail: rashidilmi@gmail.com

A. H. Bhat et al. (eds.), *Nanomaterials for Healthcare, Energy and Environment*,
Advanced Structured Materials 118, https://doi.org/10.1007/978-981-13-9833-9_2

1 Introduction

Bacterial invasion and multiple growths on healthcare materials is a serious issue to the patient's health and creates an economic burden on the society. The issue becomes more critical after bacterial adherence which leads biofilm formation where secretion as well as accumulation of extracellular products, takes place and make them less susceptible to regular antibiotic therapy because of low diffusion of antimicrobial agents into biofilm (Fig. 1) (Zaman et al. 2017; Ahonen et al. 2017). The bacterial infection damages organ functionality and therefore could be fatal for patients. Especially implant and wound care infections diagnosed at the last stage where in some case recovery is almost impossible. These dominant factors in the treatment of biomaterial-associated infections make them more difficult and challenging. Continuous efforts are in process for the development of functional bioactive agents that can act against bacterial infection and prevents biofilm formation on biomaterial surface (Anjum and Gupta 2018; Anjum et al. 2018; Tummalapalli et al. 2016a; Liu and An 2014).

To limit these infections, smart planning is needed with the help of material science, bioscience and nanoscience together (Anjum and Gupta 2018). A variety of bioactive agents are available in the literature for this purpose, however, their high cytotoxicity and low bioavailability restrict their use (Suresh et al. 2012). Therefore, nanotechnology has an emerging alternative in the healthcare sector due to their advance and fascinating features but several NPs may exhibit toxic biological effects. The challenges in the application of NPs could potentially be overcome by the designing of the *nano-within-nano* system. In which nano size bioactive agents are capped with nanodimension gels. These bioactive nanogels consist diverse structural and excellent antimicrobial properties. Moreover, it could easily penetrate into the cell wall membranes of pathogens and restrict their activity in a precise manner

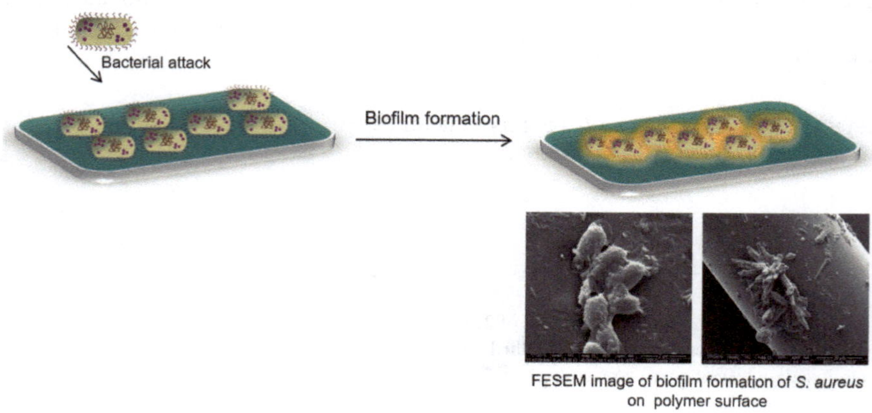

Fig. 1 Representation of bacterial growth on material surface. Reproduced with permission from Wiley (Anjum and Gupta 2018)

without direct contact to the tissue (Abed and Couvreur 2014). In recent days, bioactive nanogels are known as optimal candidates for defeating pathogens compared to the conventional bioactive agents due to their excellent bioavailability, controlled size and broad-spectrum of antimicrobial characteristics.

A wide class of bioactive agents such as antibiotics, essential oils, herbals, and metal NPs within nanogels are available for the preparation of infection resistant materials and show their effectiveness against wide range of microbes (Gayathri et al. 2017; Huang et al. 2017; Ashrafi et al. 2019). Several approaches for the preparation of bioactive nanogels such as nanoemulsion, nanoprecipitation, salt reduction, ionic gelation and radiation process have been developed (Thoniyot et al. 2015; Jin et al. 2015; Anjum et al. 2015; Li et al. 2008). Different approaches for the development of bioactive nanogels influences or controlled the morphology of NPs. For instance, nanoemulsion process leads to the formation of core-shell structure while another method supports the formation of composite structures where bioactive agents are encapsulated or stacked in the nanogels (Anjum et al. 2016). These nanogels acquired different architecture with variation in size and the resulting nanogels may be immobilized, coated, embedded and blend with the inherent material so that infection-resistant features have been designed (Fig. 2). In this book chapter, we will discuss the recent advancement in the bioengineering of functional bioactive nanogels for the preparation of infection resistant materials. Various type of bioactive agents encapsulated nanogels have been classified and discussed with suitable examples in later sections.

2 Nanoengineering of Bioactive Agents: *State of the Art*

The term nanotechnology defines engineering of materials in nanosize with diameters <100 nm and thus could be applied in healthcare applications such as infection control, wound healing, medical imaging, drug delivery, cancer treatment, and other areas have progressed rapidly (Coll Ferrer et al. 2014; Stular et al. 2017; Moya-Ortega et al. 2012; Anjum et al. 2017a; Ramos et al. 2017). Generally, metal NPs are inert in nature compared to their respective metal salts and can be directly impregnated on a material substrate where no active sites are present for the attachment of the metal NPs. However, the fast leaching of metal NPs was observed that leads cytotoxicity to the surroundings. Whereas, if the metal NPs are embedded within the hydrogel matrix by developing a nanocomposite structure which restricts the release of NPs and shows low ability to prevent bacterial growth (Liu et al. 2010; Ravindran et al. 2013). Another example of bioactive agents is essential oils and curcumin, both have low bioavailability due to poor solubility in water medium (Huang et al. 2017; Beyki et al. 2014). On this account, the main obstacle lies in developing functional nanomaterial that would easily attach to the biomaterial substrate.

These bioactive agents incorporated in nanosized gel using different approaches provide hydrogel functionality over the shell of the NPs, which generate functionality and get easily attached to the substrates. In this innovative alternate, where bio-

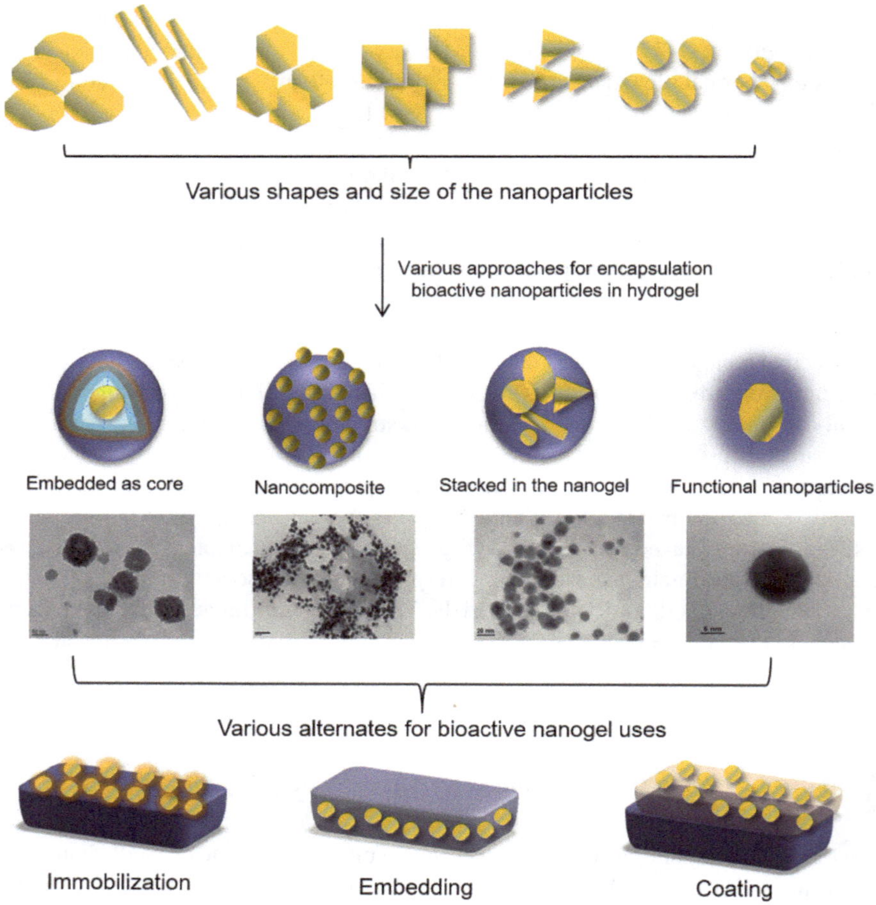

Fig. 2 Different architecture of bioactive nanogel and various alternates for their uses. Reproduced with permission from Wiley (Anjum and Gupta 2018)

compatible hydrogel sheath stays in contact with tissues and minimizes the adverse cytotoxic effect of direct contact of the conventional NPs (Jin et al. 2015; de Lima et al. 2012). These nanogels particles consist advanced features with a smart hydrogel nature. A combination of the biofunctionality and biocompatibility makes nanogel as an interesting material, thus allowing vast possibilities of designing and engineering it for healthcare materials. The state of the art in nano-engineering of gels hypothesizes that the functional groups would provide anchoring sites either for the immobilization of biomolecules or for interaction with the material surface by different bonding (covalent, ionic, coordinate etc.) while keeping safe its biocompatibility with surrounding tissue (Anjum and Gupta 2018).

3 Designing of Functional Nanogels

The crucial and important features of nanogel development are the size and chemical functionality without altering its non-toxicity and bioactivity. A plethora of functional groups e.g. –COOH, –$CONH_2$, –OH, –NH_2 and –SO_3H may be crafted on their surface by utilizing suitable monomers or polymers. A range of approaches has been employed for the development of bioactive nanogels, already discussed in the introduction section. Nanoemulsion method has exhibited great potential towards the nanogel development in which monomer/polymers were added in the water phase and subsequently mixed in the oil phase. The bioactive moiety is added in either water or oil phase depending on its solubility. This approach is favorable for the development of functional nanomaterials where bioactive moieties are embedded in nanohydrogel (Shin et al. 2018; Anjum et al. 2017b). Nanoprecipitation is a simplest method for the preparation of polymeric NPs containing bioactive agents that require two miscible solvents (Zhang et al. 2015). Both polymer and bioactive agents must dissolve in the first system (solvent), but not in the second (non-solvent). Further, the polymer solution with a bioactive agent is added to the non-solvent a rapid de-solvation of the polymer renders the precipitation of NPs. Nanoprecipitation is an easy approach which requires less energy and widely applicable without the addition of any additives for NPs designing. However, the presence of a stabilizer in the nanoprecipitation method is very important that inhibit the aggregation of the NPs and impart stability. Another approach for the preparation of NPs is ionotropic gelation and is frequently used for the synthesis of chitosan (CS) and alginate NPs by the addition of tri-polyphosphate salts as the anionic oligomeric partner, which leads the formation of colloidal NPs in the range of 100–2000 nm (Beyki et al. 2014; Zhaveh et al. 2015).

Chemical reduction method is specifically employed for the preparation of silver (Ag) or gold encapsulated nanogels by using two different approaches; (i) in situ and (ii) ex situ growth of NPs within gels (Anjum et al. 2015; Tummalapalli et al. 2015). In the former approach, both reducing and stabilizing agents work concurrently to form core-shell structure of the bioactive nanogels. While in later method Ag NPs were prepared first and then embedded into gels to form nanocomposite structure. Another smart approach is the radiation process as compared to the above-mentioned approaches since it is a chemical free process which minimizes the toxicity of bioactive nanogels (Anjum et al. 2017b; Gupta et al. 2013). Radiation initiates multiple process i.e. monomer polymerization in solution and later crosslinking of polymeric chains. Advancement in the nanogel preparation is the use of polymers instead of monomers thus will avoid the toxicity of monomer. A range of different polymers such as polyvinyl alcohol (PVA), polyvinylpyrrolidone (PVP), polyethylene glycol (PEG), CS, pectin, dextran, gelatin, alginate and cellulose are readily available for this purpose (Anjum et al. 2015; 2016, 2017a, 2017c; Tummalapalli et al. 2016b; Malzahn et al. 2014). This is an interesting approach to prepare nanogels by preselecting a polymer and hence proper control over nanogel features could be accomplished. Using these approaches different functional nanogels have been designed by encapsulating different bioactive agents for the development of healthcare products. They are classified on the basis of encapsulated bioactive agents i.e. antibiotics, nanosilver, essential oils and others, their detailed examples are discussed in the later section of the book chapter.

3.1 Antibiotic Loaded Nanogels

Antibiotics are bioactive agents that restrict bacterial growth by inhibiting bacterial functions. There are many types of antibiotics available for the treatment of bacterial infection ranging from ointments to common pills to expensive intravenous drugs for example methicillin resistant strains for *Staphylococcus aureus* (MRSA). But the major issues concern with the use of antibiotics is resistant with bacteria and restrict their functioning with low stability (Zaidi et al. 2017). Therefore, various antibiotic loaded nanogels were designed and prepared to improve bacterial resistant property (Grennhalgh et al. 2018). Different antibiotics can be either physically entrapped or encapsulated in nanogels or covalently bonded to polymeric chain. Thermoresponsive poly(*N*-isopropylacrylamide) (PNIPAAm) nanogels were prepared through emulsion polymerization for encapsulation of levofloxacin. These nanogels were added in methacrylamide CS solution to initiate free-radical crosslinking reaction of methacrylamide which yielded levofloxacin impregnated hydrogels (Shin et al. 2018). The obtained nanogel showed a maximum 27% of levofloxacin release at below lower critical temperature (LCST), whereas at higher LCST value 66% of the total drug was released.

Poly(methacrylic acid) (PMAA) nanogel was designed and prepared for loading and controlled release of metronidazole (MTZ). The MTZ/PMAA nanogel shows sustained release of MTZ in gastrointestinal medium and could be used for the prevention of mild-to-moderate Clostridium difficile infections (Chen et al. 2014). In a further study, PMAA nanogel was developed for the delivery of isoniazid (INH) and rifampin (RMP) in the digestive tract by minimizing the cytotoxicity of drugs (Chen et al. 2016). These drug loaded nanogel showed long-term antibacterial activity against multidrug-resistant bacteria in the intestine. A triple layer nanogels were designed and prepare for the effective treatment of extracellular and intracellular bacterial infections that can trigger lipase-secreting bacteria (Xiong et al. 2012). The outer shell of the nanogel consist hydrophilic PEG layer which ensures the dispersity of nanogel in aqueous medium while the inner core is composed of crosslinked polyphosphoester for the reservoir of vancomycin.

Antimicrobial nanogels were prepared by the complexation of chlorhexidine (CHX) with aqueous poly(methyl methacrylate) (PMMA) at varying content of cyclodextrin methacrylate (CD-MA) and could be used in dental applications (Kettel et al. 2011). The designing of these nanogels involves the ex situ crosslinking of the nanogel by CHX with two different β-CD-MA cones of two different nanogels. The active nanogels showed affective antimicrobial property against *Staphylococcus aureus (S. aureus)* and high effectiveness against *S. aureus* growth at low CHX concentrations 30 μg/225 cm^2. Novel surface functionalized nanocarrier for CHX loading was reported in which cationic polyelectrolyte poly(diallyldimethylammonium chloride) was used for surface functionalization of lightly-cross-linked acrylate copolymer (Carbopol) nanogel (Al-Awady et al. 2018). The antimicrobial effect of the surface functionalized nanogel carriers were investigated against *Escherichia coli (E. coli)*, *S. aureus*, *Saccharomyces cerevisiae (S. cerevisiae)* and *Chlamydomonas reinhardtii (C. reinhardtii)* and much higher activity than that of free CHX. It can be seen that the CHX-loaded nanogel reduced the microalgae viability from 90 to 0% after 1 h of incubation. SEM images of microbial cells show the influence of nanogel at different incubation period (Fig. 3). Figure 3a displays the morphology of the control sample of untreated microalgae (*C. reinhardtii*) and *E. coli* while after 1 h of incubation, cationic nanogel particles have accumulated on the cell wall of the treated *C. reinhardtii* and *E. coli*. The high magnification SEM image visualizes the hetero-coagulation of the CHX-loaded nanogel on the cells. The cluster formation and aggregation of nanogels imparts interaction between the cationic surface functionalized nanogel particles with anionic surface of the bacterial cells.

Recently, a magnetic antimicrobial nanogels were designed and prepared by solvent casting method in which CS and gelatin gels were used for the impregnation of

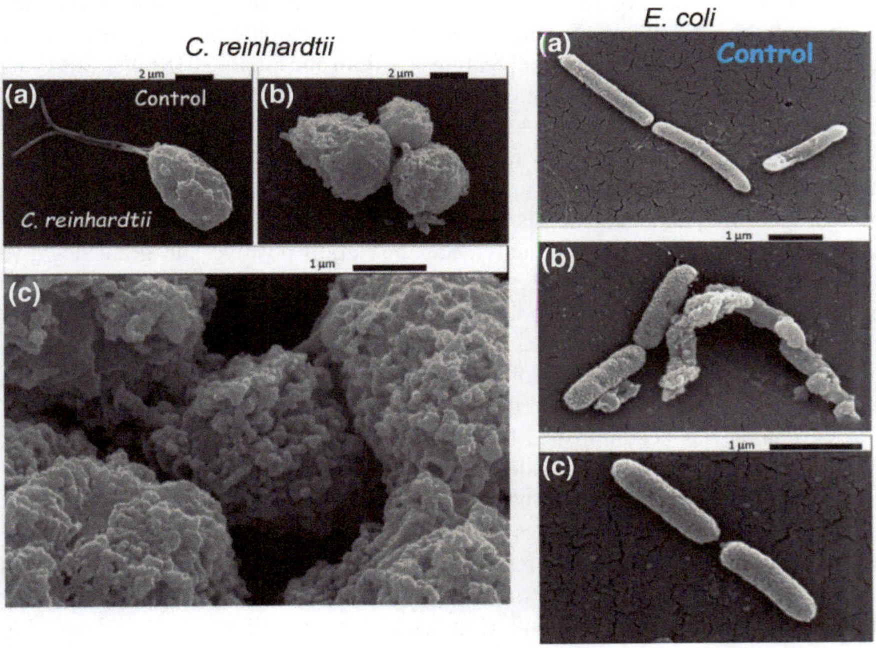

Fig. 3 SEM images of *C. reinhardtii* and *E. coli* cells (**a**) control sample of untreated *C. reinhardtii* and *E. coli* (**b**) and (**c**) samples of *C. reinhardtii* and *E. coli* after incubation in a suspension of cationic surface functionalized CHX-loaded nanogel. Reproduced with permission from RSC (Al-Awady et al. 2018)

CHX and dispersion of cobalt Iron oxide NPs using probe sonicator. The nanogels showed antibacterial property due to the presence of positively charged CHX that attacks negatively charged microbial species by disrupting their cell membrane activity (Khan et al. 2019). Still, the application of antibiotics as infection resistant materials are restricted due to its undefined mode of action against different microbes. Whereas the major concern is the resistant behavior of the antibiotics that restricts its activity against different strains. Therefore, these studies are limited and not elaborated or extended up to commercial level.

3.2 Nanosilver Loaded Nanogels

Nanosilver (nano-Ag) is a safe inorganic bioactive moiety and effective against a broad spectrum of microbes with minimum chances of bacterial resistance since it targets multiple components in bacterial cells and inhibits their activity. Nano-Ag made a significant contribution to the development of infection resistant material in comparison to other bioactive agents (Jin et al. 2017). However, the key challenge for the scientists is the limited binding ability of nano-Ag at substrate that allows the fast leaching of nano-Ag and leads toxicity to the surrounding system. Designing functional nano-Ag offers enormous potential by developing the excellent binding ability to any substrate via hydrogel features (Anjum and Gupta 2018). Various functionalized nano-Ag nanogel for infection resistant materials have been designed and developed. The synthesis of PMAA and polyacrylamide nanogels embedded with nano-Ag as core was prepared by radiation process where well dispersed core-shell nanoparticles were formed (Anjum et al. 2017b; Gupta et al. 2013). The presence of hydrogel shell, nanogel exhibits great binding ability with minimum toxicity and showed excellent antimicrobial activity against gram positive and gram negative strains. Later, these functional NPs used in combination with essential oils and coated on the air plasma functionalized polypropylene fabric for the development of antimicrobial surfaces for skin contacting materials (Anjum et al. 2017c). Alternatively, a two-step process in combination of chemical polymerization of methacrylic acid in a *water-in-oil* nanoemulsion and the reduction of silver ions with crosslinking of PMAA chains was accomplished by using radiation sources (Anjum et al. 2016). PMAA nanogels were mixed as additive with biological moieties like *Aloe vera* and curcumin which was coated on the polyester (PET) fabric for designing of herbal composite dressings. The dressings show excellent antimicrobial activity and wound healing features. Wound healing results showed interesting effects of *Aloe vera* and curcumin, in which nanogel in combination with *Aloe vera* exhibited better activity than its combination with curcumin (Fig. 4). This was attributed to the compatible chemical composition of *Aloe vera* that favored healing process.

PEG is an interesting example of a polymer that can be used for dual action in the designing of bioactive nanogel. PEG-nano-Ag particles was prepared by the growth of Ag NPs from ionic Ag and at the same time PEG also stabilized the growth of nano-Ag particles by providing a hydrogel sheath. Later, bioactive PEG- nano-Ag

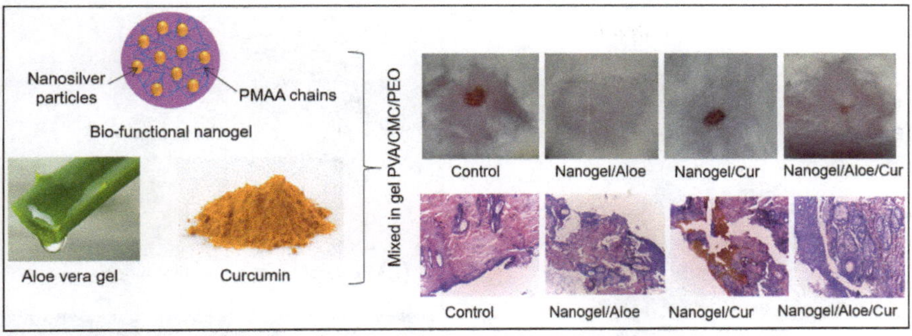

Fig. 4 Wound healing studies of bio-functionalized nanogels with *Aloe vera* and curcumin based wound dressings. Reproduced with permission from Elsevier (Anjum et al. 2016)

nanogels were immobilized by ionic interaction onto plasma functionalized PET filament surface for the development of antimicrobial suture (Anjum et al. 2017a). These antimicrobial sutures showed excellent bacteriostatic and bactericidal property against *E. coli* and *S. aureus* by inhibiting the growth of bacteria at the suture surface. These sutures showed rapid wound healing without any inflammatory response on the newly generated skin. Riham and Magdy investigated *in situ* growth of nano-Ag within epichlorohydrin crosslinked carboxymethyl CS (CMCh–Ag) nanogels (Mohamed and Sabaa 2014). The prepared gel plays dual functionality by reducing Ag ions and also stabilizes the growth of Ag NPs. The nanogels showed broad spectrum of activity against different microbes. Results concluded that the nanogel has higher antibacterial effect ($\approx$3 to 4 folds higher) toward gram-positive and gram-negative bacteria. Similarly, Ferrer and co-workers reported the development of dextran and lysozymes nanogels where nano-Ag particles were decorated as core (Coll Ferrer et al. 2014). The presence of lysozymes helped in the growth and stabilization of Ag NPs within the nano structured gel. Cryo-TEM images depict that the maximum NPs are spherical in shape with particle size in between 2 and 5 nm. These nanogels exhibited size dependent properties due to the large surface area as smaller NPs were more effective because they easily penetrate into the bacterial cell wall and restricts their activities. Whereas larger NPs shows comparatively low antimicrobial activity, their size is not allowed to penetrate the bacterial cell wall.

Recently in situ method to prepare Ag-NPs using PVA as a stabilizing gel matrix and fructose as reducing agent was reported by one of us. The synergistic effect of the bioactive nanogel with glycerol offers enhanced antimicrobial property and fast moist healing in *Swiss Albino mice* model (Fig. 5). The presence of glycerol acts as a carrier of nanogels which provides a suitable path to nano-Ag to show their activity and promoted rapid wound healing within 21 d of healing period (Anjum et al. 2015). A novel polymeric micelle was constructed where nano-Ag particles were decorated in polyaspartic acid chains in the shell. While with the help of hydrophobic interaction in the micelle curcumin was embedded into poly-Ɛ caprolactone core. Results

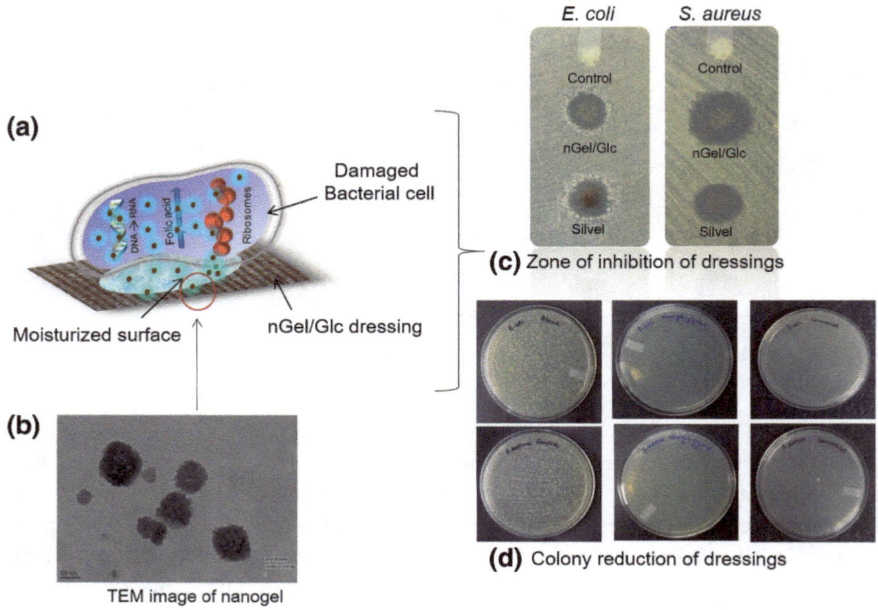

Fig. 5 **a** Schematic representation of antimicrobial action of nGel/Glc dressing and Silvel (commercial dressing), **b** TEM image of nanogel, **c** Zone inhibition analysis and **d** colony reduction analysis against *E. coli* and *S. aureus* of dressing. Reproduced with permission from Wiley (Anjum and Gupta 2018)

concluded that the synergistic effect of the nano-Ag and curcumin, nanogel exhibits improved antibacterial property against *S. aureus* and *Pseudomonas aeruginosa.*

Nano-Ag embedded PNIPAm based nanogels with methacrylic acid has been investigated for the designing of thermoresponsive fabric (Zafar et al. 2014). SEM analysis of the nanogels demonstrated the spherical morphology with the dimensions in the range of 180–200 nm. The coated fabric was effective against *Staphylococcus epidermidis (S. epidermis)* and *E. coli.* Thus it could find a potential application in textile industry as antimicrobial surgeon's care and patient drappings. A similar approach was used to prepare hybrid core-shell structure nano-Ag surrounded by NIPAm-(acrylic acid) AA copolymer shell (Wu et al. 2010). The nanogel exhibited controlled size 40–80 nm and was used in tumor cell imaging and local delivery since nanogel was capable of crosslinking the cellular barrier. As a result, it offers both optical and therapeutic functionality. Photoresponsive antimicrobial nanogel of aniline functionalized CS was prepared using complex-reduction method that stimulated the release of the silver ions in the surrounding medium. The nano-Ag particles were chelated to aniline–CS nanogel configuration. The prepared nanogel showed excellent antimicrobial behavior due to the incorporation of the Ag NPs between different lipid molecules in the bacterial cell membrane (Ballesteros et al. 2019). Microwave-assisted *in situ* growth of nano-Ag particles using sodium citrate

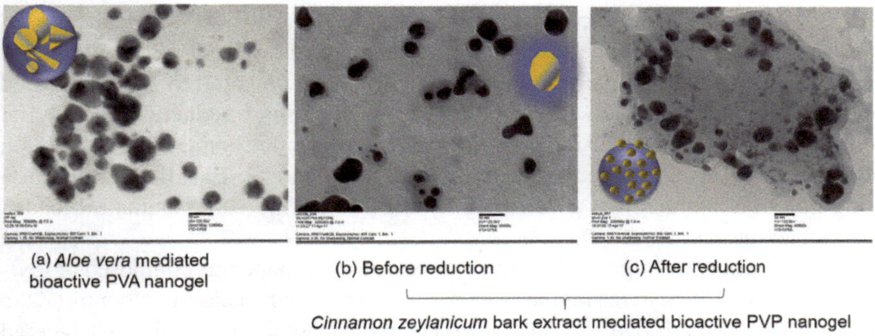

Fig. 6 TEM images of herbal extract mediated bio-reduction of nanosilver (**a**) *Aloe vera* gel (**b**) and (**c**) *Cinnamon zeylanicum* bark extract. Reproduced with permission from Wiley (Anjum and Gupta 2018)

within polyacrylic acid (PAA) PAA/PEG hydrogel was investigated (Tyliszczak and Pielichowski 2013). In this study, nano-Ag particles were prepared by the reduction of silver ions by sodium citrate salts. These NPs were stabilized by carboxylate anions of PAA chains and co-stabilized by PEG chains.

In different perspective of nano-Ag within nanogel designing different herbal extracts such as *Aloe vera* gel and *Cinnamon zeylanicum bark* extract was used for the bio-reduction of the silver ions in presence of PVA and PVP hydrogels that stabilize the growth of NPs (Fig. 6a). The influence of the pre and post addition of hydrogels matrix on the growth of NPs was analyzed. It was observed that the pre and post-condition addition of hydrogel resulted in core-shell structure in the case when the hydrogel was added in pre-condition. While the composite structure of observed when it was added in post-condition as shown in the TEM images (Fig. 6b). In this suggestion, the innovative advancement in the designing of functional nano-Ag was depicted and supports the eco-friendly approach to design nano-system prevail in future (Anjum and Gupta 2018).

3.3 *Essential Oil Loaded Nanogels*

Several recent studies have demonstrated that essential oils encapsulated nanogels have significant antimicrobial potential against multidrug-resistant pathogens. This is due to an increase in chemical stability and solubility that decreases rapid evaporation of active essential oil components (Ashrafi et al. 2019; Beyki et al. 2014; Zhaveh et al. 2015). The application of encapsulated essential oils within the nanogel also supports their controlled and sustained release thus enhances their bioavailability and efficacy against a wide class of pathogens. Different essential oils i.e. *Mentha piperita* and *Cuminum cyminum* were encapsulated in the functionalized CS nanogels and investigated the effective antimicrobial activity against *Aspergillus flavus* (Ashrafi et al. 2019; Zhaveh et al. 2015). In a similar study, *Mentha piperita* were loaded in CS nanogel and were further used as antibiofilm agent against *Streptococcus mutans (S. mutans)* to protect the dental plaque. The nanogels were prepared by sol-gel method by crosslinking CS chains with tripolyphosphate. The sharp reduction in bacterial cell adherence was observed in newly designed nanogels due to the nanoformulation of essential oil as compared with the pristine CS nanogel. Excellent antibiofilm inhibition of *S. mutans* was occurred in 50 and 400 μg/mL for essential oil loaded nanogel (Ashrafi et al. 2019).

Another study was reported for the synthesis of smart nanogels composed of PNIPAm, poly *N,N*-diethylacrylamide and AA for controlled release of thymol (Zhao et al. 2018). Thymol loading and release behavior from the nanogel was triggered by both pH and temperature variation. The antimicrobial activity of thymol loaded smart nanogel was observed against *Bacillus subtilis (B. subtilis)* and *E. coli* and result suggested that thymol loaded nanogels exhibited significantly strong antimicrobial activity (12.8 ± 0.3 to 15.0 ± 1.0 mm) against both bacteria, showing better activity against *B. subtilis* than *E. coli.* Kather et al. reported the designing of eugenol decorated anti biofouling nanogels of PEG or poly(glycidol) (PG). The *Eugenol* molecules were covalently conjugated with the terminal group of PEG or PG spacers (Kather et al. 2017). The *Eugenol* encapsulated nanogels were later coated on the dental implants and improve their antibacterial property, cell adhesion and the osteogenic differentiation of stem cells. Cell adhesion studies of the bioactive nanogel coated dental material was performed with a standard mouse cell line (L929 mouse fibroblasts), human dermal fibroblasts and human mesenchymal stem cells and observed the healthy regeneration in the tissue adjacent to dental implants. In a recent study, rosemary oil encapsulated CS with benzoic acid nanogel was reported in which both CS and benzoic acid were covalently bounded and formed self-assembled structure. Subsequently, this nanogel was added in starch-carboxymethyl cellulose blend films for the development of antimicrobial property in the blend films against *S. aureus* (Mohsenabadi et al. 2018). The free rosemary oil shows the minimum inhibitory concentration (MIC) value of 200 μg/mL, whereas the rosemary oil encapsulated CS–benzoic acid nanogel showed a sharp decrease in MIC value and observed 80 μg/mL. The enhanced inhibitory effect of these nanogels could be attributed to the larger surface area of the NPs and subsequently higher interactions with the bacterial cell walls.

3.4 *Miscellaneous*

Several other bioactive agents are also reported for the development of antimicrobial nanogels which will be discussed in the preceding section. Malzahn et al. reported the synthesis of dextran-polyacrylamide based nanogels loaded with zinc ions by miniemulsion process, in order to target MRSA. The nanogels showed system with high potential for wound care applications combining the good release of zinc ions, strong inhibition of bacterial growth with high biocompatibility (Malzahn et al.

2014). In another aspect, CS nanogels with copper ions have been developed by using physical gelation in reverse emulsion process. The core of the nanogels containing neutralized amino groups was complex with copper while the charged amino groups on the shell. Strong synergistic effect between CS and copper as fungicide or biopesticides was observed against *Fusarium graminearum* (Brunel et al. 2013). Surface functionalized PAA-based nanogels (Carbopol Aqua SF1) coated with a cationic polyelectrolyte were designed for berberine (BRB) encapsulation (Al-Awady et al. 2017). BRB is an isoquinoline alkaloid which can be found in number of commonly therapeutic plant species and possess strong antimicrobial and other biological activities. The cationic polyelectrolyte coated nanogel showed strong antimicrobial activity against *C. reinhardtii* and *E. coli* even for very short incubation times. The increase in the antimicrobial activity was attributed to the favorable electrostatic adhesion between the cationic polyelectrolyte coated nanogel and the anionic cell membranes allowing the accumulation of BRB particles on the cell walls and lead to the cell death. In a further study, highly efficient antimicrobial nanocarriers for BRB on surface-functionalized polymer and the cationic surfactant octadecyltrimethylammonium bromide. The antimicrobial action of functional nanocarrier was investigated on three different microbes, *C. reinhardtii*, *Saccharomyces cerevisiae* and *E. coli* and compared to the activity with the effect of free BRB as well as non-coated BRB-loaded nanocarriers (Al-Obaidy et al. 2019). The result reflects the interactions between the functional nanogel and the anionic surface of the bacterial cell walls which delivers the high concentration of BRB. These nanogels shows effective antimicrobial and disinfecting activities and could be used as specifically for dental formulations, wound dressings and antifungal agents.

Recently, the inherent antimicrobial nanogels were prepared by copolymerization of styrene, polycaprolactone-hydroxyethyl methacrylate, and polyhexamethylene guanidine hydrochloride methacrylate (Han et al. 2018). The nanogels were finished on the surface of cotton fabric that decreases the bacterial adhesion by altering the hydrophobicity and the bacteriostatic rate. The antibacterial activity of the functionalized fabric was investigated against *E. coli* and *S. aureus* and showed more than 86% activity even after 50 times of mechanical washing of the fabric. Similarly, physically crosslinked of multifunctional poly(ethylene imine) was designed that comprises hydrophobic groups. The nanogels were used for the preparation of thin film coatings and displayed excellent antimicrobial activity against a wide range of different bacteria due the presence of azetidinium groups in the poly(ethylene imine) nanogel (Chattopadhyay et al. 2016). Rectorite encapsulated lysozyme-alginate nanogels were designed and used as a nanofiller for the development of cellulose nanofibers. The presence of lysozyme in the nanogel leads the damage of bacterial cell wall whereas the variation of rectorite in the nanogel synergized the bacterial inhibition effect. Rectorite is a kind of layered silicate structures with similar characteristics to montmorillonite also shows an absorptive capacity on bacteria and inhibit the bacterial proliferation. The developed nanofibrous material showed excellent antibacterial property against *E. coli* and *S. aureus* (Zhang et al. 2015). Cell viability test was performed to measure the cytotoxicity of nanogel immobilized mats with a standard mouse cell line (L929 mouse fibroblasts). SEM

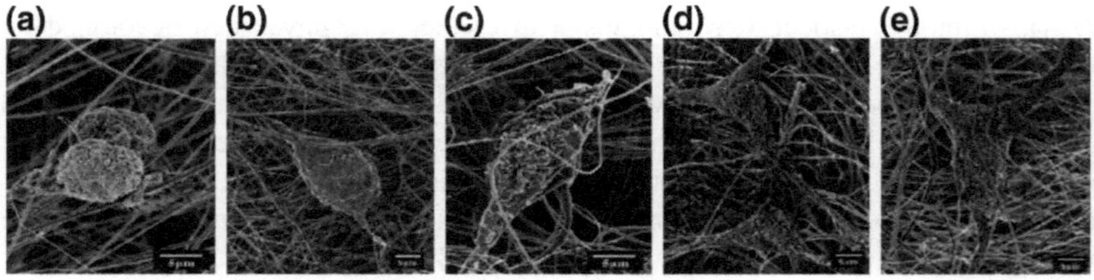

Fig. 7 SEM images of L929 cells attached on (a) cellulose nanofibrous mats, and nanogels immobilized mats with different rectorite content after 72 h of incubation (b) (NG0/ALG)0.5, (c) (NG61/ALG)0.5, (d) (NG31/ALG)0.5 and (e) (NG31/ALG)1. Reproduced with permission from Elsevier (Zhang et al. 2015)

images were taken after L929 cells proliferated on nanofibrous mats with or without assembling of nanogels. In Fig. 7a, round cells were observed on cellulose nanofibrous mats whereas, nanogel immobilized mats show that the cells were intensively attached to the surfaces and tended to grow along the orientation of nanofibers with fusiform shape. It can be found that the increased content of rectorite in nanogel correlated with the better condition of cell morphology and adhesion onto mats.

Antimicrobial peptides (AMP) are also promising antimicrobial agents with high efficacy in the treatment of infections associated with biofilm formation (Klodzinska et al. 2018). Bioactive nanogel composed of octenyl succinic anhydride-modified hyaluronic acid nanogels with embedded lysine-based α-peptide/β-peptoid. The obtained nanogels have excellent biocompatibility and biodegradability, as well as the shielding of AMP and minimize side effects associated with direct exposure of peptide to cells, such as cytotoxicity towards non-targeted cells (Klodzinska et al. 2018). After inspiration with AMP, in a further investigation, the antimicrobial nanodots were designed using a one-step hydrothermal process. These antimicrobial nanodots possessed unique properties such as amphiphilic and zwitterionic structure, having excellent antibacterial properties due to its small size of which leads the bacterial membrane permeabilization and rapid bactericidal kinetics (Xu et al. 2018). An injectable thermosensitive nanogels with encapsulation of graphene oxide (GO) was designed where the pH-dependent hydrogen bonding interactions between the carboxyl and oxethyl groups induce the reversible assembly of GO and nanogels. The hybrid nanogel possessed high antibacterial activity and demonstrated thermo responsive drug release behavior with high viability of three-dimensional encapsulated cells. The presence of the GO in the nanogel is responsible for the inherent antimicrobial nature against *E. coli* (Cheng et al. 2018).

4 Conclusion and Future Perspectives

Nanoengineering of bioactive gels is the fascinating approach for the development of functional nanomaterials with novel architecture towards the development of healthcare materials. Nanosized bioactive agents have been a suitable option for the development of antimicrobial materials. Due to the inert nature of NPs, they have poor ability to interact to any surface, there is a strong need to find a suitable alternative. The encapsulation of bioactive agents within a nanogel seems to an innovative approach which may lead to the design of functional NPs, where bioactive agents are surrounded by a biocompatible hydrogel layer. The aim of this book chapter is to explore the designing of functional nanomaterials with unique features e.g. inherent antimicrobial, highly active and many more. In different sections of the chapter, we discuss the various routes for the designing and development of bioactive nanogels and their applications in specific areas. These bioactive agents may be encapsulated in the gel matrix during the pre and post-polymerization steps of the gels to provide suitable dimensions to the nanogel. Among all process, the radiation process is very effective in developing nanogels because of chemical-free approach which reduces the toxicity of the by-products. The future direction to develop functional nanomaterials could be the use of different plant extracts or bioproducts in presence of hydrogels. Another interesting and useful approach is the addition of gamma sources that synergized the growth of NPs without using a bunch of chemicals and thus will reduce the toxicity of NPs as well. The functional nanomaterials have controlled architecture with specific functionality which creates the binding ability with any substrate and make them antimicrobial in nature. In conclusion, this is the collaborative efforts of scientists, engineers and policymakers would continue to develop functional nanomaterials with much wider domains of applications. Such bio-interactive nanogels would certainly open up a revolution in the world of antimicrobial medical devices leading to enormous control over healthcare expenditures.

References

Abed N, Couvreur P (2014) Nanocarriers for antibiotics: a promising solution to treat intracellular bacterial infections. Int J Antimicrob Agents 43(6):485–496. https://doi.org/10.1016/j.ijantimicag.2014.02.009

Ahonen M, Kahru A, Ivask A, Kasemets K, Koljalg S, Mantecca P, Vinkovic Vrcek I, Keinanen-Toivola MM, Crijns F (2017) Proactive approach for safe use of antimicrobial coatings in healthcare settings: opinion of the COST action network AMiCI. Int J Environ Res Public Health 14(4):366. https://doi.org/10.3390/ijerph14040366

Al-Awady MJ, Fauchet A, Greenway GM, Paunov VN (2017) Enhanced antimicrobial effect of berberine in nanogel carriers with cationic surface functionality. J Mater Chem B 5(38):7885–7897. https://doi.org/10.1039/C7TB02262J

Al-Awady MJ, Weldrick PJ, Hardman MJ, Greenway GM, Paunov VN (2018) Amplified antimicrobial action of chlorhexidine encapsulated in PDAC-functionalized acrylate copolymer nanogel carriers. Mater Chem Front 2(11):2032–2044. https://doi.org/10.1039/c8qm00343b

Al-Obaidy SSM, Greenway GM, Paunov VN (2019) Dual-functionalised shellac nanocarriers give a super-boost of the antimicrobial action of berberine. Nanoscale Advances 1(2):858–872. https://doi.org/10.1039/c8na00121a

Anjum S, Gupta B (2018) Bioengineering of functional nanosilver nanogels for smart healthcare systems. Glob Chall 2(10):1800044. https://doi.org/10.1002/gch2.201800044

Anjum S, Sharma A, Tummalapalli M, Joy J, Bhan S, Gupta B (2015) A novel route for the preparation of silver loaded polyvinyl alcohol nanogels for wound care systems. Int J Polym Mater Polym Biomater 64(17):894–905. https://doi.org/10.1080/00914037.2015.1030660

Anjum S, Gupta A, Sharma D, Gautam D, Bhan S, Sharma A, Kapil A, Gupta B (2016) Development of novel wound care systems based on nanosilver nanohydrogels of polymethacrylic acid with Aloe vera and curcumin. Mater Sci Eng C Mater Biol Appl 64:157–166. https://doi.org/10.1016/j.msec.2016.03.069

Anjum S, Gupta A, Sharma D, Kumari S, Sahariah P, Bora J, Bhan S, Gupta B (2017a) Antimicrobial nature and healing behavior of plasma functionalized polyester sutures. J Bioact Compat Polym 32(3):263–279. https://doi.org/10.1177/0883911516668665

Anjum S, Patra S, Gupta B (2017b) Preparation and bactericidal action of biofunctional polyacrylamide nanogels. Adv Mater Lett 8(1):13–18

Anjum S, Gupta A, Sharma D, Dalal P, Gupta B (2017c) Skin compatibility and antimicrobial studies on biofunctionalized polypropylene fabric. Mater Sci Eng C Mater Biol Appl 69:1043–1050. https://doi.org/10.1016/j.msec.2016.08.009

Anjum S, Singh S, Benedicte L, Roger P, Panigrahi M, Gupta B (2018) Biomodification strategies for the development of antimicrobial urinary catheters: overview and advances. Glob Chall 2(1):1700068. https://doi.org/10.1002/gch2.201700068

Ashrafi B, Rashidipour M, Marzban A, Soroush S, Azadpour M, Delfani S, Ramak P (2019) Mentha piperita essential oils loaded in a chitosan nanogel with inhibitory effect on biofilm formation against *S. mutans* on the dental surface. Carbohydr Polym 212:142–149. https://doi.org/10.1016/j.carbpol.2019.02.018

Ballesteros CAS, Bernardi JC, Correa DS, Zucolotto V (2019) Controlled release of silver nanoparticles contained in photoresponsive nanogels. ACS Appl Bio Mater 2(2):644–653. https://doi.org/10.1021/acsabm.8b00366

Beyki M, Zhaveh S, Khalili ST, Rahmani-Cherati T, Abollahi A, Bayat M, Tabatabaei M, Mohsenifar A (2014) Encapsulation of Mentha piperita essential oils in chitosan–cinnamic acid nanogel with enhanced antimicrobial activity against Aspergillus flavus. Ind Crops Prod 54:310–319. https://doi.org/10.1016/j.indcrop.2014.01.033

Brunel F, El Gueddari NE, Moerschbacher BM (2013) Complexation of copper(II) with chitosan nanogels: toward control of microbial growth. Carbohydr Polym 92(2):1348–1356. https://doi.org/10.1016/j.carbpol.2012.10.025

Chattopadhyay S, Heine E, Mourran A, Richtering W, Keul H, Möller M (2016) Waterborne physically crosslinked antimicrobial nanogels. Polym Chem 7(2):364–369. https://doi.org/10.1039/C5PY01566A

Chen T, Chen L, Li H, Chen Y, Guo H, Shu Y, Chen Z, Cai C, Guo L, Zhang X (2014) Design and in vitro evaluation of a novel poly (methacrylic acid)/metronidazole antibacterial nanogel as an oral dosage form. Colloids Surf, B 118:65–71

Chen T, Li Q, Guo L, Yu L, Li Z, Guo H, Li H, Zhao M, Chen L, Chen X, Zhong Q, Zhou L, Wu T (2016) Lower cytotoxicity, high stability, and long-term antibacterial activity of a poly(methacrylic acid)/isoniazid/rifampin nanogel against multidrug-resistant intestinal *Mycobacterium tuberculosis*. Mater Sci Eng C Mater Biol Appl 58:659–665. https://doi.org/10.1016/j.msec.2015.08.055

Cheng W, Chen Y, Teng L, Lu B, Ren L, Wang Y (2018) Antimicrobial colloidal hydrogels assembled by graphene oxide and thermo-sensitive nanogels for cell encapsulation. J Colloid Interface Sci 513:314–323. https://doi.org/10.1016/j.jcis.2017.11.018

Coll Ferrer MC, Dastgheyb S, Hickok NJ, Eckmann DM, Composto RJ (2014) Designing nanogel carriers for antibacterial applications. Acta Biomater 10(5):2105–2111. https://doi.org/10.1016/j.actbio.2014.01.009

de Lima R, Seabra AB, Duran N (2012) Silver nanoparticles: a brief review of cytotoxicity and genotoxicity of chemically and biogenically synthesized nanoparticles. J Appl Toxicol 32(11):867–879. https://doi.org/10.1002/jat.2780

Gayathri NK, Aparna V, Maya S, Biswas R, Jayakumar R, Mohan CG (2017) Preparation, characterization, drug release and computational modelling studies of antibiotics loaded amorphous chitin nanoparticles. Carbohydr Polym 177:67–76. https://doi.org/10.1016/j.carbpol.2017.08.112

Grennhalgh R, Dempsey-Hibbert N, Vagg-Whitehead K (2018) Antimicrobial strategies and economic considerations for polymeric medical implants. Int Biodeterior Biodegrad 136

Gupta B, Gautam D, Anjum S, Saxena S, Kapil A (2013) Radiation synthesis of nanosilver nanohydrogels of poly(methacrylic acid). Radiat Phys Chem 92:54–60. https://doi.org/10.1016/j.radphyschem.2013.07.020

Han H, Zhu J, Wu DQ, Li FX, Wang XL, Yu JY, Qin XH (2018) Inherent guanidine nanogels with durable antibacterial and bacterially antiadhesive properties. Adv Funct Mater 1806594. https://doi.org/10.1002/adfm.201806594

Huang F, Gao Y, Zhang Y, Cheng T, Ou H, Yang L, Liu J, Shi L, Liu J (2017) Silver-decorated polymeric micelles combined with curcumin for enhanced antibacterial activity. ACS Appl Mater Interfaces 9(20):16880–16889. https://doi.org/10.1021/acsami.7b03347

Jin J-C, Xu Z-Q, Dong P, Lai L, Lan J-Y, Jiang F-L, Liu Y (2015) One-step synthesis of silver nanoparticles using carbon dots as reducing and stabilizing agents and their antibacterial mechanisms. Carbon 94:129–141. https://doi.org/10.1016/j.carbon.2015.05.084

Jin JC, Wu XJ, Xu J, Wang BB, Jiang FL, Liu Y (2017) Ultrasmall silver nanoclusters: highly efficient antibacterial activity and their mechanisms. Biomater Sci 5(2):247–257. https://doi.org/10.1039/c6bm00717a

Kather M, Skischus M, Kandt P, Pich A, Conrads G, Neuss S (2017) functional isoeugenol-modified nanogel coatings for the design of biointerfaces. Angew Chem Int Ed 56(9):2497–2502. https://doi.org/10.1002/anie.201609180

Kettel MJ, Dierkes F, Schaefer K, Moeller M, Pich A (2011) Aqueous nanogels modified with cyclodextrin. Polymer 52(9):1917–1924

Khan MA, Mujahid M, Gul IH, Hussain A (2019) Electrochemical study of magnetic nanogel designed for controlled release of chlorhexidine gluconate. Electrochim Acta 295:113–123. https://doi.org/10.1016/j.electacta.2018.10.098

Klodzinska SN, Molchanova N, Franzyk H, Hansen PR, Damborg P, Nielsen HM (2018) Biopolymer nanogels improve antibacterial activity and safety profile of a novel lysine-based alpha-peptide/beta-peptoid peptidomimetic. Eur J Pharm Biopharm 128:1–9. https://doi.org/10.1016/j.ejpb.2018.03.012

Li J, Yu S, Yao P, Jiang M (2008) Lysozyme−dextran core−shell nanogels prepared via a green process. Langmuir 24(7):3486–3492. https://doi.org/10.1021/la702785b

Liu G, An Z (2014) Frontiers in the design and synthesis of advanced nanogels for nanomedicine. Polym Chem 5(5):1559–1565. https://doi.org/10.1039/C3PY01502E

Liu J, Sonshine DA, Shervani S, Hurt RH (2010) Controlled release of biologically active silver from nanosilver surfaces. ACS Nano 4(11):6903–6913. https://doi.org/10.1021/nn102272n

Malzahn K, Jamieson WD, Dröge M, Mailänder V, Jenkins ATA, Weiss CK, Landfester K (2014) Advanced dextran based nanogels for fighting *Staphylococcus aureus* infections by sustained zinc release. J Mater Chem B 2(15):2175–2183

Mohamed RR, Sabaa MW (2014) Synthesis and characterization of antimicrobial crosslinked carboxymethyl chitosan nanoparticles loaded with silver. Int J Biol Macromol 69:95–99. https://doi.org/10.1016/j.ijbiomac.2014.05.025

Mohsenabadi N, Rajaei A, Tabatabaei M, Mohsenifar A (2018) Physical and antimicrobial properties of starch-carboxy methyl cellulose film containing rosemary essential oils encapsulated in chitosan nanogel. Int J Biol Macromol 112:148–155. https://doi.org/10.1016/j.ijbiomac.2018.01.034

Moya-Ortega MD, Alvarez-Lorenzo C, Concheiro A, Loftsson T (2012) Cyclodextrin-based nanogels for pharmaceutical and biomedical applications. Int J Pharm 428(1–2):152–163. https://doi.org/10.1016/j.ijpharm.2012.02.038

Ramos AP, Cruz MAE, Tovani CB, Ciancaglini P (2017) Biomedical applications of nanotechnology. Biophys Rev 9(2):79–89. https://doi.org/10.1007/s12551-016-0246-2

Ravindran A, Chandran P, Khan SS (2013) Biofunctionalized silver nanoparticles: advances and prospects. Colloids Surf B Biointerfaces 105:342–352. https://doi.org/10.1016/j.colsurfb.2012.07.036

Shin B, Kim J, Vales TP, Yang SK, Kim J-K, Sohn H, Kim H-J (2018) Thermoresponsive drug controlled release from chitosan-based hydrogel embedded with poly(N-isopropylacrylamide) nanogels. J Polym Sci Part A Polym Chem 56(17):1907–1914. https://doi.org/10.1002/pola.29073

Stular D, Jerman I, Naglic I, Simoncic B, Tomsic B (2017) Embedment of silver into temperature- and pH-responsive microgel for the development of smart textiles with simultaneous moisture management and controlled antimicrobial activities. Carbohydr Polym 159:161–170. https://doi.org/10.1016/j.carbpol.2016.12.030

Suresh AK, Pelletier DA, Wang W, Morrell-Falvey JL, Gu B, Doktycz MJ (2012) Cytotoxicity induced by engineered silver nanocrystallites is dependent on surface coatings and cell types. Langmuir 28(5):2727–2735. https://doi.org/10.1021/la2042058

Thoniyot P, Tan MJ, Karim AA, Young DJ, Loh XJ (2015) Nanoparticle-hydrogel composites: concept, design, and applications of these promising, multi-functional materials. Adv Sci (Weinh) 2(1–2):1400010. https://doi.org/10.1002/advs.201400010

Tummalapalli M, Deopura BL, Alam MS, Gupta B (2015) Facile and green synthesis of silver nanoparticles using oxidized pectin. Mater Sci Eng C Mater Biol Appl 50:31–36. https://doi.org/10.1016/j.msec.2015.01.055

Tummalapalli M, Anjum S, Kumari S, Gupta B (2016a) Antimicrobial surgical sutures: recent developments and strategies. Polym Rev 56(4):607–630. https://doi.org/10.1080/15583724.2015.1119163

Tummalapalli M, Berthet M, Verrier B, Deopura BL, Alam MS, Gupta B (2016b) Drug loaded composite oxidized pectin and gelatin networks for accelerated wound healing. Int J Pharm 505(1–2):234–245. https://doi.org/10.1016/j.ijpharm.2016.04.007

Tyliszczak B, Pielichowski K (2013) Novel hydrogels containing nanosilver for biomedical applications—synthesis and characterization. J Polym Res 20(7):191. https://doi.org/10.1007/s10965-013-0191-8

Wu W, Zhou T, Berliner A, Banerjee P, Zhou S (2010) Smart core−shell hybrid nanogels with Ag nanoparticle core for cancer cell imaging and gel shell for pH-regulated drug delivery. Chem Mater 22(6):1966–1976. https://doi.org/10.1021/cm903357q

Xiong MH, Li YJ, Bao Y, Yang XZ, Hu B, Wang J (2012) Bacteria-responsive multifunctional nanogel for targeted antibiotic delivery. Adv Mater 24(46):6175–6180. https://doi.org/10.1002/adma.201202847

Xu HV, Zheng XT, Wang C, Zhao Y, Tan YN (2018) Bioinspired antimicrobial nanodots with amphiphilic and zwitterionic-like characteristics for combating multidrug-resistant bacteria and biofilm removal. ACS Appl Nano Mater 1(5):2062–2068. https://doi.org/10.1021/acsanm.8b00465

Zafar M, Shah T, Rawal A, Siores E (2014) Preparation and characterisation of thermoresponsive nanogels for smart antibacterial fabrics. Mater Sci Eng C Mater Biol Appl 40:135–141. https://doi.org/10.1016/j.msec.2014.03.033

Zaidi S, Misba L, Khan AU (2017) Nano-therapeutics: a revolution in infection control in post antibiotic era. Nanomedicine 13(7):2281–2301. https://doi.org/10.1016/j.nano.2017.06.015

Zaman SB, Hussain MA, Nye R, Mehta V, Mamun KT, Hossain N (2017) A review on antibiotic resistance: alarm bells are ringing. Cureus 9(6):e1403. https://doi.org/10.7759/cureus.1403

Zhang X, Malhotra S, Molina M, Haag R (2015a) Micro-and nanogels with labile crosslinks–from synthesis to biomedical applications. Chem Soc Rev 44(7):1948–1973

Zhang J, Li X, Tian J, Lu Y, Shi X, Zhan Y, Du Y, Liu H, Deng H (2015b) Antimicrobial activity and cytotoxicity of nanofibrous mats immobilized with polysaccharides-rectorite based nanogels. Colloids Surf B Biointerfaces 133:370–377. https://doi.org/10.1016/j.colsurfb.2015.04.051

Zhao Y, Sun W, Saldaña MD (2018) Nanogels of poly-N-isopropylacrylamide, poly-N, N-diethylacrylamide and acrylic acid for controlled release of thymol. J Polym Res 25(12):253. https://doi.org/10.1007/s10965-018-1644-x

Zhaveh S, Mohsenifar A, Beiki M, Khalili ST, Abdollahi A, Rahmani-Cherati T, Tabatabaei M (2015) Encapsulation of *Cuminum cyminum* essential oils in chitosan-caffeic acid nanogel with enhanced antimicrobial activity against *Aspergillus flavus*. Ind Crops Prod 69:251–256

An Overview of Unique Metal Oxide Nanostructures for Biosensor Applications

Leonard Sean Anthony, Veeradasan Perumal, Norani Muti Mohamed, Mohamed Shuaib Mohamed Saheed and Subash C. B. Gopinath

Abstract Metal oxide nanostructures have recently garnered much recognition because of their exceptional electrical, optical and molecular properties that provide a fascinating platform for the development of biosensors. Till recent years, various methods and techniques such as chemical precipitation, electrodeposition and hydrothermal treatment have been utilised to fabricate metal oxide-based biosensors with high sensitivities and low detection limits. However, a number of these fabricated devices are irreversible and have low sensitivities. Therefore, more efficient and reliable biosensing technologies need to be developed. Biosensors are usually defined as an analytical equipment that consists of a biological recognition component integrated to a transducer and signal processing unit that converts, amplifies and displays biochemical activity into a measurable signal. The goal of this bioelectrical combination is to detect pathogenic and physiological molecules in the food or body for the early detection as well as treatment of diseases using the high sensitivity and selectivity of biological sensing. We review here some of the main strategies used in the fabrication of metal oxide nanostructures, including the advancements in biosensors based on metal oxides and evaluate their properties with the aim of motivating greater interest in improving the development of biosensors for medical diagnosis.

L. S. Anthony · V. Perumal (✉) · N. M. Mohamed · M. S. M. Saheed
Centre of Innovative Nanostructures and Nanodevices (COINN), Universiti Teknologi PETRONAS, 32610 Seri Iskandar, Perak Darul Ridzuan, Malaysia
e-mail: veeradasan.perumal@utp.edu.my

V. Perumal
Department of Mechanical Engineering, Universiti Teknologi PETRONAS, 32610 Seri Iskandar, Perak Darul Ridzuan, Malaysia

N. M. Mohamed · M. S. M. Saheed
Department of Fundamental and Applied Sciences, Universiti Teknologi PETRONAS, 32610 Seri Iskandar, Perak Darul Ridzuan, Malaysia

S. C. B. Gopinath
Institute of Nano Electronic Engineering, Universiti Malaysia Perlis, 01000 Kangar, Perlis, Malaysia

School of Bioprocess Engineering, Universiti Malaysia Perlis, 02600 Arau, Perlis, Malaysia

A. H. Bhat et al. (eds.), *Nanomaterials for Healthcare, Energy and Environment*, Advanced Structured Materials 118, https://doi.org/10.1007/978-981-13-9833-9_3

Keywords Transition metal · Metal oxides · Synthesis · Nanostructures and biosensor

1 Introduction

Nanotechnology plays a vital role in almost every scientific field today. The major aspect of nanotechnology is the capability to work at the atomic and molecular level and to develop various tuneable structures with varying configurations such as nanorods, nanofibers, nanowires, nanosheets and nanoflowers (Malwal and Packirisamy 2018). Within the past three decades, the growing need for advanced and efficient materials have brought about the development of different types of nanomaterials. Among these, metal oxides have been identified as one of the more promising functional materials and continue to be intensively studied because of their distinct properties. Metal oxides are a chemical compound formed between positive metals ions, called cations and negative oxygen ions called anions. These compounds require a minimum of one oxygen atom, and two metallic atoms. They have a special electronic structure that enables them to exhibit metallic, semiconducting and even insulative characteristics. In the field of nanomaterials, metal oxide nanostructures demonstrate unique electrochemical properties. This is due to their high density of corner or edge surface sites and their limited size (Fernandez-Garcia and Rodriguez 2007). Their small particle size induces modifications in the cell parameters and structural transformations, improving the thermodynamic stability, conductivity and chemical reactivity of the nanostructure. They also have tuneable mechanical, magnetic, optical and thermal properties along with an extremely high surface area and sensitivity. All these properties allow them to be widely used in various applications such as photovoltaic cells, lithium ion batteries, catalysts, gas sensors, biosensors and so on.

On the other hand, biosensors are analytical devices that consist of a biological recognition component integrated to a transducer and signal processing unit that converts, amplifies and displays biochemical activity into a measurable signal (Jha 2016). Clark and Lyons pioneered the development of biosensor in 1962. They suggested that enzyme electrodes can be formed by immobilizing enzymes on electrical detectors and then went on to devise the first-generation biosensor for monitoring glucose concentrations in blood. A typical biosensor consists of two components; the biological recognition component and the physical component comprising the transducer and signal conditioners. The biological component, which may be an enzyme or antibody, binds with the testing analyte to produce a biological response. Sometimes the analyte releases electrons, gas, heat, or hydrogen ions. The transducer then converts those biological responses into an electrical output, which can be measured (Chakraborty 2017). The output signal generated by the transducer can be digitized to show real-time measurements on a display with the help of a signal conditioning circuit and a microcontroller. Figure 1 shows the schematic diagram of a typical biosensor.

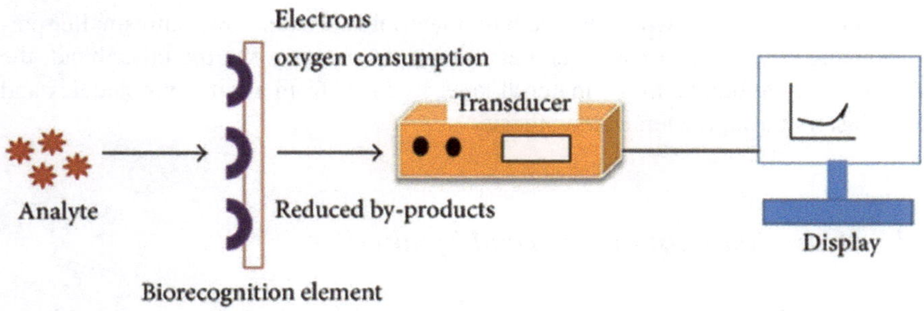

Fig. 1 Structure of a biosensor. Reproduced with permission from Dhull et al. (2013)

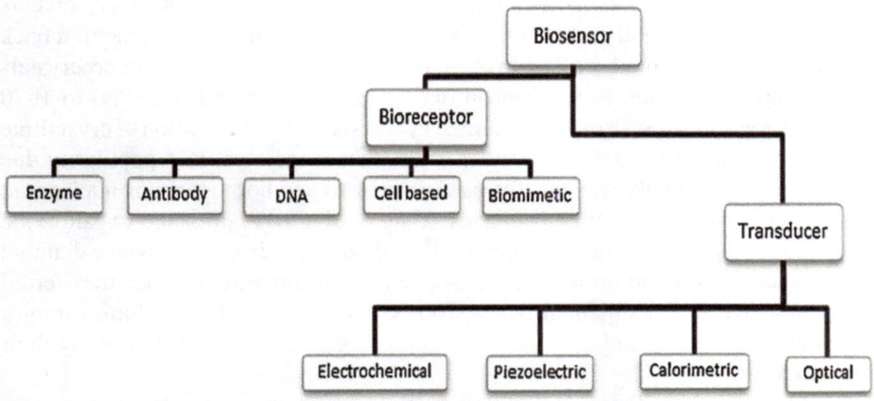

Fig. 2 Classification of biosensors. Reproduced with permission from Perumal and Uda (2014)

Numerous types of biosensors are also available as shown in Fig. 2 (Perumal and Hashim 2013). They are categorised into biosensors based on their biological recognition elements and biosensors based on their transduction methods. This book chapter will first provide an overview on the fabrication methods of metal oxide nanostructures and then present a comprehensive review on currently available metal oxide-based biosensors within the past decade.

2 Nanostructure Synthesis Methods

The fabrication of metal oxide nanostructures with a specific morphology is a challenging task for researchers today. The selection of a synthesis technique depends on the material of interest and can be a significant factor in determining the size, shape and composition of the metal oxide nanostructure desired. In general, there are two main routes for the synthesis of nanostructures, namely the top-down and bottom-up approach. In the top-down route, the bulk material is broken down into smaller particles using physical processes such as milling or grinding. On the other hand, the bottom-up route builds up the material layer by layer from atoms to molecules and then clusters through chemical synthesis.

2.1 *Hydrothermal/Solvothermal Synthesis*

Hydrothermal synthesis is generally defined as using high temperature and high-pressure water conditions to synthesise or grow crystals (Hayashi and Hakuta 2010). It is typically performed at ambient pressure below 2 bar and at temperatures between 200–300 °C. Similarly, the solvothermal method uses a variety of other solvents besides water. Schafhautl reported the earliest hydrothermal synthesis method back in 1845. He transformed freshly precipitated silicic acid to quartz microcrystals using a Papin's digester. From then on out, scientific journals from 1960 to 1970 frequently reported the use of hydrothermal synthesis for the fabrication of crystalline structures (Ashik et al. 2018). Over the years, this process grew in popularity due to its use of eco-friendly solvents for particle growth and homogeneous nucleation, which is also one of its main advantages. Yang et al. (2017) prepared cobalt oxide microspheres using a one-step hydrothermal method. They dropped ethylene diamine into a cobalt nitrate solution while being sonicated. The mixture was then transferred to an autoclave and maintained below 100 °C. After 12 h, the resulting samples were washed, freeze dried and annealed to remove water without destroying their nanostructures.

On the other hand, Zhang et al. (2018) utilised a solvothermal method to surface-engineer a multiple stages of nickel oxide rod-on-sheet nanoarrays on nickel foam. In a typical procedure, they dissolved a piece of nickel foam in a solution of ammonium fluoride, nickel nitrate and urea. The resulting mixture was then transferred into an autoclave and maintained at high temperature before being calcined in an argon filled tube furnace. Hosseini et al. (2018) used the hydrothermal method to fabricate a DNA biosensor to determine menadione. They dissolved graphene oxide and subsequently added iron chloride and ascorbic acid, forming a homogenous solution which was then sonicated. After that, hydrazine hydrate was added, and the solution was treated hydrothermally at 180 °C for 8 h. The hydrothermal synthesis method provides efficient control of the uniformity and rate of material growth, nucleation and aging through the immediate crystallization of nanomaterials from the solution as well as regulating experimental perimeters (Ashik et al. 2018). Therefore, the morphology, material properties, size and structure are easily controlled. The simple hydrothermal synthesis setup also makes the overall process more cost saving compared to other physical methods. Figure 3 shows the typical hydrothermal process.

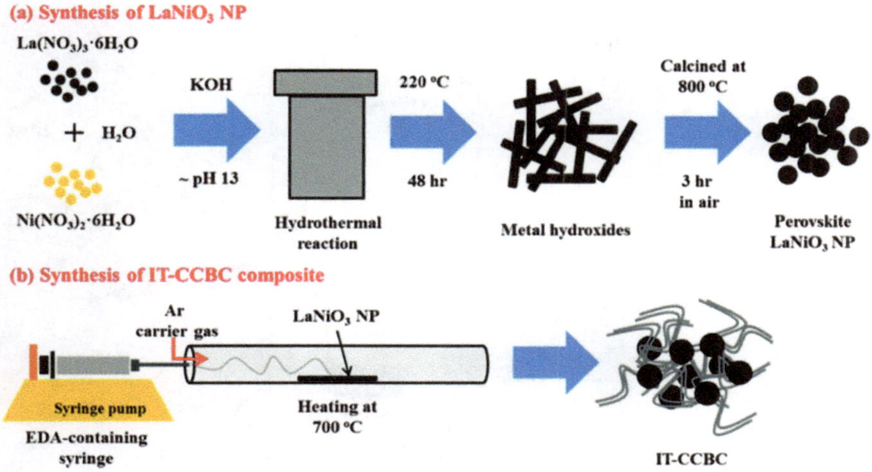

Fig. 3 Typical hydrothermal synthesis process. Reproduced with permission from Acharya et al. (2015)

2.2 *Chemical Precipitation Synthesis*

In the chemical precipitation process, metals in the hydroxide form are precipitated from an aqueous precursor solution through the combined action of a base in the solvent and then consequently decomposed to form metal oxide nanostructures. When the concentration of one solid is above the solubility limit of the other, precipitation transpires (Xia et al. 2014). The release of anions and cations are regulated to control particle growth and nucleation as well as synthesize monodispersed nanoparticles. Mohamad et al. (2016) used the precipitation method to synthesise iron oxide nanoparticles. They pumped sodium hydroxide and iron chloride into water under a nitrogen gas atmosphere while continuously stirring. After two hours of crystallisation, the precipitates were collected. Chen et al. (2017) also used to precipitation method to explore manganese dioxide microspheres as a biosensing platform for effective detection of mycoplasma ovipneumoniae. They dissolved manganese sulphate and nitric acid in water and stirred. The solid products were washed and mixed with potassium permanganate. The resultant solids were diluted with hydrochloric acid and vacuum dried at 180 °C for 4 h.

Besides that, Almeida et al. (2017) utilised chemical precipitation to create an enzyme-modified metal oxide electrode for the development of a biosensor using iron oxide nanoparticles. They synthesised the nanoparticles by dissolving iron sulphate and iron chloride in an acidic medium and then added ammonium hydroxide under a nitrogen environment to precipitate the nanoparticles. Moreover, Tiwari et al. (2015) precipitated iron oxide nanoparticles for the fabrication of a nuclei acid sensor to immobilize E-coli. They synthesised the nanoparticles by mixing iron acetylacetonate, hexadecanediol and phenyl ether under nitrogen atmosphere. The

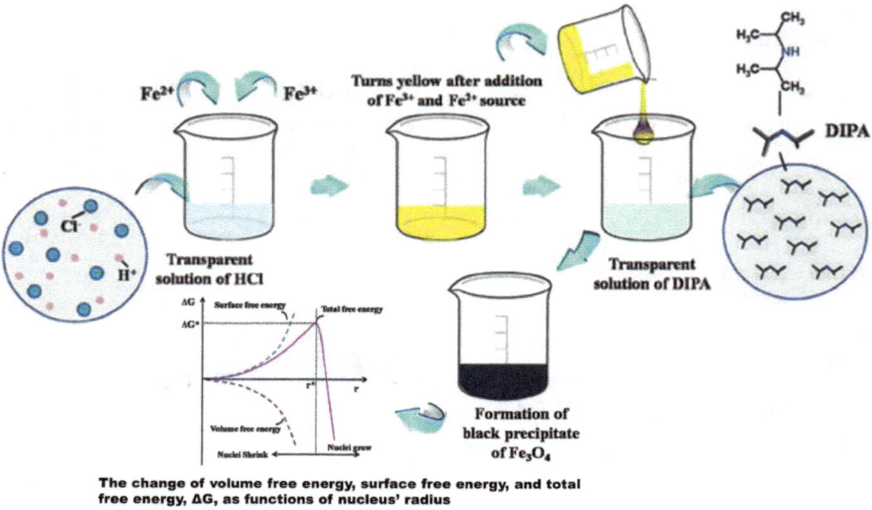

Fig. 4 Typical chemical precipitation technique. Reproduced with permission from Tarafdar and Adhikari (2016)

obtain precipitate was then washed and centrifugated. In another research, Thandavan et al. (2015) used the precipitation method to develop a combined interface using iron oxide nanoparticles and CNTs to detect hydrogen peroxide. They dissolved iron chloride in ammonium hydroxide solution. The resultant precipitates were then heated, washed and dried. The major advantage of this approach is its fast precipitation process and ease of scalability for commercial synthesis. The process does not need high temperatures since it is conducted at room temperature. Even so, it is a challenge to precisely manipulate the chemical homogeneity, morphology and particle size of the metal oxides due to the fast precipitation process. Figure 4 shows the standard chemical precipitation process.

2.3 Chemical Vapour Deposition (CVD) Synthesis

The CVD process is generally defined as passing volatile precursor gases over several substrates which reacts to produce a thin film on the surface. The entire process is performed in a reaction chamber using different energy sources such as electrons, hot filaments, lasers, or plasmas at temperatures between 200 and 1200 °C (Fernandez-Garcia and Rodriguez 2007). Any by-products are removed through gaseous flow. The reaction temperature, reaction rate and concentration of the precursors significantly affects the quality, crystallinity and physical properties of the deposited film. There are several important components in the CVD process, namely the precursor supply and the carrier gas. The initial precursor must be in the gaseous form at suffi-

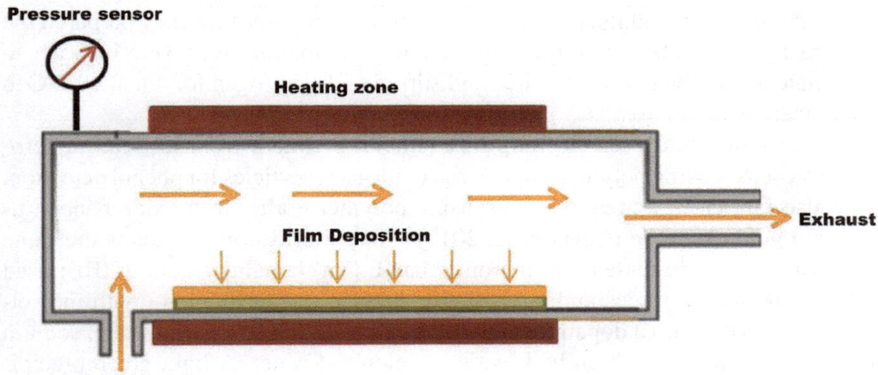

Fig. 5 Chemical vapor deposition process. Reproduced with permission from Zhang et al. (2016a, b)

ciently high vapour pressure for the reaction to take place. Process parameters for the carrier gas such as flow rate, pressure and rate of convection also affect the growth of the nanostructures (Ashik et al. 2018). Usually, the CVD synthesis process results in highly stable solid materials which have been used extensively to synthesize various types of nanocarbons. Guo et al. (2014) synthesised Ni nanosheets using a plasma arc CVD apparatus. They dissolved nickel nitrate in ethanol and sprayed the mixture onto molybdenum substrate, which was then reacted the presence of argon and hydrogen gasses. The major advantage is in the results where pure, uniform and reproduceable nanoparticles as well as films are fabricated. The semiconductor industry often uses this technique to produce high performance thin films due to its uniform coating of nanoparticles. However, the complexity of the process and the price of the precursors had led to limited research conducted (Ashik et al. 2018). Figure 5 shows the CVD process.

2.4 Sol-Gel Synthesis

The sol-gel technique is a flexible process used for synthesizing different metal oxides. This technique generally enables significant control of the crystallographic and morphological properties of the nanostructures. The term sol-gel is derived from the action of micro particles in a solution (sol) that accumulate under specific conditions and eventually combine together to create a coherent network (gel) (Xia et al. 2014). Wahab et al. (2018) prepared zinc oxide nanorod thin films by firstly dissolving zinc acetate dehydrate in a mixed solution of monoethanolamine and methoxyethanol at room temperature. After aging for 24 h, the solution was used to spin coat several substrates and after uniform coating, the nanorods were grown by annealing at high temperature. Besides that, Cevik et al. (2012) fabricated an amperometric biosensor to determine phenol and its derivatives using the sol-gel process. They prepared the particles by firstly dissolving iron chloride and ammonium hydroxide in water to precipitate a gel. The gel was washed and stirred under nitrogen for 4 h at 100 °C to ensure inert atmosphere.

Similarly, the same research group did another experiment to construct an amperometric biosensor using polymer grafted iron oxide nanoparticles for phenol detection. They also fabricated a urea biosensor using polymer grafter iron oxide nanoparticles on a gold electrode (Cevik et al. 2012). Their fabrication process is the same as the first report discussed. On the other hand, Roychoudhury et al. (2016) used nickel oxide nanoparticles and a tyrosinase enzyme conjugate along with the sol-gel method to develop a dopamine biosensor. They dissolved nickel nitrate, sodium dodecyl sulphate and sodium hydroxide in deionised water. A light green precipitate was acquired, separated by centrifugation, rinsed with ethanol and then oven dried. Besides that, Nesakumar et al. (2016) used a zinc oxide nanocuboid modified platinum electrode to develop a sensitive acetylcholinesterase biosensor based on the sol-gel method. They dissolved zinc acetate into water and the dropped sodium hydroxide to the mixture while heating. Later, the solution was cooled, the precipitate separated by centrifugation and collected. One of the advantages of this method is that it can be performed in both an aqueous and non-aqueous medium. Oxygen for oxide formation is usually provided by water during the aqueous sol-gel process. However, for the non-aqueous route, oxygen is contributed by solvents like ethers, alcohol, ketones and aldehydes (Ashik et al. 2018). Figure 6 shows the typical sol-gel process.

2.5 Thermal Decomposition Synthesis

The thermal decomposition method is a simple process that involves the use of a high temperature thermal furnace to vaporize precursor metal substrates at temperatures between 300 and 800 °C in the presence of air to facilitate the deposition of nanostructures. Chen et al. (2016) fabricated copper oxide nanowires through thermal decomposition. The sequentially mixed copper nitrate, sodium hydroxide and diazine at 60 °C. The residue was washed and then treated in a furnace for 5 h at 400 °C. Wang and Zhang (2010) utilised thermal decomposition to fabricate an electrochemical DNA biosensor to detect specific target DNA. They reduced a blend of zinc oxide and graphite powder at 900 °C for 30 min. The advantages of this method is that it is convenient and allows for the rapid and large-scale production of nanostructures without the need for hazardous chemicals. However, other aspects like the requirement of additional materials such as auxiliary for vacuum pressure, source containers, substrates, and process monitoring add on to the cost of the method and makes it difficult in controlling the synthesized nanomaterials morphology. Figure 7 shows the thermal decomposition process.

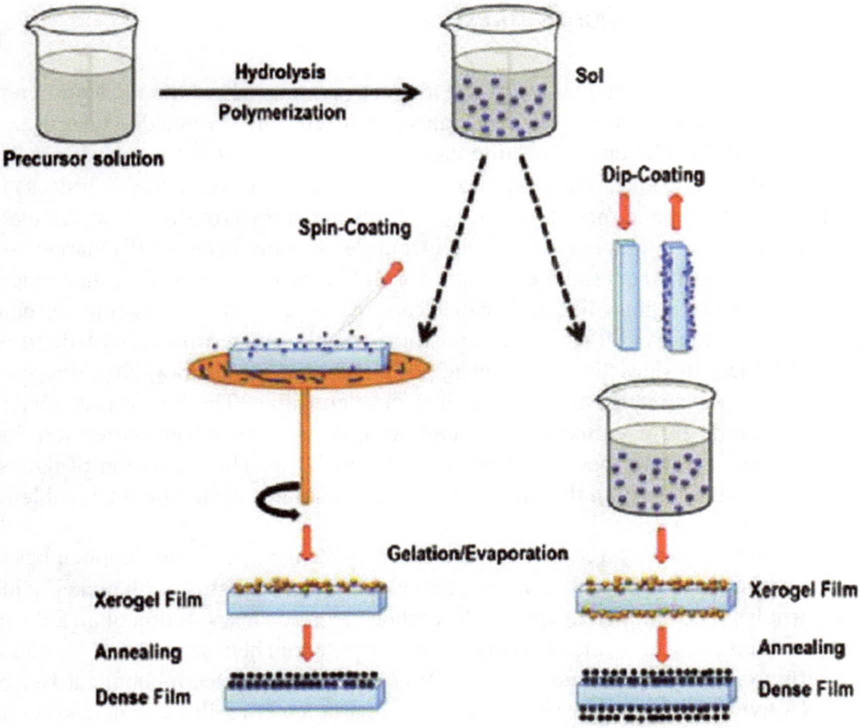

Fig. 6 Typical sol-gel process. Reproduced with permission from Mahmood and Naeem (2017)

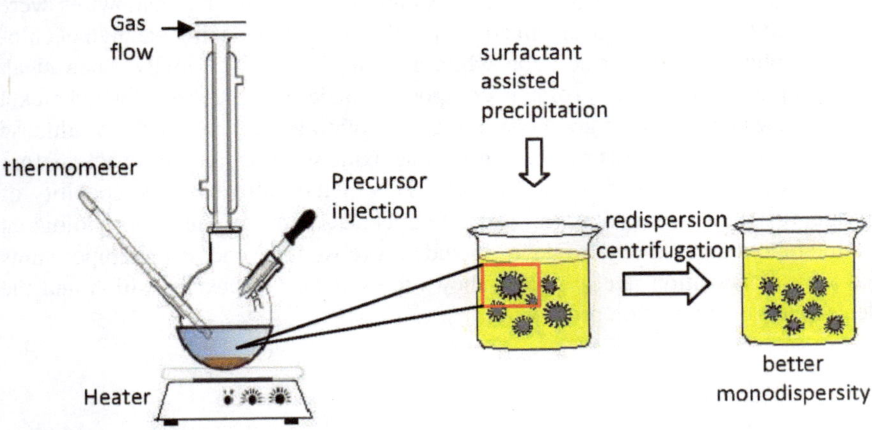

Fig. 7 Thermal decomposition process. Reproduced with permission from Evrim (2013)

2.6 Electrodeposition Synthesis

The electrodeposition method is similar to electroplating where the nucleation and growth of crystallites are done using an electric current (Fernandez-Garcia and Rodriguez 2007). The entire reaction can be broken down into two steps: precipitation of hydroxide and formation of oxide. Reetz (2008) developed this process in 1994. In this technique, heterogeneous electrons are transferred between the electroactive species in the medium and the electrode. The metal atoms discharged are then converted to their crystalline states. Yu et al. (2016) electrodeposited manganese oxide thin films on a glass-like carbon electrode in the solution of potassium permanganate and sulphuric acid. After the deposition, the films were washed with distilled water and dried. Besides that, Salimi et al. (2007) fabricated a haemoglobin biosensor using cobalt oxide nanoparticles through electrodeposition. The film was deposited on the electrode from a mixture of cobalt chloride and phosphate buffer solution using potential cycling repeating between 0.5 and 1.1 V. The formation of nanostructures occurs directly in the final state which eliminates weak bonding problems among the nanoparticles.

Liu et al. (2014) prepared a bio-electrochemical sensor for acetaminophen based on electrochemically reduced graphene that were electrodeposited with nickel oxide nanoparticles. They dipped the glass-like carbon electrode in a solution of nickel sulphate, sodium sulphate while applying a constant potential between −1.5 V for 150 s. Furthermore, Noorbakhsh and Salimi (2011) developed an electrochemical sensor for DNA hybridisation using nickel oxide nanoparticles. The film was deposited on the GCE from nickel chloride and phosphate buffer solution using a repetitive cycled from 0 to 1 V. Next, Tettamanti et al. (2018) used electrodeposited nickel nanowires build a nanostructured platform for the oxidation of ethanol. The nanowires were synthesised by deposition using an aqueous solution of nickel sulphate, nickel chloride and phosphate buffer solution while applying −1.20 V. Finally, Shen et al. (2016) reported the synthesis of nanocomposites made from electrodeposited nickel nanoparticles and reduced graphene oxide to develop a GCE electrode for glucose sensing. They added nickel nitrate to graphene oxide solution, and nitric acid before electrodepositing at 0–1.5 V for 40 cycle. This method also has the capability of producing 3D characteristics with accurate thickness and consistent morphology at room temperature. Other advantages include low cost, low processing temperatures and higher deposition rate. Figure 8 shows the standard electrodeposition and the electrophoretic deposition processes.

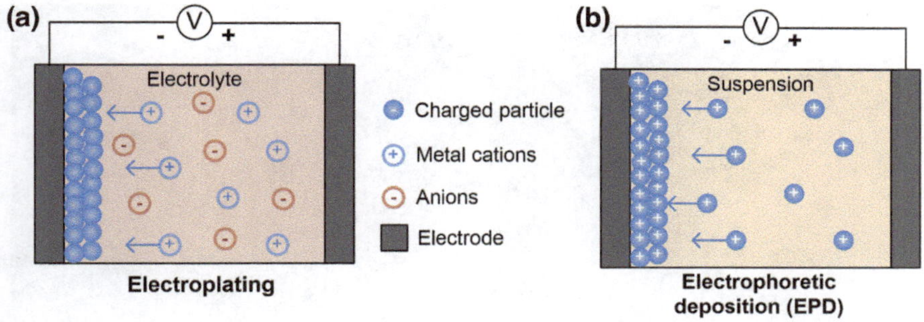

Fig. 8 Electrodeposition process. Reproduced with permission from Santos et al. (2015)

3 Metal Oxide Composites

3.1 Manganese Oxide Based Composites

As a well-known transition metal, manganese oxide is regarded as one of the most attractive candidates to construct biosensors due to its high stability, low cost and excellent catalytic capabilities (Fernandez-Garcia and Rodriguez 2007). Manganese oxide nanostructures have various shapes and morphologies. Depending on its form, the material can have a band gap between 0.26 and 0.7 eV. They have been extensively utilised to construct chemical sensors or biosensors in recent years. Chinnasamy et al. (2015) synthesised manganese oxide nanoparticles by the chemical reduction route at different growth temperatures. The increase in growth temperature lead to a morphology change from clusters into a mixture of nanorods and nanoflakes. These nanoparticles were then investigated as a chronoamperometric biosensor by immobilising them on a platinum electrode to test for hydrogen peroxide. Compared to other samples prepared at 40 and 80 °C, the sample prepared at 100 °C showed more prominent sensing capabilities. The chronoamperometric signals showed linear responses to the concentration of hydrogen peroxide within a range of 1×10^{-3} to 5×10^{-3} M. They discovered that the morphology of the manganese oxide nanostructures influences the performance of the sensor. There are a greater number of adsorption sites on the surface of the platinum electrode due to the morphological combination of nanorods and nanoflakes, thus increasing the electrocatalytic behaviour for the oxidation of hydrogen peroxide. Figure 9 shows the results of Chinnasamy's work.

Moreover, Zhang et al. (2014) created a hydrogen peroxide sensor using ultra-thin manganese oxide nanosheets that were synthesised from a simple hydrothermal approach. The nanosheets displayed high electrochemical activity when detecting hydrogen peroxide in phosphate buffer solution. They reported sensors which showed a wide linear range from 5×10^{-6} to 3.5×10^{-3} M and a low limit of detection of 1.5×10^{-6} M. The sensitivity of the electrode was also high at 130.56 μA mM^{-1} cm^{-2}. These characteristics are due excellent electrocatalytic activity of the nanosheets because of its large surface are. Finally, Xu and Zhang (2017) used a novel two-step

Fig. 9 FESEM image of pure manganese oxide particles prepared at 100 °C. Inset figure: Mixture of nanorods and nanoflakes. Reproduced with permission from Chinnasamy et al. (2015)

chemical precipitation method to develop a nanocomposite consisting of manganese dioxide nanowires and graphene oxide. They used the nanocomposite as an electrode to determine hydrogen peroxide. The sensor showed good electrocatalytic performance. It had a wide linear range from 4.90 to 4.50 mmol L^{-1} and a low limit of detection of 0.48 mmol L as well as a high sensitivity of 191.22 mA (mmol L^{-1}) $^{-1}$ cm^{-2}. The graphene oxide showcased strong adsorption ability which enabled the manganese oxide nanowires to attach easily. The rapid diffusion of analytes was improved by the large surface area of the nanowires, resulting in enhanced efficiency and better electrocatalytic performance.

3.2 *Cobalt Oxide Based Composites*

Among the different kinds of transition metals, cobalt oxide is one of the lesser researched materials for biosensors. The few studies done however indicated that cobalt oxide has excellent electrocatalytic properties, low cost, good biocompatibility and chemical stability as compared with noble metal-based materials. Depending on its form, the material can have a band gap between 2.0 and 2.6 eV. Yang et al. (2017) prepared porous cobalt oxide—graphene microspheres using a facile hydrothermal method and then tested the nanostructures as an enzyme free glucose biosensor. They obtained uniformly shaped flower-like structure microspheres with an average radius of 4.1 μm. The microspheres were formed from interconnected networks of cobalt oxide and graphene sheets. This provided wide active sites through the porous structure and offered efficient pathways for electron transfer and electrocatalytic reactions. The sensor had a wide range between 0 and 18 mM, high sensitivity of 628 μA mM^{-1} cm^{-2} as well as a low detection limit of 38 nM, which shows a high electrocatalytic performance.

On the other hand, Ali et al. (2015) synthesised cobalt oxide nanostructures through aqueous hydrothermal growth and applied them to develop an non-enzyme electrochemical glucose biosensor. The cobalt oxide nanostructures obtained exhibited a nanodisc like morphology with sizes between 300 and 500 nm. The sensor also showed a wide detection range between 0.5 and 5.0 mM, a sensitivity of 27.33 μA mM^{-1} cm^{-2} as well as a low detection limit of 0.8 μM. One unique aspect to note is that although there were common interferents like ascorbic acid, dopamine and uric acid, the sensor still displayed a significant selectivity towards glucose. Figure 10 shows the results of Ali's work.

Furthermore, Zhang et al. (2016a, b) synthesised cobalt oxide hollow nanododecahendra by a simple thermal decomposition of cobalt for the application of a glucose biosensor. The sensor developed exhibited an excellent performance for detecting glucose with a wide linear range of 2.0 μM to 6.06 mM, a low limit of detection of 0.58 μM as well as a high sensitivity of 708.4 μA mM^{-1} cm^{-2}. Using the cobalt oxide nanododecahendra, they also developed a non-enzyme glucose fuel cell which can produce steady milliamps of electricity for a significant period.

3.3 Nickel Oxide Based Composites

Nickel oxide is one of the most typical and important materials belonging to the family of transition metal oxides, which has been widely used in fabricating biosensors due to the presence of fast electron transport property, high electro-catalytic activity, high surface energy, low cost, chemical stability and biocompatibility (Fernandez-Garcia and Rodriguez 2007). Depending on the form, the material can have a band gap around 3.4 eV. Many nickel oxide-based nanostructures have been fabricated by different methods and excellent electrochemical performances have been demonstrated. Zhang et al. (2018) proposed a simple strategy to surface-engineer multiple stages of nickel oxide rod-on-sheet nanoarrays on nickel foam using the hydrothermal method. Their biosensor showed a superior sensing performance with a fast response time of less than 4 s, a wide linear response range of 0.75–3837 μM, a high sensitivity of 2739.5 μA mM^{-1} cm^{-2} and a low limit of detection of 0.75 μM. This

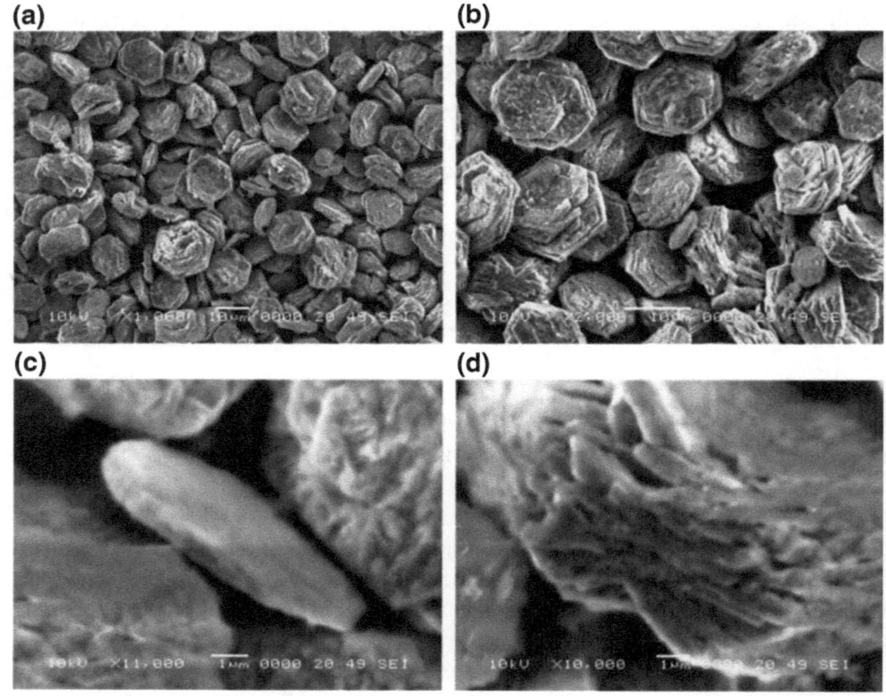

Fig. 10 SEM images of cobalt oxide nanodiscs. Reproduced with permission from Ali et al. (2015)

is due to the 3D architecture of the electrode. The nickel oxide nanoarray architectures provides a high surface area which enables a high usage of active materials and effective facilitation of charge/ions and mass transfer for electrocatalysis.

Furthermore, Sattarahmady et al. (2014) investigated different nickel-based composites for the catalytic oxidation of acetylcholine. Among the different nanocomposites made, the one from nickel nanoshells-carbon microparticles-Nafion showed the highest performance. It had been irreversibly oxidised by acetylcholine with excellent catalytic activity. Their nanocomposite showed a high sensing performance with detection limit of 49.33 nM and a sensitivity of 48.58 mAM^{-1} cm^{-2}. The sensitivity shown by the biosensor was better than previously reported literature. Finally, Zhao et al. (2016) used a simple aqueous deposition to develop uniform nickel oxide mesoporous nanowalls on 3D nickel foams coated with nanosheets made of reduced graphene oxide. As a biosensor, the electrode can detect glucose with a remarkable low limit of detection of 10 mM and an ultrahigh sensitivity of 3.23 mA mM^{-1} cm^{-2}. Figure 11 shows Zhao's work.

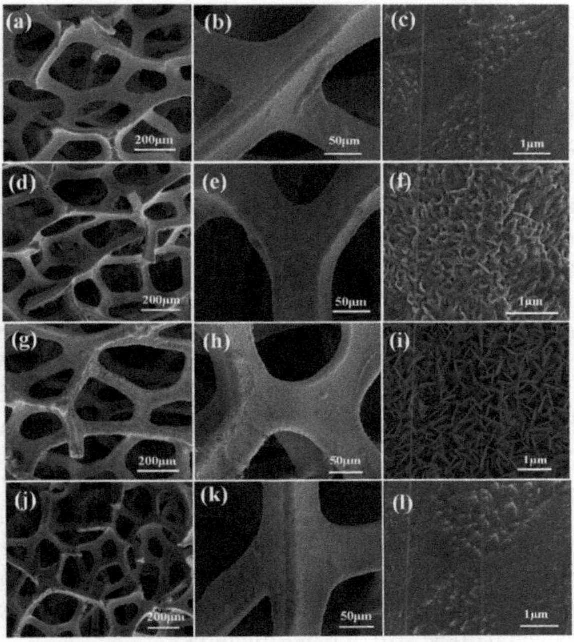

Fig. 11 SEM images of 3D nickel foam-graphene oxide-nickel oxide nanowall composites. Reproduced with permission from Zhao et al. (2016)

3.4 *Zinc Oxide Based Composites*

Zinc oxide is one of the transition metals that have been extensively studied over the last decade for biosensors because of its biocompatibility, chemical stability, high catalytic properties, non-toxicity, unique morphology, and sensitivity (Fernandez-Garcia and Rodriguez 2007). Depending on the form, the material can have a band gap between 3.3 and 4.0 eV. Many zinc oxide-based nanostructures have been produced using various methods and enhanced biosensing performances have been reported in these systems with promising applications. Nesakumar et al. (2016) developed zinc oxide nanocuboids on their platinum electrode for a highly sensitive acetylcholinesterase cyclic voltammetric biosensor. The electrode was tested to detect the presence of carbosulfan. Their biosensor displayed a linearity of 5–30 nM and a detection limit of 0.24 nM. Besides that, Perumal et al. (2016) used an aqueous hydrothermal method to fabricate a zinc oxide-gold hybrid worm-like nanostructure. The worm-like structure was formed by the doping of gold nanoparticles on the ZnO nanorods. The sensor was then used to test for the presence of cholera. The nanocomposite displayed an enhancement in its sensitivity and limit of detection because of its larger surface area caused by the gold nanoparticles. Besides that, Mazaheri et al. (2017) fabricated an electrode made from a combination of reduced graphene oxide and nickel nanoparticles decorated on zinc oxide nanoarrays created by a one-step hydrothermal method for glucose biosensors. The prepared biosensor possessed a low limit of detection of 0.15 μM, wide linear range over 0.5 μM to 1.11 mM, and a sensitivity of 2030 $\mu A\ cm^{-2}\ mM^{-1}$. Other electroactive substances such as ascorbic acid and uric acid did not produce any noticeable interference. These properties are due to the 3D structure with a large surface area (Fig. 12).

Over the past several years, there has been a number of exciting developments in the fabrication of unique metal oxide nanostructures. However, no particular group or institution has established a complete and detailed fabrication template of these nanostructures and the mechanism that governs them. Based on the four focus metals of this research paper, it can also be noted various techniques can be utilised to produce similar nanostructures with varying qualities and properties. Another importance is that the discovery of new novel metal oxide nanostructures can help improve the design and functionality of the biosensors used in the electronics and biomedical industry. These new nanostructures may display enhanced properties and their combination with constantly improving biosensing technology could lead to a wealth of unexpected applications. Hence, a need arises to complete the above missing research as well as work towards creating a finalized template for the detailed synthesis of metal oxide nanostructures and nanocomposites. This will help researchers develop a better understanding of the structural and property relationship, degree of control over morphology and particle size of nanostructures.

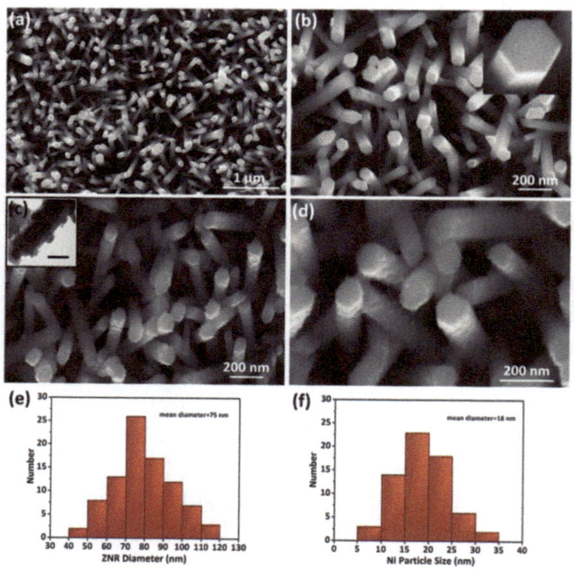

Fig. 12 SEM image of synthesised zinc oxide nanoarrays. Reproduced with permission from Mazaheri et al. (2017)

4 Conclusion

The development of metal oxide nanostructures has travelled a long journey since the last decade with the same noble objective of improving our sensing technologies. This dynamic material has been applied in many areas of research; from engineering, medical science, physics, chemistry and biology. It is a promising candidate for lowering detection limits and analysis time with greater accuracy in medical diagnostic systems. Throughout this chapter, the use of metal oxide nanostructures and their distinct improvement of biosensor characteristics have been well explored. However, absolute control over the fabrication of metal oxide nanostructures remains a difficulty. Some of the main issues that require the attention of researchers involve, creating cost effective, more sustainable ways of manufacturing these metal oxides and enabling a greater degree of control over the fabrication of the size and shape of the metal oxide nanostructures. Further progress in the area of metal oxide-based biosensing technology will boost up the future of medical diagnostics and health care.

References

Acharya S, Martha S, Sahoo PC, Parida K (2015) Glimpses towards the modification of perovskite with graphene-analogous materials in photocatalytic applications INORGANIC. https://doi.org/10.1039/c5qi00124b

Ali R, Nafady A, Hussain Z et al (2015) Development of sensitive non-enzymatic glucose sensor using complex nanostructures of cobalt oxide. Mater Sci Semicond Process 34:373–381. https://doi.org/10.1016/j.mssp.2015.02.055

Almeida I, Henriques F, Carvalho MD, Viana AS (2017) Carbon disulfide mediated self-assembly of laccase and iron oxide nanoparticles on gold surfaces for biosensing applications. J Colloid Interface Sci 485:242–250. https://doi.org/10.1016/j.jcis.2016.09.042

Ashik UPM, Kudo S, Hayashi J (2018) An overview of metal oxide nanostructures. Elsevier Ltd

Cevik E, Mehmet S, Baykal A (2012) A novel amperometric phenol biosensor based on immobilized HRP on poly (glycidylmethacrylate)-grafted iron oxide nanoparticles for the determination of phenol derivatives. Sens Actuators B. 173:396–405. https://doi.org/10.1016/j.snb.2012.07.026

Chakraborty M (2017) An overview of biosensors and devices. pp 1–24. https://doi.org/10.1016/b978-0-12-803581-8.10316-9

Chen M, Hou C, Huo D et al (2016) An ultrasensitive electrochemical DNA biosensor based on a copper oxide nanowires/ single-walled carbon nanotubes nanocomposite. Appl Surf Sci 364:703–709. https://doi.org/10.1016/j.apsusc.2015.12.203

Chen Q, Zhang L, Jiang F et al (2017) Chemical MnO_2 microsphere absorbing Cy5-labeled single strand DNA probe serving as powerful biosensor for effective detection of mycoplasma ovipneumoniae. Sens Actuators B Chem 244:1138–1144. https://doi.org/10.1016/j.snb.2017.01.104

Chinnasamy R, Rao GM, Thangavelu R (2015) Synthesis and electrocatalytic properties of manganese dioxide for non-enzymatic hydrogen peroxide sensing. Mater Sci Semicond Process 31:709–714. https://doi.org/10.1016/j.mssp.2014.12.054

Dhull V, Gahlaut A, Dilbaghi N, Hooda V (2013) Acetylcholinesterase biosensors for electrochemical detection of organophosphorus compounds : a review. https://doi.org/10.1155/2013/731501

Evrim U (2013) Surface modification of nanoparticles used in biomedical applications. Mod Surf Eng Treat 185–207. http://dx.doi.org/10.5772/55746

Fernández-Garcia M, Rodriguez J (2007) Metal oxide nanoparticles. doi:BNL-79479-2007-BC

Guo W, Li H, Li M (2014) Synthesis of nickel nanosheet/graphene composites for biosensor applications. Carbon N Y 79:636–645. https://doi.org/10.1016/j.carbon.2014.08.043

Hayashi H, Hakuta Y (2010) Hydrothermal synthesis of metal oxide nanoparticles in supercritical water. Mater (Basel) 3:3794–3817. https://doi.org/10.3390/ma3073794

Hosseini M, Mirzaei M, Torkzadeh-mahani M (2018) Double strand DNA-based determination of menadione using a Fe_3O_4 nanoparticle decorated reduced graphene oxide modified carbon paste electrode. Bioelectrochemistry 124:165–171. https://doi.org/10.1016/j.bioelechem.2018.07.014

Jha SN (2016) Biosensor. Rapid Detect Food Adulterants Contaminants, pp 125–145. https://doi.org/10.1016/b978-0-12-420084-5.00005-6

Liu G, Chen H, Lin G et al (2014) One-step electrodeposition of graphene loaded nickel oxides nanoparticles for acetaminophen detection. Biosens Bioelectron 56:26–32. https://doi.org/10.1016/j.bios.2014.01.005

Mahmood A, Naeem A (2017) Sol-gel-derived doped ZnO thin films. InTECH. https://doi.org/10.5772/67857

Malwal D, Packirisamy G (2018) Recent advances in the synthesis of metal oxide (MO) nanostructures. Elsevier Ltd

Mazaheri M, Aashuri H, Simchi A (2017) Three-dimensional hybrid graphene/nickel electrodes on zinc oxide nanorod arrays as non-enzymatic glucose biosensors. Sens Actuators B: Chem 251:462–471. https://doi.org/10.1016/j.snb.2017.05.062

Mohamad N, Lockman Z, Abdul K (2016) Study of ITO glass electrode modified with iron oxide nanoparticles and nafion for glucose biosensor application. Procedia Chem 19:50–56. https://doi.org/10.1016/j.proche.2016.03.116

Nesakumar N, Sethuraman S, Maheswari U (2016) Electrochemical acetylcholinesterase biosensor based on ZnO nanocuboids modified platinum electrode for the detection of carbosulfan in rice. Biosens Bioelectron 77:1070–1077. https://doi.org/10.1016/j.bios.2015.11.010

Noorbakhsh A, Salimi A (2011) Development of DNA electrochemical biosensor based on immobilization of SSDNA on the surface of nickel oxide nanoparticles modified glassy carbon electrode. Biosens Bioelectron 30:188–196. https://doi.org/10.1016/j.bios.2011.09.010

Perumal V, Hashim U (2013) Advances in biosensors: principle, architecture and applications. J Appl Biomed 11:1–34. https://doi.org/10.1016/j.jab.2013.02.001

Perumal V, Hashim U, Gopinath SCB et al (2016) A new nano-worm structure from gold-nanoparticle mediated random curving of zinc oxide nanorods. Biosens Bioelectron 78:14–22. https://doi.org/10.1016/j.bios.2015.10.083

Reetz MT (2008) Size—selective synthesis of nanostructured metal and metal oxide colloids and their use as catalysts

Roychoudhury A, Basu S, Jha SK (2016) Dopamine biosensor based on surface functionalized nanostructured nickel oxide platform. Biosens Bioelectron 84:72–81. https://doi.org/10.1016/j.bios.2015.11.061

Salimi A, Hallaj R, Soltanian S (2007) Immobilization of hemoglobin on electrodeposited cobalt-oxide nanoparticles: direct voltammetry and electrocatalytic activity. Biophy Chem 130:122–131. https://doi.org/10.1016/j.bpc.2007.08.004

Santos L, Neto JP, Crespo A et al (2015) Electrodeposition of WO3 nanoparticles for sensing applications

Sattarahmady N, Heli H, Vais RD (2014) A flower-like nickel oxide nanostructure: synthesis and application for choline sensing. Talanta 119:207–213. https://doi.org/10.1016/j.talanta.2013.11.002

Shen Z, Gao W, Li P et al (2016) Highly sensitive nonenzymatic glucose sensor based on nickel nanoparticle—attapulgite-reduced graphene oxide-modified glassy carbon electrode. Talanta 159:194–199. https://doi.org/10.1016/j.talanta.2016.06.016

Tarafdar JC, Adhikari T (2016) Nanotechnology in soil science

Tettamanti CS, Ramírez ML, Gutierrez FA et al (2018) Nickel nanowires-based composite material applied to the highly enhanced non-enzymatic electro-oxidation of ethanol. Microchem J 142:159–166. https://doi.org/10.1016/j.microc.2018.06.023

Thandavan K, Gandhi S, Nesakumar N et al (2015) Hydrogen peroxide biosensor utilizing a hybrid nano-interface of iron oxide nanoparticles and carbon nanotubes to assess the quality of milk. Sens Actuators B Chem 215:166–173. https://doi.org/10.1016/j.snb.2015.03.041

Tiwari I, Singh M, Mouli C, Sumana G (2015) Electrochemical genosensor based on graphene oxide modified iron oxide—chitosan hybrid nanocomposite for pathogen detection. Sens Actuators B Chem 206:276–283. https://doi.org/10.1016/j.snb.2014.09.056

Wahab HA, Salama AA, El Saeid AA et al (2018) Zinc oxide nano-rods based glucose biosensor devices fabrication. Results Phys 9:809–814. https://doi.org/10.1016/j.rinp.2018.02.077

Wang J, Li S, Zhang Y (2010) A sensitive DNA biosensor fabricated from gold nanoparticles, carbon nanotubes, and zinc oxide nanowires on a glassy carbon electrode. Electrochimica Acta 55(15): 4436–4440

Xia X, Zhang Y, Chao D et al (2014) Solution Synthesis of metal oxides for electrochemical energy storage applications. Nanoscale 6:5008–5048. https://doi.org/10.1039/c4nr00024b

Xu H, Zhang W (2017) Graphene oxide—Mno2 nanocomposite-modified glassy carbon electrode as an efficient sensor for H_2O_2. 28:143–148. https://doi.org/10.1016/j.cclet.2016.10.008

Yang M, Jeong J, Lee KG et al (2017) Biosensors and bioelectronics hierarchical porous microspheres of the Co_3 O_4 @ graphene with enhanced electrocatalytic performance for electrochemical biosensors. Biosens Bioelectron 89:612–619. https://doi.org/10.1016/j.bios.2016.01.075

Yu G, Zhao Q, Wu W et al (2016) A facile and practical biosensor for choline based on manganese dioxide nanoparticles synthesized in-situ at the surface of electrode by one-step electrodeposition. Talanta 146:707–713. https://doi.org/10.1016/j.talanta.2015.06.037

Zhang P, Guo D, Li Q (2014) Manganese oxide ultrathin nanosheets sensors for non-enzymatic detection of H_2O_2. Mater Lett 125:202–205. https://doi.org/10.1016/j.matlet.2014.03.172

Zhang E, Xie Y, Ci S et al (2016b) Porous Co_3O_4 hollow nanododecahedra for nonenzymatic glucose biosensor and biofuel cell. Biosens Bioelectron 81:46–53. https://doi.org/10.1016/j.bios.2016.02.027

Zhang Y, Zhao D, Zhu W et al (2018) Multi-stage nickel oxide rod-on-sheet nanoarrays on Ni foam: a superior catalytic electrode for ultrahigh-performance electrochemical sensing of glucose. Sens Actuators B Chem 255:416–423. https://doi.org/10.1016/j.snb.2017.08.078

Zhang Q, Sando D, Nagarajan V (2016a) Chemical route derived bismuth ferrite thin films and nanomaterials. J Mater Chem C 4:4092–4124. https://doi.org/10.1039/C6TC00243A

Zhao B, Wang T, Jiang L et al (2016) NiO mesoporous nanowalls grown on RGO coated nickel foam as high-performance electrodes for supercapacitors and biosensors. Electrochim Acta 192:205–215. https://doi.org/10.1016/j.electacta.2016.01.211

Nanoparticles; Their Use as Antibacterial and DNA Cleaving Agents

Irshad Ul Haq Bhat, Sayyed Jaheera Anwar, Emamalar Subramaniam and Aabid Hussain Shalla

Abstract Nanoparticles have revolutionized the field of nanotechnology, owing to their specialized characteristics. The immense use of nanoparticles in various fields has led to the great demand in terms of production as well as the fabrications methods. Apart from conventional methods, the use of environmental friendly methods had gained much impetus, particularly, using biological material, including plant extracts. The obtained metals based nanoparticles are applied in innumerable areas including drug discovery, antibacterial and DNA cleaving agents, while as DNA cleaving ability of nanoparticles is to be explored extensively. The various nanoparticles generated have been characterized by different spectroscopic, optical and thermal techniques to establish their precise structure and size. Thus, this chapter reports the importance of nanotechnology in the form of nanoparticles based on their various modes of synthesis, their biological significance in terms of targeting microbes and DNA.

Keywords Biological targets · Nanoparticles · Antimicrobial · DNA

1 Introduction

Nanotechnology is a broad arena of applied sciences in which the preparation, characterization and exploration of nano-scale materials ranging from 1 to 100 nm is studied (Shanmugaiah et al. 2015). The nanotechnology has also received tremendous attention in the field of life science, most importantly in biomedical science and biotechnology (Savithramma et al. 2011). This is because of its ability to deal with

I. U. H. Bhat (✉) · S. J. Anwar
School of Fundamental Science, Universiti Malaysia Terengganu, 21030 Kuala Terengganu, Malaysia
e-mail: irshadbhat78@gmail.com

E. Subramaniam
Faculty of Earth Science, Universiti Malaysia Kelantan, Campus Jeli, 17600 Jeli, Kelantan, Malaysia

A. H. Shalla
Department of Chemistry, Islamic University of Science and Technology, Awantipora 192122, Jammu and Kashmir, India

A. H. Bhat et al. (eds.), *Nanomaterials for Healthcare, Energy and Environment*, Advanced Structured Materials 118, https://doi.org/10.1007/978-981-13-9833-9_4

matter on the atomic and molecular level (Govindaraju et al. 2010). Nanotechnology is further directly and indirectly applied in the areas viz chemistry, physics, biology, medicine and material sciences (Das et al. 2013). Nanotechnology has successfully attained reputation in the broad spectrum areas of Paints, Textiles, cosmetics, pharma industry, food science, optics, mechanics, electronics and many more (Gurjas et al. 2012; Iravani 2011). The advanced application in nanotechnology has offered many new developments in life sciences, industrial biotechnology and health care (Gnanajobitha et al. 2013). The increasing progress in nanotechnology also promotes the betterment of other interdisciplinary field such as smart materials, therapeutics, and alternate source of energy, environmental services and also in other fields (Firdhouse et al. 2014). The advanced technology has derived many improvements and yet, current therapies have some negative effects such as existence of some serious diseases such as drug-induced disease and multiple-drug-resistant tuberculosis that urge more researches and development of nanotechnology (Zhou et al. 2012). Nanotechnology is now rising rapidly, dealing with devices production and construction. The nanoparticles synthesis accompanied by varied sizes, chemical compositions and morphologies are major part of nanotechnology. The great thing about nanotechnology is the development of vital technique in the synthesis and congregation of nanoparticles in a cleaner and eco-friendly way (Iravani et al. 2014). Nanotechnology has developed biosensing appliances for the detection of protein biomarkers, infectious agents, nucleic acids, cancer cells, drugs, etc. Recently developing tests based on nanotechnology aimed at timely detection, and subsequently controlling foodborne ailments have been reported (Krishna et al. 2018).

2 Nanoparticles

Nanoparticles have received a great attention as the fundamental building blocks of nanotechnology, a field of applied science and technology that regulates the matter over atomic and molecular scale (Govindaraju et al. 2010). This is because of their distinctive properties that differ related to bulk materials such as optical, electrical and catalytic properties. These unique properties exhibited by nanoparticles are contributed by the deviations in their size, structure, shape, distribution and composition of particles in it (Jena et al. 2015). These changes are made at the nanometer scale, this can eventually manipulate the properties of the nanoparticles by reducing or altering their size. The special features attributed to nanoparticles are the use of nano-sized inorganic and organic particles produced for extensive application in medical field including metal oxides of iron, zinc and copper in biomedical research (Azam et al. 2012; Chokkareddy and Redhi 2018).

Generally, there are numerous methods that have reported the nanostructure material synthesis, comprising physical, chemical, and hybrid methods (Salari et al. 2016). Based on the previous reports on the preparation methods of the nanoparticles, the extensively used methods are micelles process, microwave, salt reduction, radiolysis, ultrasonic, solvothermal synthesis, (Guzman et al. 2009), chemical reduction, thermal decomposition, sonochemical and photochemical (Shameli et al. 2012).

The application of physical method for nanoparticle synthesis require expensive equipment, elevated temperature, pressure and also, large space area for setting machineries; whereas, the use of chemical methods require harmful chemicals that may inherent in the synthesized nanoparticles and would prove to be hazardous to be used for medical purpose. Because of these reasons green approach towards nanoparticle synthesis have come into light which uses plants, fungus, bacteria, and algae instead of chemicals (Agarwal et al. 2017). The necessity of high energy, temperature, pressure and toxic chemicals are not needed in this method (Pattanayak and Nayak 2013). Out of a number of methods in green approach, plant-mediated nanoparticle synthesis is found to be more trustworthy and cost effective route due to the biodiversity of plant biomasses and provide single step synthesis (Pramila et al. 2018). In addition, the plant materials can be used as capping as well as reducing agents for the preparation of nanoparticles (Chokkareddy and Redhi 2018). Thus, scientists are concentrating on plant material assisted nanoparticle synthesis.

The plant based nanoparticles has achieved relatively a broad spectrum applications in the fields of medical, agricultural, catalytic, cosmetic, food, pollution control etc. (Vijayaraghavan and Ashokkumara 2017). There are huge number of problems in living conditions caused by different types of microorganisms through microbial contamination of air, water and soil (Ahamed et al. 2014). The paramount that fight infectious diseases are known as antimicrobial agents, any compound that locally kill microorganisms or lowered their growth, without creating any harm or toxic to surrounding tissues (Hajipour et al. 2012).

Hence, the researchers are trending towards the use of nanoparticles as antibacterial materials due to their excellent antimicrobial activities (Anandgaonker et al. 2015; Agarwal et al. 2017). The antibacterial activity of nanoparticles is highly influenced by their composition, intrinsic properties and the type of microorganism. It has been also reported that nanoparticles with smaller particle size show a good antimicrobial activity (Dizaj et al. 2015). Generally, the frequently used microorganisms to access the antimicrobial property of nanoparticles are *Escherichia coli* and *Staphylococcus aureus* (Azam et al. 2012). This is because it is most non-fastidious organisms and shows good bacteria culture reproducibility (Shameli et al. 2012).

Recent studies have revealed that, plant mediated synthesized metal nanoparticles like Fe_3O_4, Cu, Zn and, Ag are known to possess excellent antibacterial activities (Saif et al. 2016; Velvizhi and Ravi 2016).

Apart from that, there is ever rising interest towards studies done to inspect the interaction between nanoparticles and DNA. This interaction can cause the cleaving activity of the DNA. The cleavage of DNA and generation of oxidation-induced DNA adducts in contact with metal oxide nanoparticles have been proved by recent studies (Puja et al. 2015). However, the cleaving capability of the DNA by nanoparticles also has been not explored in detail. The high attention on the DNA cleavage activity can lead to the production of many useful drugs.

In the present scenario the term "nanoparticle" is well known and its popularity is increasing throughout the globe due to its vast applications. The term "*nano*" is derived from the Greek word "*nanos*," which means dwarf, and indicates a dimension on the scale of one-billionth (10–9) of a meter in size (Narayanan and Sakthivel 2010). Nanoparticles are a form of structure that can be found as materials with submicron dimension that ranging from 1 to 100 nm (Chandrasekar et al. 2013). Recently, nanoparticles have been consistently studied by researches owing to their uniqueness in terms characteristics, physical, chemical, mechanical, electronic, electrical, magnetic, thermal, dielectric, optical and biological properties that are much more differ from bulk materials (Abboud et al. 2014). The characteristics that exhibited by nanoparticles are highly influenced by their size, shape and composition. These properties can be modified easily by reducing or changing their size and structure, particularly when alteration occurs at nanometer scale (Azam et al. 2012). These unique properties contribute to the wide application of nanoparticles in catalysis, plasmonics, optoelectronics, biological sensor and pharmaceutical (Govindaraju et al. 2010). Nanoparticles are said to be more reactive than other molecules due to its greater surface area per weight than larger particles. On that account, nanoparticles have been utilized in many fields such as biosensors, biomedicine, and bio nanotechnology (Shanmugaiah et al. 2015). Thus, nanoparticles can be well thought-out as the building blocks for the generation of technology with application in various industrial segments (Abboud et al. 2014). Some nanoparticles have been designed in such way that is being widely applied to replace the usage of harmful chemicals (Sharma et al. 2014). Till now, there are various types of nanoparticles that have been successfully synthesized and applied in many fields.

3 Overview of Synthesis Methods of Nanoparticles

There are different synthetic methods of nanoparticles that have been reported previously and can be categorized as physical, chemical and biological methods (El-Aziz 2014). All this while, these methods have been applied successfully for the synthesizing various types of nanoparticles. However, there are some disadvantages by using these methods. Firstly, it is considered as hazardous to the environment because it involves production of dangerous byproducts in great quantity. Next, these methods involve monotonous and environmentally challenging procedures (Mariam et al. 2014). Moreover, the synthesis of nanoparticles by using physical and chemical methods have very low production rate compared to that of biological method (Kanchana et al. 2011). This leads to the researchers' efforts to develop simple methods of synthesis which consist environmentally acceptable solvent system (Jena et al. 2015). This encourages the biological mediators for the synthesis of nanoparticles using biological agents such as bacteria, fungus, yeast and plant extract as an alternatives for those hazardous chemical methods (El-Aziz 2014). Such methods are environmentally acceptable and ecofriendly as well as cheap and rapid (Ponarulselvam et al. 2012).

3.1 Physical Methods

Plasma arcing, thermal evaporate, ball milling, pulsed laser desorption, spray pyrolysis, ultrathin films, lithographic techniques, layer by layer growth, sputter deposition, diffusion flame and molecular beam epistaxis synthesis are some of the physical methods of nanoparticles synthesis (Joerger et al. 2000). Evaporation-condensation and laser ablation are the most widely used approaches in physical methods for synthesis nanoparticles. Some of the benefits of the physical methods are the preparation of thin films without solvent contamination and production of evenly distributed nanoparticles. However, there are some disadvantages of this method. This is because of the tube furnace that have been used at atmospheric pressure which takes up a large space, requires high amount of energy and longer time to become thermally stable (Iravani et al. 2014).

3.2 Chemical Methods

There are many different approaches of synthesizing nanoparticles through chemical methods. Electrodeposition, chemical solution deposition, sol–gel process, chemical vapour deposition (Panigrahi et al. 2004; Oliveira et al. 2005), soft chemical method, catalytic route, Langmuir Blodgett method, hydrolysis (Pileni 1997), wet chemical method and co-precipitation method (Gan et al. 2012) are the chemical methods are used for Nanoparticle synthesis. Some of them are chemical reduction with different organic or inorganic reducing agents, physicochemical reduction, electrochemical techniques, and radiolysis (Iravani et al. 2014). However, the application of chemical method to synthesis nanoparticles are becoming limited day by day because the chemicals that are toxic which are used get adsorbed onto the nanoparticles surface (Gannimani et al. 2014). However, there are some limitations in the application of this method in clinical field as it use toxic substances on the nanoparticles surface and also non-polar solvents (Abboud et al. 2014).

3.3 Biological Methods

Biological methods in preparing nanoparticles are more advantageous and has high prospect in current nanotechnology field. Based on a study by Kanchana et al. (2011), biological way for the nanoparticles synthesis is vital because it is believed to create a new platform to prevent and fight diseases. Some illustrations of green methods are mixed-valence polyoxometalates, polysaccharides, tollens, biological and irradiation method. The reducing agents and stabilizing agents that used in these methods are environment friendly and non-toxic (Iravani et al. 2014). According to El-Aziz (2014), the nanoparticles can be synthesized by using biological agents for example bacteria, fungi, yeast and plants. Moreover, synthesizing nanoparticles using green method is more advantageous compared to physical and chemical methods as it is cheaper, basically simple, and can be produced in a large scale without consuming toxic and harmful chemicals (Das et al. 2013). The stability of the synthesized nanoparticles is also high (Kanchana et al. 2011). Putting the advantages aside, microbial intermediated synthesis of nanoparticles is not recommended as it is less feasible due to its expensive and high maintenance of the aseptic conditions (Das et al. 2013). As studies revealed much valuable information on the advantages of the using plants in the synthesis of nanoparticles, biological method of synthesizing nanoparticles is highly interested by the potential factories (Das et al. 2013). Generally, plants contain various types of accumulated metals in the form of heavy-metals. Thus, the production of metal nanoparticles from the plant extracts has attained some need due to the hovering of economic and environmental alarms of the chemical and physical methods. Up to date, there are many plants that have been put to use for the synthesis of nanoparticles. Example of the plants used are *Azadirachta indica, Chenopodium album, Murraya koenigii, Ocimum sanctum, Anacardium occidentale, Hibiscus rosa-sinensis, Acalypha indica, Gliricidia sepium, Mangifera indica, Magnolia kobus, Diopyros kaki, Nelumbo nucifera* and *Helianthus annus* (Yilmaz et al. 2011). Typically, there are extracellular and intracellular methods of synthesizing nanoparticles such as leaf broth, fruits, stem, seed, flowers (Das et al. 2011), sun dried leaves and growing plants on various medium (Kanchana et al. 2011).

4 Characterization of Nanoparticles

Generally, the synthesized nanoparticles are characterized by various analysis methods. The regularly utlized techniques for the characterization are UV-Visible spectroscopy, Transmission Electron Miscroscopy (TEM), X-Ray Diffraction (XRD), scanning electron microscopy (SEM), Fourier transform infrared spectroscopy (FTIR), Energy dispersive X-ray (EDX), Atomic Force Microscopy (AFM) (Logeswari et al. 2015) and Thermogravimetric Analysis (TGA). These analyses are carried out to monitor the formation of nanoparticles and to determine the size, shape, morphology, composition, dielectric environment and structure of the nanoparticles (Shameli et al. 2012). Generally, UV-Visible spectroscopy is used to determine the size and shape of nanoparticles in aqueous form. Dynamic light scattering (DLS) is used to measure the average size distribution of metal nanoparticles dispersed in aqueous media (Arshad et al. 2018). It is also used to learn the formation of the nanoparticles, given that there is existence of surface plasmon resonance for the metal (Mahmoud et al. 2013). Specifically, X-Ray diffraction characterization is done to identify intensity of peak, position and width of the nanoparticles (Dubey et al. 2009). The morphology and shape of the synthesized nanoparticles can be identifies using the TEM and SEM analysis (Al-Othman et al. 2014). Besides, FTIR analysis is able to reveal the details on the kind of functional group that exist in the nanoparticles

(Gannimani et al. 2014). The thermal stability of the newly synthesized nanoparticles can be identified through the Thermogravimetric Analysis (TGA).

The magnetic density studies are done using Gauss or Tesla meter, Vibrating sample magnetometer (VSM) for the measurement of magnetic properties; Hysteresis loop, field-cooled and zero-field-cooled curves are deliberated by superconducting quantum interference magnetometer. X-ray photoelectron spectroscopy (XPS) has been bought into service to determine oxidation state of produced nanoparticle (Paramita and Rashmi 2018).

5 Types of Nanoparticles

Classification of nanoparticles is usually depends on their dimensionality, morphology, alignment, accumulation and regularity. The functional properties and toxicology of nanoparticles is mainly based upon their shape and morphology (Buzea et al. 2007). Nanoparticles are ordered as one-, two- and three-dimensional nanoparticles on the basis of dimensionality. One-dimensional nanoparticles comprise thin films that are applicable in electronics and sensor devices; two-dimensional nanoparticles include carbon nanotubes of high adsorption ability and stability where, three-dimensional nanoparticles contain dendrimers, quantum dots, etc. (Pal et al. 2011). Nanoparticles are differentiated into flat, spherical and crystalline depending upon the morphology. Based on number of nanoparticles, it can exist in single form or in the nanocomposites form (Chokkareddy and Redhi 2018). Nanoparticles can also be categorized as carbon-based, polymer-based, lipid-based, metal oxide-based, metal-based, and magnetic nanomaterials based on the ingredients used in their synthesis (Ibrahim and Ehab 2018).

Generally, particles of an element are exists in many forms which may be reactive or unreactive. Similarly, nanoparticles that are synthesis through nantechnology may available as single metal nanoparticles or oxides of metal nanoparticles. Some of the metal nanoparticles include gold, silver, nickel, cobalt, zinc, platinum, palladium and copper (Shaik et al. 2014); whereas that of metal oxides include ZnO, CuO, Fe_3O_4, MgO (Shamaila et al. 2018). Both types of the nanoparticles reveal different physical and chemical properties (Fernández-García and Rodriguez 2007).

5.1 *Metal Oxide Nanoparticles*

As our world is developing with the presence advanced technology, fields such as chemistry, physics and material science are accompanying the development. Metal oxides hold a significant role in those areas. This is due to the ability of metal elements to form large variety of oxide compounds. When focusing to metal oxide nanoparticles, they can show physical and chemical properties that are much unique when compared to bulk materials. This is due to its smaller size and higher density of corner surface sites (Fernández-García and Rodriguez 2007). The metal oxide nanoparticles of transition metals are essential division of semiconductors that have broad application in magnetic storage media, gas sensors, solar energy transformation, and electronics (Abboud et al. 2014). Generally, metal oxides nanoparticles are highly ionic in nature. They can be organized with very high surface areas as well as with crystal morphologies that exhibits various corners and other reactive sites (Stoimenov et al. 2002). Such properties allow metal nanoparticles such as TiO_2 and ZnO have used in many polymeric materials. The firmness, toughness, and service life of polymeric materials can be improved by adding those types of nanoparticles in it (Jeng and Swanson 2006). However, there are studies that revealed that the metal oxides nanoparticles could be toxic to certain organisms. They stated that the presence of these particles in sunscreens could be more dangerous than the UV radiation and presence in solar batteries can leads to adverse environmental effects than carbon dioxide release from conventional energy sources (Puzyn et al. 2011).

5.2 *Metal Nanoparticles*

Nowadays, there is high demand for metal nanoparticles in the field of nanotechnology. The focus is high in the metal nanoparticles production because they are well-known for their uniqueness in terms of optical, electronic, magnetic, mechanical and chemical properties which distinguish them from bulk materials (Gnanajobitha et al. 2013). Importantly, the role of metal nanoparticles in the production of newfangled biosensors and in improvement of prevailing biosensing techniques so that it can fulfill the highly sensitive biomolecular diagnostics requirement is noticeable. Metal nanoparticles also widely used for the expansion of biosensors variety such as nanobiosensors, nanoprobes and other nanotechnology-based tools which can be advantageous in scientific research on basic biology. This is because it has unique physicochemical properties. This property also leads to its additional application in biosensors such as infrared, fluorescence and Raman spectroscopy (Doria et al. 2012).

5.3 *Cadmium (Cd)*

Cadmium (Cd) is a metal which is lustrous, silver-white, ductile and very malleable. There are many cadmium based nanoparticles that have been synthesized over the past few years. Among them, CdS nanoparticles is the most widely synthesized and used.

Its application is very promising as it has high photosensitivity. Thus, they are used to detect visible radiations, improving effectiveness of solar cells, in LEDs and as sample photoconductor in optoelectronic devices (Bansal et al. 2012). Other application that utilizes CdS are in transistors, photovoltaics, photocatalysis, environmental, photoluminescence, infrared photodetectors and biological sensors (El-Shanshoury et al. 2012). Another form of cadmium based nanoparticles is CdO which has some unique properties such as having direct band gap of 2.3 eV. This property of CdO nanoparticles allows it be applied in works of cadmium-coated baths and paint pigments manufacturing (Manickathai et al. 2008). These cadmium based nanoparticles are defined as appropriate emitting 'core' materials they can emit brightly in the visible range and near the region of infrared in the electromagnetic spectrum (Rodríguez-Fragoso et al. 2012).

5.4 Selenium (Se)

Naturally, selenium can be found in various oxidation states such as elemental selenium, selenide, selenite and selenate with oxidation states of 0, −2, +4 and +6 respectively. Selenium is said to be analogous to sulphur in terms of structure (Sasidharan and Balakrishnaraja 2014). Selenium (Se) nanoparticles have many unique properties and great potential applications. A report by Tam et al. (2010) shows that these properties allow its wide application for various applications. Selenium has numbers of roles in many different applications such as photovoltaic cells, photographic exposure meters, rectifiers, and xenography. Selenium is also well-known for its function as one of the trace minerals that is essential which helps to prevent cellular damage, thyroid gland function regulation and in immune system functioning (Sharma et al. 2014).

Another importance of selenium is as an anti-oxidant in human body. Thus, selenium nanoparticles also draw high attention other than its uniqueness in mechanical, optical, electrical, biologic and chemical properties. There are many methods of synthesis of Se nanoparticles have been reported before, that includes physical vapor deposition, wet chemical method and vapor phase diffusion (Ingole et al. 2010). The application of selenium in pharmaceutical and industrial sectors such as pesticides and algaecides is undeniable (Sasidharan and Balakrishnaraja 2014).

5.5 Copper (Cu)

Copper is among one of the metal nanoparticles that has received great attention due to its cheap and high availability likened to other nanoparticles. Besides that, copper nanoparticles have high thermal conductivity in nature, high surface area to volume ratio, cheaper manufacture cost, catalytic activity, antibacterial potency, optical and magnetic properties than any other nanoparticles (Umer et al. 2014). There are numerous means for nanoparticles synthesis that have been reported so far, such as chemical reduction, colloidal synthesis with both reduction and extraction steps as well as evaporation and condensation of metal vapor. Rapid oxidation of copper leads to the difficulty during the process of the stable, uniformly-shaped and monodisperse copper nanoparticles synthesis (Kathad and Gajera 2014). The main function that copper holds is as a biocatalyst that assists the biological reaction in human body (Chityala et al. 2014).

6 Application of Nanoparticles

The applications of nanoparticles are promisingly high as it is used in many fields based on their properties and capability. Generally, silver nanoparticles is amongst the extensively used nanoparticles as it is highly applicable as catalysts and opto-electronics, detection, diagnosis, antimicrobials and therapeutics in medical fields (Sulaiman et al. 2013). Silver nanoparticles are also famous for their disinfecting property and application in traditional medicines culinary items (Dastager 2010). It also has demand in the production of medicinal products such as skin ointments, creams for wound infection, medical devices as well as silver impregnated polymers (Sharma et al. 2014). Next is zinc oxide nanoparticle which is a good antibacterial agent as well as has high chemical stability and electronic properties (Satheeskumar et al. 2014). Gold nanoparticles, also known as nanogold is another type of nanoparticle that has provided benefits in many fields such as medicine, nursing, cosmetology, dentistry and the construction industry Pulit and Banach (2013). It has been used as capable agent in thermo ablation, radiotherapy and drug and gene delivery (Firdhouse et al. 2014). Nevertheless, iron oxide nanoparticles also has proved it's extend through the application in cell separator, contrast agent in MRI imaging system, drug delivery in cancer therapy and hyperthermia as it is biocompatible and cheap (Palanisamy et al. 2013). Zero valent nanoparticle is also one of the nanoparticles that have been used in medical procedures, storage and usage of food materials (Chandrasekar et al. 2013). Subsequently, copper oxide nanoparticles have grabbed higher attention as it is the simplest member of copper compound which present with variety of beneficial physical properties. This includes electron correlation effects, high temperature; superconductivity, and spin dynamics. These properties allow it to be used as in catalysis, batteries, heat transfer fluids, gas sensors, and solar energy. Moreover, its crystal structures acquire a narrowband gap, which is responsible for the photocatalytic and photovoltaic properties (Ahamed et al. 2014). Last but not least, nickel nanoparticles received great focus due to their contribution as catalysts and as magnetic materials (Mariam et al. 2014). Nanomaterials have expanded its application in the field of food safety through food processing. Nanomaterials have been used for detecting and removing food contaminants as antimicrobials and antioxidants (Kuswandi et al. 2017; Naghdi et al. 2016).

6.1 Antimicrobial Activity of Nanoparticles

One of the major reasons for nanoparticles have been receiving great attention is the potent antimicrobial properties on microorganisms. A previous research has proven that several categories of nanoparticles with antimicrobial activity have shown their efficiency towards infectious diseases (Hajipour et al. 2012). The antimicrobial property of the nanoparticles is a subject to its composition, intrinsic properties, surface modification and the type of microorganisms (Dizaj et al. 2015). The antimicrobial activity of nanoparticles has been proven through various studies with human pathogenic bacteria such as *Escherichia coli* and *Staphylococcus aureus* (Azam et al. 2012). It has been outlined that the fact behind the antimicrobial property of nanoparticles is that the reaction that takes place between nanoparticles and bacteria which includes biosorption, nanoparticles breakdown or aggregation and cellular uptake (Zhou et al. 2012). Theoretically, it is stated that the antimicrobial activity takes place when the nanoparticles attaches and cause the permeability of the cell membrane, leading to the effects on the proper transport through the membrane, which will eventually cause cell damages and death (Gornicka et al. 2014).

Senthil and Ramesh investigated the antibacterial activity of *Tridax procumbens* synthesised iron oxide (Fe_3O_4) nanoparticles hostile to gram negative bacteria *Pseudomonas aeruginos* (Senthil and Ramesh 2012). The biosynthesized Ag NPs using plant extract have been studied that show excellent antimicrobial behavior in opposition to both *E. coli* and *S. aureus* (Velvizhi and Ravi 2016). Pectin-capped Pt nanoparticles showed toxicity against Gram-positive as well as Gram-negative bacteria in both in vitro and in vivo studies (Niranjan et al. 2018).

6.2 DNA Cleavage Activity

On the basis of literature survey, it can be clearly noticed that the DNA has been the main target of the study. For an example, in antitumor drugs, the small molecules and DNA interaction could damage the cancer cells, which will eventually prevent the cancer cell division and cause cell death. This occurs when the metal ion binds to the specific site of DNA and makes some changes on the DNA. The changes are due to the reduction and oxidation reaction of the metal ion that are bound to the DNA (Babu et al. 2011). Recent studies have demonstrated DNA cleavage properties of biosynthesized CuO nanoparticle using Chamomile flower extract (Duman et al. 2016). Another study has shown antibacterial effect of green Ag nanoparticle through DNA digestion (Mohammad et al. 2018).

7 Conclusion

Several synthetic approaches have been discussed for the metal nanoparticles synthesis, out of which the plant based metal nanoparticle synthesis, proves to be more successful. These methods provide perfect morphology to nanoparticles, act as a stabilizing agent, toxic free reactions and easy access which lead to prefer to other conventional methods. The present chapter focuses on the antibacterial and emerging DNA cleavage activity of green metal nanoparticles. These activities help nanotechnology to extend its wings in the food processing for removing food contaminants and in the medical field to treat diseases like cancer respectively. Special focus on non-noble metal nanoparticles namely Cd, Se and, Cu have been made as they are less expensive compared to noble metals.

References

Abboud Y, Saffaj T, Chagraoui A et al (2014) Biosynthesis, characterization and antimicrobial activity of copper oxide nanoparticles (Conps) produced using brown alga extract (*Bifurcaria bifurcata*). Appl Nanosci 4(5):571–576

Agarwal H, Venkat Kumar S, Rajeshkumar S (2017) A review on green synthesis of zinc oxide nanoparticles—an eco-friendly approach. Resour Effic Technol 3:406–413

Ahamed M, Alhadlaq HA, Khan MAM et al (2014) Synthesis, characterization, and antimicrobial activity of copper oxide nanoparticles. J Nanomater. https://doi.org/10.1155/2014/637858

Al-Othman MR, El-Aziz ARMABD, Mahmoud MA et al (2014) Application of silver nanoparticles as antifungal and antialatoxin B1 produced by *Aspergillus flavus*. Dig J Nanomater Bios 9(1):151–157

Anandgaonker P, Kulkarni G, Gaikwad S et al (2015) Synthesis of TiO_2 nanoparticles by electrochemical method and their antibacterial application. Arab J Chem. https://doi.org/10.1016/j.arabjc.2014.12.015

Arshad M, Khan A, Farooqi ZH et al (2018) Green synthesis, characterization and biological activities of silver nanoparticles using the bark extract of *Ailanthus altissima*. Mater Sci-Poland 36(1):21–26

Azam A, Ahmed AS, Oves M et al (2012) Antimicrobia activity of metal oxide nanoparticles against Gram-positive and Gram-negative bacteria: a comparative study. Int J Nanomed 7:6003–6009

Babu MSS, Patrudu TB, Reddy KH (2011) DNA binding and cleavage activity of binuclear metal complexes with benzil-α-monoxime thiosemicarbzone. Eur J Chem 8(S1):S309–S317

Bansal P, Jaggi N, Rohilla SK (2012) "Green" synthesis of Cds nanoparticles and effect of capping agent concentration on crystallite size. Res J Chem Sci 2(8):69–71

Buzea C, Pacheco II, Robbie K (2007) Nanomaterials and nanoparticles: sources and toxicity. Biointerphases 2(4):17–71

Chandrasekar N, Kumar KMM, Balasubramnian KS et al (2013) Facile synthesis of iron oxide, iron-cobalt and zero valent iron nanoparticles and evaluation of their anti microbial activity, free radicle scavenginging activity and antioxidant assay. Dig J Nanomater Bios 8(2):765–775

Chityala VK, Kumar KS, Macha R et al (2014) DNA cleavage, cytotoxic activities, and antimicrobial studies of ternary copper (ii) complexes of isoxazole schiff base and heterocyclic compounds. Bioinorg Chem Appl. https://doi.org/10.1155/2014/691260

Chokkareddy R, Redhi G (2018) Green synthesis of metal nanoparticles and its reaction mechanisms: synthesis, characterization and their applications. In: Green metal nanoparticles. https://doi.org/10.1002/9781119418900.ch4

Das RK, Gogoi N, Bora U (2011) Green synthesis of gold nanoparticles using *Nyctanthes arbor-tristis* flower extract. Bioprocess Biosyst Eng 34(5):615–619

Das J, Paul DM, Velusamy P (2013) *Sesbania grandiflora* leaf extract mediated green synthesis of antibacterial silver nanoparticles against selected human pathogens. Spectrochim Acta A 104:265–270

Dastager SG (2010) Antimicrobial activity of silver nanoparticles. Dig J Nanomater Bios 5(2):447–451

Dizaj SM, Mennati A, Jafari S et al (2015) Antimicrobial activity of carbon-based nanoparticles. Adv Pharm Bull 5(1):19–23

Doria G, Conde J, Veigas B et al (2012) Noble metal nanoparticles for biosensing applications. Sensors 12(2):1657–1687

Dubey M, Bhadauria S, Kushwah BS (2009) Green synthesis of nanosilver particles from extract of *Eucalyptus hybrida* (Safeda) leaf. Dig J Nanomater Bios 4(3):537–543

Duman F, Ocsoy I, Kup FO et al (2016) Chamomile flower extract-directed CuO nanoparticle formation for its antioxidant and DNA cleavage properties. Mater Sci Eng, C 60:333–338

El-Aziz ARA (2014) Eco-friendly biosynthesis of silver nanoparticles by *Aspergillus parasiticus*. Dig J Nanomater Bios 9(4):1485–1492

El-Shanshoury AER, Sobhy EE, Mohamed EE (2012) Rapid biosynthesis of cadmium sulfide (CdS) nanoparticles using culture supernatants of *Escherichia coli* ATCC 8739, *Bacillus subtilis* ATCC 6633 and *Lactobacillus acidophilus* DSMZ 20079T. Afr J Biotechnol 11(31):7957–7965

Fernández-García M, Rodriguez JA (2007) Metal oxide nanoparticles. Nanomaterials: inorganic and bioinorganic perspectives. Brookhaven National Laboratory, Upton

Firdhouse MJ, Lalitha P, Sripathi SK (2014) An undemanding method of synthesis of gold nanoparticles using *Pisonia grandis* (R. Br.). Dig J Nanomater Bios 9(1):385–393

Gan PP, Ng SH, Huang Y et al (2012) Green synthesis of gold nanoparticles using palm oil mill effluent (POME): a low-cost and eco-friendly viable approach. Bioresour Technol 113:132–135

Gannimani R, Perumal A, Krishna SB et al (2014) Synthesis and antibacterial activity of silver and gold nanoparticles produced using aqueous seed extract of *Protorhus longifolia* as a reducing agent. Dig J Nanomater Bios 9(4):1669–1679

Gnanajobitha G, Paulkumar K, Vanaja M et al (2013) Fruit-mediated synthesis of silver nanoparticles using *Vitis vinifera* and evaluation of their antimicrobial efficacy. J Nanostruct Chem 3(1):1–6

Gornicka E, Mikiciuk J, Wronska A et al (2014) The influence of silver nanoparticles on fecal bacteria susceptibility. Dig J Nanomater Bios 9(1):347–354

Govindaraju K, Tamilselvan S, Kiruthiga V et al (2010) biogenic silver nanoparticles by *Solanum torvum* and their promising antimicrobial activity. J Biopestic 3(1):394–399

Gurjas K, Tanvir S, Amit (2012) Nanotechnology: a review. Int J Edu App Res 2(1)

Guzman MG, Dille J, Godet S (2009) Synthesis of silver nanoparticles by chemical reduction method and their antibacterial activity. Int J Biomol Eng 2(3):04–111

Hajipour MJ, Fromm KM, Ashkarran AA et al (2012) Antibacterial properties of nanoparticles. Trends Biotechnol 30(10):499–511

Ibrahim ME, Ehab S (2018) Green synthesis of metallic nanoparticles using biopolymers and plant extracts. In: Green metal nanoparticles. Scrivener Publishing LLC, pp 293–319

Ingole AR, Thakare SR, Khati NT et al (2010) Green synthesis of selenium nanoparticles under ambient condition. Chalcogenide Lett 7(7):485–489

Iravani S (2011) Green synthesis of metal nanoparticles using plants. Green Chem 13:2638–2650

Iravani S, Korbekandi H, Mirmohammadi SV et al (2014) Synthesis of silver nanoparticles: chemical, physical and biological methods. Res Pharm Sci 9(6):385–406

Jena J, Pradhan N, Dash BP et al (2015) Pigment mediated biogenic synthesis of silver nanoparticles using diatom *Amphora* Sp. and its antimicrobial activity. J Saudi Chem Soc 19(6):661–666

Jeng HA, Swanson J (2006) Toxicity of metal oxide nanoparticles in mammalian cells. J Environ Sci Health A 41(12):2699–2711

Joerger R, Klaus T, Granqvist CG (2000) Biologically produced silver–carbon composite materials for optically functional thin-film coatings. Adv Mater. https://doi.org/10.1002/(sici)1521-4095(200003)12:6%3c407::aid-adma407%3e3.0.co;2-o

Kanchana A, Agarwal I, Sunkar S et al (2011) Biogenic silver nanoparticles from *Spinacia oleracea* and *Lactuca sativa* and their potential antimicrobial activity. Dig J Nanomater Bios 6(4):1741–1750

Kathad U, Gajera HP (2014) Synthesis of copper nanoparticles by two different methods and size comparision. Int J Pharma Bio Sci 5(3):533–540

Krishna VD, Wu K, Su D, Cheeran MCJ, Wang J-P, Perez A et al (2018) Nanotechnology: review of concepts and potential application of sensing platforms in food safety. Food Microbiol 75:47–54

Kuswandi B, Futra D, Heng LY (2017) Nanosensors for the detection of food contaminants A2-Oprea, Alexandra Elena. In: Grumezescu AM (ed) Nanotechnology applications in food. Academic Press, pp 307–333

Logeswari P, Silambarasan S, Abraham J (2015) Synthesis of silver nanoparticles using plants extract and analysis of their antimicrobial property. J Saudi Chem Soc 19(3):311–317

Mahmoud M, Al-Sohaibani SA, Al-Othman MR et al (2013) Synthesis of extracellular silver nanoparticles using *Fusarium semitectum* (KSU-4) isolated from Saudi Arabia. Dig J Nanomater Bios 8(2):589–596

Manickathai K, Viswanathan SK, Alagar M (2008) Synthesis and characterization of Cdo and Cds nanoparticles. Indian J Pure Appl Phys 46(8):561–564

Mariam AA, Kashif M, Arokiyaraj S et al (2014) Bio-synthesis of NiO and Ni nanoparticles and their characterization. Dig J Nanomater Bios 9(3):1007–1019

Mohammad MK, Mehrnaz K, Amir R (2018) A comparative study of stability, antioxidant, DNA cleavage and antibacterial activities of green and chemically synthesized silver nanoparticles. Artif Cells Nanomed Biotechnol 46(S3):S1022–S1031

Naghdi M, Taheran M, Sarma SJ et al (2016) Nanotechnology to remove contaminants. In: Eric (eds) Nanoscience in food and agriculture 1. Springer International Publishing, Cham, pp 101–128

Narayanan KB, Sakthivel N (2010) Biological synthesis of metal nanoparticles by microbes. Adv Colloid Interface Sci 156(1–2):1–13

Niranjan T, Rajasekar C, Gan G. Redhi (2018) Green synthesis of platinum nanoparticles and their biomedical applications. In: Green metal nanoparticles. Scrivener Publishing LLC, pp 603–627

Oliveira MM, Ugarte D, Zanchet D et al (2005) Influence of synthetic parameters on the size, structure, and stability of dodecanethiol-stabilized silver nanoparticles. J Colloid Interface Sci 292:429–435

Pal SL, Jana U, Manna PK et al (2011) Nanoparticle: an overview of preparation and characterization. J Appl Pharm Sci 01:228–234

Palanisamy KL, Sundaram NM, Devabharathi V et al (2013) Synthesis and characterization of olive oil mediated iron oxide nanoparticles. Dig J Nanomater Bios 8(2):607–612

Panigrahi S, Kundu S, Ghosh SK et al (2004) General method of synthesis for metal nanoparticles. J Nanopart Res 6:411–414

Paramita K, Rashmi (2018) Green tiny magnets: an economic and eco-friendly remedy for environmental damage. In: Green metal nanoparticles. Scrivener Publishing LLC, pp 245–292

Pattanayak M, Nayak PL (2013) Ecofriendly green synthesis of iron nanoparticles from various plants and spices extract. Int J Plant Anim Environ Sci 3(1):68–78

Pileni MP (1997) Nanosized particles made in colloidal assemblies. Langmuir 13(13):3266–3276

Ponarulselvam S, Paneerselvam C, Murugan A (2012) Synthesis of silver nanoparticles using leaves of *Catharanthus roseus* Linn. G. Don and their antiplasmodial activities. Asian Pac J Trop Biomed 2(7):574–580

Pramila K, Ravi KY, Deepak KS et al (2018) Biogenesis of metal nanoparticles and their pharmacological applications: present status and application prospects. J Nanostruct Chem 8(3):217–254

Puja K, Cynthia O, Boon HB et al (2015) Nanotoxicity: an interplay of oxidative stress, inflammation and cell death. Nanomaterials 5(3):1163–1180

Pulit J, Banach M (2013) Environment-friendly method for obtaining gold nanoparticles based on plant extract. Dig J Nanomater Bios 8(3):1295–1300

Puzyn Tomasz, Bakhtiyor R et al (2011) Using nano-QSAR to predict the cytotoxicity of metal oxide nanoparticles. Nat Nanotechnol 6(3):175–178

Rodríguez-Fragoso P, Reyes-Esparza J, León-Buitimea A et al (2012) Synthesis, characterization and toxicological evaluation of maltodextrin capped cadmium sulfide nanoparticles in human cell lines and chicken embryos. J Nanobiotechnol. https://doi.org/10.1186/1477-3155-10-47

Saif S, Tahir A, Chen Y (2016) Green synthesis of iron nanoparticles and their environmental applications and implications. Nanomaterials 5(209):1–26

Salari Z, Danafar F, Dabaghi S et al (2016) Sustainable synthesis of silver nanoparticles using macroalgae Spirogyra varians and analysis of their antibacterial activity. J Saudi Chem Soc 20(4):459–464

Sasidharan S, Balakrishnaraja R (2014) Comparison studies on the synthesis of selenium nanoparticles by various micro-organisms. Int J Pure App Biosci 2(1):112–117

Satheeskumar S, Ramesh K, Srinivasan N (2014) Exploration Of synthesis, structural, morphology and antibacterial activity of Zn_{1-x-y}, Mg_x, Al_yO nano particles. Dig J Nanomater Bios 9(4):1323–1330

Savithramma N, Rao ML, Rukmini K (2011) Antimicrobial activity of silver nanoparticles synthesized by using medicinal plants. Int J Chem Tech Res 3(3):1394–1402

Senthil M, Ramesh C (2012) Biogenic synthesis of Fe_3O_4 nanoparticles using *Tridax procumbens* leaf extract and its antibacterial activity on *Pseudomonas aeruginosa*. Dig J Nanomater Bios 7(3):1655–1660

Shaik S, Mkize L, Khumalo M et al (2014) *Tetradenia riparia*-mediated synthesis of nano-gold particles. Dig J Nanomater Bios 9(2):567–573

Shamaila S, Sajjad AKL, Najam-Ul-Athar R et al (2018) Green synthesis of metal-based nanoparticles and their applications. In: Green metal nanoparticles. Scrivener Publishing LLC, pp 23–77

Shameli K, Ahmad MB, Jazayeri SD (2012) Investigation of antibacterial properties silver nanoparticles prepared via green method. Chem Cent J 6(1):1–10

Shanmugaiah V, Harikrishnan H, Al-Harbi NS et al (2015) Facile synthesis of silver nanoparticles using *Streptomyces* Sp. VSMGT1014 and their antimicrobial efficiency. Dig J Nanomater Bios 10(1):179–187

Sharma G, Sharma AR, Kurian M et al (2014) Green synthesis of silver nanoparticle using *Myristica fragrans* (NUTMEG) seed extract and its biological activity. Dig J Nanomater Bios 9(1):325–332

Stoimenov PK, Klinger RL, Marchin GL et al (2002) Metal oxide nanoparticles as bactericidal agents. Langmuir 18(17):6679–6686

Sulaiman GM, Mohammad AAW, Abdul-Wahed HE et al (2013) Biosynthesis, antimicrobial and cytotoxic effects of silver nanoaparticles using *Rosmarinus officinalis* extract. Dig J Nanomater Bios 8(1):273–280

Tam K, Ho CT, Lee JH et al (2010) Growth mechanism of amorphous selenium nanoparticles synthesized by *Shewanella* sp. HN-41. Biosci Biotechnol Biochem 74(4):696–700

Umer A, Naveed S, Ramzan N et al (2014) A green method for the synthesis of copper nanoparticles using L-ascorbic acid. Materia (Rio J) 19(3):197–203

Velvizhi K, Ravi S (2016) Green synthesize and characterization of the plant mediated silver nanoparticles using *Solanum nigrum* leaf extract. Int J Sci Res (IJSR) Index Copern Value 79:57

Vijayaraghavan K, Ashokkumara T (2017) Plant-mediated biosynthesis of metallic nanoparticles: a review of literature, factors affecting synthesis, characterization techniques and applications. J Environ Chem Eng 5(5):4866–4883

Yilmaz M, Turkdemir H, Kilic MA et al (2011) Biosynthesis of silver nanoparticles using leaves of *Stevia rebaudiana*. Mater Chem Phys 130(3):1195–1202

Zhou Y, Kong Y, Kundu S et al (2012) Antibacterial activities of gold and silver nanoparticles against *Escherichia coli* and Bacillus Calmette Guerin. J Nanobiotechnol 10(19):1–9

Antimicrobial Property of Biosynthesized Silver Nanoparticles

Santheraleka Ramanathan, Subash C. B. Gopinath, M. K. Md Arshad, Prabakaran Poopalan, Veeradasan Perumal and Mohamed Shuaib Mohamed Saheed

Abstract The reduction of silver metal through the headway of nanotechnology results in the production of silver nanoparticle (AgNP). It is an apparent metallic nanoparticle widely recognized in nanotechnology for its application in various platforms. Physical, chemical and biological methods have been studied and developed in the past decade to synthesize crystalline AgNP. Among these three, biological synthesis known as green synthesis has been extensively accepted as the most eco-accommodating and an efficient technique for the production of AgNP. Biological method consumes bio-reducing agents present in the bio-extract used for the synthesis of AgNP, such as microorganisms (bacteria and fungi) and plants. Extract from the part of the plant such as, stem, leaf, flower and bark is utilized as bioreducing agent, where its extractions ready to incorporate AgNP. Besides, bioreducing agents/stabilizing agents, commonly known as capping agents present in the biological extraction for the green synthesis of AgNP omits the consumption of additional chemical agents. The size and structure of AgNP from biosynthesis could be portrayed by electron microscopes and structural analysis. In the present chapter, the

S. Ramanathan · S. C. B. Gopinath (✉)
Institute of Nano Electronic Engineering, Universiti Malaysia Perlis,
01000 Kangar, Perlis, Malaysia
e-mail: subash@unimap.edu.my

S. C. B. Gopinath
School of Bioprocess Engineering, Universiti Malaysia Perlis,
02600 Arau, Perlis, Malaysia

M. K. M. Arshad · P. Poopalan
School of Microelectronic Engineering, Universiti Malaysia Perlis, Pauh Putra,
02600 Arau, Perlis, Malaysia

V. Perumal
Department of Mechanical Engineering, Universiti Teknologi PETRONAS,
32610 Seri Iskandar, Perak Darul Ridzuan, Malaysia

M. S. M. Saheed
Centre of Innovative Nanostructures and Nanodevices (COINN), Universiti Teknologi PETRONAS,
32610 Seri Iskandar, Perak Darul Ridzuan, Malaysia

A. H. Bhat et al. (eds.), *Nanomaterials for Healthcare, Energy and Environment*,
Advanced Structured Materials 118, https://doi.org/10.1007/978-981-13-9833-9_5

biosynthesized AgNPs from various biological extractions were presented with their significant application as antimicrobial agent.

Keywords Silver nanoparticle · Green synthesis · Bioreducing agents · Antimicrobial activity

1 Introduction

Nanotechnology is an essential emerging area of study that has been expanding interest, focuses on the development of synthetic and biological methods of engineering nanoparticles. Due to the enhanced properties of nanoparticles, their application grows swiftly on various scopes such as biomedical, pharmaceutical, antimicrobial, and drug-deliveries (Fujimaki et al. 2010; Gopinath et al. 2013b, c; Lakshmipriya et al. 2013a, b; Suk et al. 2017; Perumal et al. 2018; Sumitha et al. 2018; Theivasanthi et al. 2018; Letchumanan et al. 2019). Nanoparticles consist of metallic compound purposely considered as promising as they carry antibacterial properties due to their layer large surface area to volume ratio. Because of this, there is a remarkable interest for other researchers to study about antimicrobial activity of metallic nanoparticles (Suominen et al. 2016; Bhattarai et al. 2018; Amiri et al. 2019; Dong et al. 2019). The various metallic nanoparticles have gained deliberation in recent times because of their exquisite importance, and they include gold, silver, platinum (Gopinath et al. 2012; Lakshmipriya et al. 2016a; Logaranjan et al. 2016; Velusamy et al. 2018; Wang et al. 2018).

The emergence of nanoparticle conjugation with biomolecules has received high attention in recent technologies (Fujimaki et al. 2010; Lakshmipriya et al. 2013a, 2014, 2016b; Cheen et al. 2017). Among different nano-metallic materials, silver evidence the prominent contribution in the nanotechnology industries. Silver nanoparticles (AgNPs) are applied for electronic industries, material and solvent treatments, optical engineering and not ignoring biomedicine and pharmaceutical fields. Health care and medicinal aspects use AgNPs in almost all services and applications due to its anti-infection properties (Arasu 2010; Adhya et al. 2015; Akram et al. 2016). AgNPs are known to be the generic medicine with its potentials. The properties of AgNPs against pathogens extend its usage in medicines, equipment, and devices. It mimics the role of medicines and drugs to save a human from several illnesses and justifies AgNP to be a generic medicine (Ramanathan and Gopinath 2017).

Several methods have been established with AgNP synthesis. The physical method of synthesizing AgNP has advantageous as it prevents the solvent contamination of nanoparticle but the consumption of a high amount of energy for particle condensation and evaporation are not negligible. In addition, extremely high temperature and pressure modulation are involved, which causes a longer time for AgNP production, indirectly increases the cost of nanoparticle synthesis. In the chemical method, reducing agents, and protective agents are used to synthesizing AgNP and avoid agglomeration in order to synthesise high purity and stable AgNP. The consumption of strong chemicals causes contamination to the AgNP synthesized. Even though the reaction is quick and cost-effective compared to the physical method, it is difficult to control the variation in size and shape of AgNP without strong reducing and stabilizing agents (Iravani et al. 2014; Firdhouse and Lalitha 2015). So as to fix these difficulties, two-methods for AgNP synthesis are found, where the reduction is done using a strong reducing agent at first, followed by stabilization using weak chemical reagent. Yet, the usage of more than one chemical reagent in AgNP synthesis encounters risk in the toxicity carried by the synthesized nanoparticles. Thus, biosynthesis is observed to be an ecologically inviting method for stable AgNP synthesis and overwhelms among different methods because of its capability of incorporating AgNP utilizing just bio-based reagents. The ideal conditions for biosynthesis, for example, the temperature and humidity are manageable, thusly, it highly welcomed in a large framework of AgNP synthesis (Ghorbani et al. 2011; Anbu et al. 2018). Biosynthesis of AgNP is straightforward and known as a one-advance procedure which ready to produce stable nanoparticles.

2 Biosynthesis of Nanoparticles

Biosynthesis of AgNP is the bioreduction of silver ions by utilizing natural reducing agents, such as microorganisms, biodegradable polymers and plants. It is known as an ecologically friendly technique since the reducing and stabilizers agents are bio-based agents, for example, microorganism, peptides, proteins and starch comprise of different types of microscopic organisms, parasites, yeast, and green growth, finally plants and its concentrates. In biosynthesis, the utilization of bio-based reducing and capping agents in AgNP synthesis eliminates the incorporation of organic solvents and chemical reagents. Biosynthesis of AgNP is classified as enzymatic and non-enzymatic relying upon the sort of reducing agents. Biodegradable polymers, starch, peptides and proteins experience enzymatic nanoparticle reduction. Natural reducing sources such as plants and microorganism are considered as non-enzymatic reduction. The non-enzymatic reduction is a quick procedure compared with enzymatic one, and able to manage high-temperature and pH conditioned synthesizes (Ge et al. 2014; Ahmed et al. 2016).

Several herbal plants have been used against microbial infections. Herbal plants are natural without involves chemical, so it will not give any extra side-effects; and it is effective with native components. So far, all products using herbal plants have been recognized all over the world, this is why herbal plants participate in improving the current antibiotics. The improvement in medicine itself can help to develop society's research on the effectiveness of plant extract and improves medicine quality. Herbal plants are cheaper because it is natural and has no chemical, eventually decreases the price of the antibiotics. In addition, there are no side effects, so no extra investment has to be added. This will help less fortunate people who have a limited source. At the same time, the cost to make antibiotics is increasing; it uses expensive chemicals and needs money to be acquired. The government and society will also need money for their development and basic needs. If it is all wasted on the development of antibiotics, society will not be able to develop. Herbal plants do not cost much and can be obtained easily. The cost of making will decrease tremendously using herbal plants. There will be enough money left for the development of society, including the basic needs, education and opportunities for employment. This is how herbal plants can decrease the source to be used in medicine. Most of the diseases and illnesses involve bacterial infection, and antibiotics (Gopinath et al. 2014). If antibiotics become cheaper, unfortunate people will be healthy and society will be able to develop. Therefore, the biosynthesis of AgNP is highly encouraged by using herbal plants mediates high-performance antimicrobial property and enable the application in medicinal and pharmaceutical fields (Sulaiman et al. 2013; Elgorban et al. 2016).

3 Stabilizing Agent

The stabilizing agent alludes as a capping agent, one of the imperative components required in the AgNP synthesis. The essential capacity of stabilizing agent is to balance out the nanoparticle shaped after the reduction (Neena et al. 2012; Tanner et al. 2015; Ajitha et al. 2016). It keeps the agglomeration and aggregation of AgNPs. Common stabilizing agents utilized in AgNP synthesis are, polyvinylpyrrolidone (PVP), sodium citrate, triethylenetetramine (TEA) and polyvinyl liquor (PVA) depending on the type of metal ion used. The role of the stabilizing agent in a variety of shape, size, and encompassing interaction in a solvent are noteworthy. Notwithstanding, in the biosynthesis of AgNP, the microorganism and plants are able to execute as both reducing and stabilizing agents. The constituents present in cells of plant and microorganisms extricates are acts as both reducing and stabilizing agents which eliminates the utilization of additional stabilizing agent in the AgNP synthesis. Kelsey Beyer states that stabilizing agents on the AgNP surface influences the morphology and gives toxic effect when it is in contact with pathogens or foreign materials (Ajitha et al. 2014; Srikar et al. 2016).

4 Biosynthesis Methods

4.1 *Bacteria*

AgNP biosynthesis by using microorganisms is reported in 2000 for the first time by utilizing *Pseudomonas stutzeri* to synthesis AgNP with less than 200 nm in size. Then, *Bacillus licheniformis* is studied to synthesis smaller AgNP, which is about 50 nm in range. The visible color change of the reaction mixture to yellowish brown affirms the formation of AgNP. The examined reducing enzyme in bacteria is the nitrate reductase compound plays the role in reducing silver ions to AgNP. The

AgNP synthesis by microscopic organisms is considered stable as the AgNP could be synthesized in 24 h. The culture of *Escherichia coli*, *Enterobacter cloacae*, and (*Enterobacteriaceae*) *Klebsiella pneumonia* requires a very short time, about 5 min to perform the quick reaction in AgNP synthesis. On 2009, AgNP from silver nitrate utilizing *K. pneumonia* with the submersion of noticeable light illumination is examined. The investigation effectively structures AgNP with 1–6 nm size and very much characterized shape (Ghorbani et al. 2011; Iravani et al. 2014). The biosynthesis of AgNP happens extracellularly where the catalysts and sugars help the reduction of silver ions to AgNP on the cell surface.

4.2 Fungi

The fungal extracellular resources have been widely exposed for a range of downstream applications (Gopinath et al. 2002, 2003, 2013a, 2017; Anbu et al. 2005; Musa et al. 2018a, b, c). Literature have evidence the biosynthesis of nanoparticles from fungal strains. Fungi are increasingly effective in AgNP synthesis because of its temperament of discharging enzymes and it is easy to manage in research and laboratories. A study researched in 2001 in the transformation of fluid silver ions to AgNP using *Verticillium* and AgNP were synthesized extracellularly, resulting in 13–37 nm size of AgNPs. *Fusarium oxysporum* found to be a suitable fungal strain in synthesizing stable AgNPs with response to the enzymatic reaction, where the enzyme is secreted by the fungal strain itself. A few strains of *F. oxysoprum* are resulting in diverse efficiencies of biosynthesis of nanoparticles. *S. aureus* prompts antimicrobial movement of the contagious strain in cotton textures. Ghorbani et al. (2011) state that in the presence of a-NADPH, *F. phytochelatin* and *F. oxysporum* reduce nanoparticles with the guide of nitrate reductase in the absence of oxygen. The biosynthesis of the nanoparticle by fungi is known to be quicker and produce stable AgNP because of the nearness or nature of incorporating different catalysts so as to catalyze the procedure and balance out nanoparticle (Iravani et al. 2014; Firdhouse and Lalitha 2015; Ibrahim 2015).

4.3 Plants

The biosynthesis of nanoparticles from plant extricates is more engaging than microscopic organisms and different microorganisms (Anbu et al. 2018; Sumitha et al. 2018; Wang et al. 2019). Biosynthesis of AgNP from plants is categorized as green synthesis since the source of synthetic substances to produce and stabilize nanoparticles are only from green plants (Chung et al. 2016; Pattanayak et al. 2017). *Camella Sinensis*, regularly known as green tea plant remove orchestrate AgNPs with the use of theophylline and caffeine as characteristic of reduction and stabilization substances. Additionally, *Ocimum sanctum* extracts consist of ascorbic which go about as the reducing specialist of silver ions to AgNPs. The biosynthesis of AgNP from plants is a lot less expensive and ecological amicable because of the absence of any outer utilization of compound in the synthesis. *Datura metal* leaf is equipped for integrating AgNP since the plant is exceedingly comprehensive with biomolecules, for example, alcoholic mixes, amino acids, catalyst, polysaccharides, and proteins. The biomolecules go as reducing substances, which encourages the immediate change of salts to nanoparticles in the solvent arrangement. For a few plants extricates, the aging rate of nanoparticles can be controlled with the response temperature. *Diospyros kaki* and *Magnolia kobus* plants extricate accelerate the development of AgNPs as the response temperature is expanded. The investigation represented that an about 11 min is expected to create the AgNP by using the plants extricates, at higher temperature condition. *Capsicum annuum* is known for changing its auxiliary protein structure with response to silver ions in the AgNP synthesis. A study explores the number of particles incorporated with the bio-plant extract concentrations. High centralization of plant extricates gives the high substance of biomolecules in the response therefore delivers a higher number of nanoparticles (Ahmed et al. 2015; Elgorban et al. 2016; Małgorzata and Depciuch 2016; Srikar et al. 2016). The biosynthesis of AgNP from microorganisms, plants, and fungi with the size range and reaction time are tabulated in Table 1.

5 Biosynthesized AgNP-Mediated Antimicrobial Activity

Disc diffusion technique is a standout amongst the most widely recognized strategy used to decide the antimicrobial activity of numerous components (Foo et al. 2018). The technique is financially cheap since it can lead effectively in research facilities and the hardware and materials required are reasonable. Disc diffusion technique can be performed with different types of microorganisms to distinguish the antimicrobial action. The strategy to measure the result is clear and efficient. When the sample disperses at the circumference of the disc, it frames a locale with no bacterial contamination, where the antimicrobial action of the sample is examined. A disc is vaccinated with the biosynthesized nanoparticle and put it on a microorganism agar plate. The plates are incubated for 24 h at 35–37 °C. At that point the plates are inspected for bacterial restraint area on the agar plate. Type of microorganism, inoculating media and agar plate, the density of sample on the disc, and examination conditions are the variables that may influence the result and elucidation of disc diffusion technique (Driscoll et al. 2012). Next, the well diffusion which has been found after the disc diffusion technique, since circle dissemination inapplicable to analyze a wide range of antifungal actions. The strategy for well diffusion is that a well is made on the agar plate of the testing organism. The biosynthesized nanoparticle sample is dropped into the wells instead of adding on the disc as in disc diffusion. The plates are incubated for medium-term and inspected for the restraint zone which speaks to the antimicrobial action of the sample (Natheer et al. 2015).

Table 1 Biosynthesis of silver nanoparticles from plants and microorganisms (Ghorbani et al. 2011)

Organism		Size range (nm)	Reaction time
Bacteria	*Bacillus licheniformis*	40–50	24 h
	Klebsiella pneumonia	28.3–122	5 min
	Escherichia coli	28.2–122	5 min
	Enterobacter cloacae	28.2–122	5 min
	P. stutzeri	<200	48 h
	Klebsiella pneumonia	1–6	20 min
	Staphylococcus aureus	120–180	5 min
Fungi	*Fusarium oxysporum*	20–50, 2–5, 5–15	28 h
	Aspergillus fumigatus	5–25	10 min
	Verticillium	25 ± 12	72 h
	Aspergillus flavus	8.92 ± 1.61	72 h
	Nitrate reductase (enzyme purified from *Fusarium oxysporum*)	10–25	5 h
	Cladosporium cladosporioides	10–100	78 h
	Penicillium fellutanus	5–25	24 h
Plants	Alfalfa sprouts	2–20	24 h
	Pelargonium graveolens	16–40	24 h
	Azadirachta indica (neem)	5–35	4 h
	Jatropha curcas (latex)	20–40	4 h
	Aloe vera	15.2 ± 4.2	–
	Cinnamomum camphora	55–80	–
	Emblica officinalis	10–25	–
	Apiin (Henna leaves)	21–39	1 min
	Ocimum	3–20	–
	Cassia auriculata	20–40	48 h

6 Antimicrobial Property of Biosynthesized Silver Nanoparticles

Silver compounds are proven to exhibit resistance to microorganisms due to its natural antimicrobial effect. AgNP with nanoscale size and shape conveys unique physical and chemical properties in term of catalytic effect, electrical and magnetic properties. Due to its excellent physiochemical properties, AgNP is highly welcomed in the application of controlling and preventing microbial growth (Zhang et al. 2016). The biosynthesis of AgNP using herbal plants and microorganisms that carries antimicrobial property in nature further enhance the AgNP synthesized in the antimicrobial action against pathogenic organisms. The consumption of herbal green plants not only enhances the antimicrobial property, but also ensure that AgNP synthesized has no toxicity towards the human and it is favorable to be used in medicinal applications (Velusamy et al. 2016). The absence of any chemical reagents and high energy equipment in the biosynthesis of AgNP save the cost and time of nanoparticle synthesis providing an ecofriendly synthesis procedure. Figure 1 shows the schematic illustration of biosynthesis of AgNP using herbal extracts and its enhancement in antimicrobial effect (Prabhu and Poulose 2012; Gurunathan et al. 2014; Adhya et al. 2015; Elgorban et al. 2016; Gopinath et al. 2018).

In the analysis of antimicrobial property of biosynthesized of AgNP, gram-positive and gram-negative bacteria and fungi are usually examined. Both disc and well diffusion assays are applicable for bacteria examination, however, to analyze the antifungal property of AgNP, well diffusion assay is preferred. The efficiency of biosynthesized AgNP in inhibiting microbial growth is analyzed based on the zone of inhibition formed around the disc or well on the inoculated agar plate. The measurements are taken by using a scientific ruler in millimeter. The bigger the zone of microbial inhibition shown by the tested analyte, the higher the antimicrobial property. Figure 2 shows the antibacterial properties shown by herbal plants extracts and biosynthesized AgNP using the herbal plant extracts against *Escherichia coli* (gram-negative bacteria) and *Bacillus* sp. (gram-positive bacteria). It is obvious that the microbial inhibition zone shown by biosynthesized AgNP is larger than the herbal plant extract. The nanoscale AgNP with a large surface area/volume has high ability in penetrating the cell wall of bacteria and disintegrates it. However, herbal extract alone is unable to show prominent inhibition zone without the aid of nanoparticles (Latha and Kannabiran 2006; Nagarajan 2009; Pratheeba et al. 2014; Gopinath et al. 2018; Ramanathan et al. 2018).

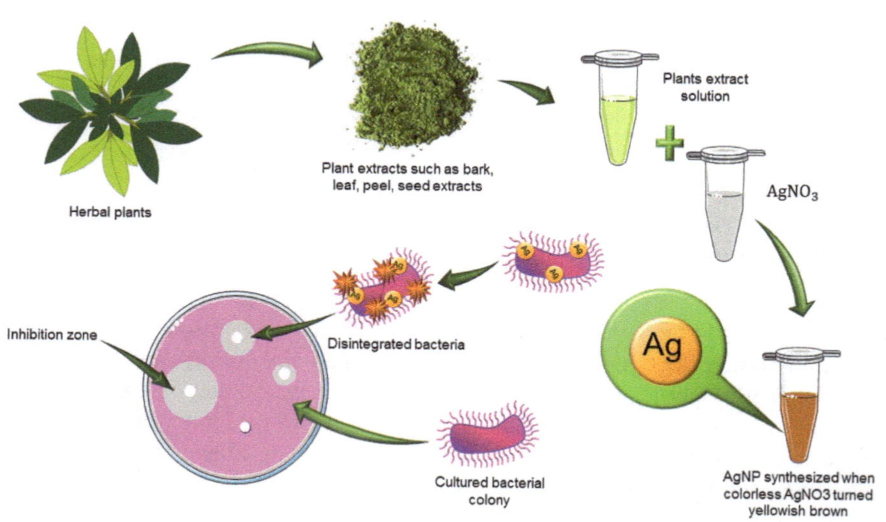

Fig. 1 Schematic illustration of biosynthesized AgNP and its effect on the antimicrobial property

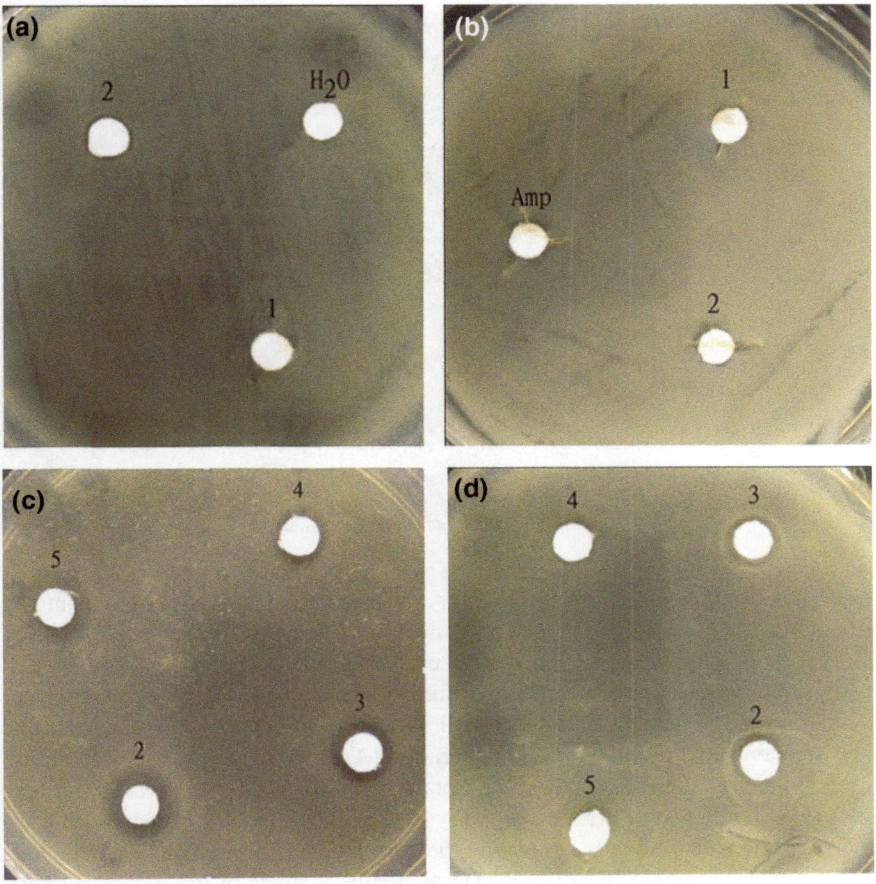

Fig. 2 Antibacterial effect is shown by herbal plant extract against **a** *E. coli* and **b** *Bacillus* sp. and antibacterial effect shown by biosynthesized AgNP against **c** *E. coli* and **d** *Bacillus* sp.

Aspergillus niger is a commonly used fungus in the analysis of antifungal property of biosynthesized AgNP. Figure 3 shows the image of *A. niger* grown plate where the antifungal property of AgNP was analyzed using disc diffusion technique. The antifungal effect shown by AgNP is evidenced by the apparent zone of antifungal inhibition implies the strength of biosynthesized AgNP in inhibiting the fungal growth (Pereira et al. 2008; Abdehgah et al. 2017; Bahrami-Teimoori et al. 2017; Jalal et al. 2018; Rónavári et al. 2018).

The mechanism of AgNP in antimicrobial movement is as yet equivocal. In any case, it is generally acknowledged that AgNP can attach and diffuse through the bacterial cell wall through the periplasmic space of cell wall into the cytoplasm thus causes arrangement of 'pits' on the structure of cell layer. The activity expands the cell porousness for more AgNPs diffuse into the cell and leads the disintegration of

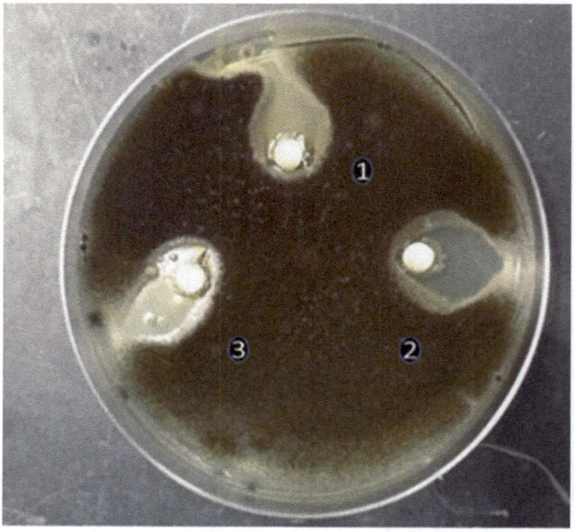

Fig. 3 Antifungal effect against *A. niger* on selective agar plate shown by biosynthesized AgNP using herbal extracts. The numbers 1, 2, and 3 refer to the experimental set-up with triplicate

bacterial cells (Suk and Gopinath 2017). Another component of AgNP in changing the cell mass of microscopic organisms is the arrangement of free radicals by AgNP. The AgNP has the capacity to discharge free radicals when it is in contact with the cell mass of microorganism. The capacity of free radicals in harming the cell membrane of microscopic organisms causes the cell surface to be increasingly permeable to outer segments and prompts cell disintegration (Anbu et al. 2018). The mechanism of AgNP in resisting the growth of microbial is likewise clarified by acid and base idea. Silver has delicate corrosive property in nature. The AgNP decreased from the silver particle is additionally conveying the corrosive property when it is in contact with different segments. Then again, DNA of cells is made of bases, for example, sulfur and phosphorus as their significant parts. Here, the AgNP diffuse through the cell layer of the cell and may follow up on the base segments of DNA and end the DNA replication of the cell, which results in the end of microscopic organisms (Chamakura et al. 2011; Reidy et al. 2013).

7 Conclusion and Future Prospective

The necessity for the inexpensive and eco-accommodating synthesis of AgNP is imperative in the developing innovation and technology of nanoparticles. Physical and chemical combinations of AgNP have a few imperfections, for example, high utilization of vitality energy, perilous toxins and synthetic reagents, which makes it troublesome for the synthesis of AgNP in expansive scale. These could lead to the development of harmful and toxic AgNP. In light of the survey, the natural strategy utilizing plant extricates under green synthesis of AgNP is observed to be the most inviting technique. It is more propitious than other natural techniques, have a few disadvantages. AgNP incorporated utilizing plant extricates has noteworthy application in biomedicinal and pharmaceutical fields. It can possibly be utilized in hardware and apparatuses related to heath appliances. Hence, the biosynthesis of AgNP utilizing plant separates is the engaging strategy for the massive synthesis of AgNP.

References

Abdehgah IB, Khodavandi A, Shamsazar A et al (2017) In vitro antifungal effects of biosynthesized silver nanoparticle by *Candida albicans* against *Candida glabrata*. Biomed Res (India) 28:2870–2876

Ahmed S, Ahmad M, Swami BL (2015) Green synthesis of silver nanoparticles using *Azadirachta indica* aqueous leaf extract. J Radiat Res Appl Sci 9:1–7. https://doi.org/10.1016/j.jrras.2015.06.006

Ahmed S, Ahmad M, Swami BL, Ikram S (2016) A review on plants extract mediated synthesis of silver nanoparticles for antimicrobial applications: a green expertise. J Adv Res 7:17–28. https://doi.org/10.1016/j.jare.2015.02.007

Ajitha B, Ashok Kumar Reddy Y, Reddy PS (2014) Biogenic nano-scale silver particles by *Tephrosia purpurea* leaf extract and their inborn antimicrobial activity. Spectrochim Acta Part A Mol Biomol Spectrosc 121:164–172. https://doi.org/10.1016/j.saa.2013.10.077

Ajitha B, Kumar Reddy YA, Reddy PS et al (2016) Role of capping agents in controlling silver nanoparticles size, antibacterial activity and potential application as optical hydrogen peroxide sensor. RSC Adv 6:36171–36179. https://doi.org/10.1039/c6ra03766f

Akram FE, El-Tayeb T, Abou-Aisha K, El-Azizi M (2016) A combination of silver nanoparticles and visible blue light enhances the antibacterial efficacy of ineffective antibiotics against methicillin-resistant *Staphylococcus aureus* (MRSA). Ann Clin Microbiol Antimicrobials 15:48. https://doi.org/10.1186/s12941-016-0164-y

Amiri M, Salavati-Niasari M, Akbari A (2019) Magnetic nanocarriers: evolution of spinel ferrites for medical applications. Adv Coll Interface Sci 265:29–44. https://doi.org/10.1016/j.cis.2019.01.003

Anbu P, Gopinath SCB, Hilda A et al (2005) Purification of keratinase from poultry farm isolate-*Scopulariopsis brevicaulis* and statistical optimization of enzyme activity. Enzyme Microb Technol 36:639–647. https://doi.org/10.1016/j.enzmictec.2004.07.019

Anbu P, Gopinath SCB, Yun HS, Lee C-G (2018) Temperature-dependent green biosynthesis and characterization of silver nanoparticles using balloon flower plants and their antibacterial potential. J Mol Struct 1177:302–309. https://doi.org/10.1016/j.molstruc.2018.09.075

Arasu T (2010) Stable silver nanoparticle synthesizing methods and its applications. J Biosci Res 1:259–270

Adhya A, Bain J, Ray O, Hazra A, Adhikari S, Dutta G, Ray S, Majumdar BK (2015) Healing of burn wounds by topical treatment: a randomized controlled comparison between silver sulfadiazine and nano-crystalline silver. J Basic Clin Pharm 6:29–34. https://doi.org/10.4103/0976-0105.145776

Bahrami-Teimoori B, Nikparast Y, Hojatianfar M et al (2017) Characterisation and antifungal activity of silver nanoparticles biologically synthesised by *Amaranthus retroflexus* leaf extract. J Exp Nanosci 12:129–139. https://doi.org/10.1080/17458080.2017.1279355

Bhattarai R, Bachu R, Boddu S, Bhaduri S (2018) Biomedical applications of electrospun nanofibers: drug and nanoparticle delivery. Pharmaceutics 11:5. https://doi.org/10.3390/pharmaceutics11010005

Chamakura K, Perez-ballestero R, Luo Z et al (2011) Comparison of bactericidal activities of silver nanoparticles with common chemical disinfectants. Colloids Surf B 84:88–96. https://doi.org/10.1016/j.colsurfb.2010.12.020

Cheen OC, Gopinath SCB, Perumal V et al (2017) Aptamer-based impedimetric determination of the human blood clotting factor IX in serum using an interdigitated electrode modified with a ZnO nanolayer. Microchim Acta 184:117–125. https://doi.org/10.1007/s00604-016-2001-6

Chung IM, Park I, Seung-Hyun K et al (2016) Plant-mediated synthesis of silver nanoparticles: their characteristic properties and therapeutic applications. Nanoscale Res Lett 11:1–14. https://doi.org/10.1186/s11671-016-1257-4

Dong P, Rakesh KP, Manukumar HM et al (2019) Innovative nano-carriers in anticancer drug delivery—a comprehensive review. Bioorg Chem 85:325–336. https://doi.org/10.1016/j.bioorg.2019.01.019

Driscoll AJ, Bhat N, Karron RA et al (2012) Disk diffusion bioassays for the detection of antibiotic activity in body fluids: applications for the pneumonia etiology research for child health project. Clin Infect Dis 54:159–164. https://doi.org/10.1093/cid/cir1061

Elgorban AM, Al-rahmah AN, Rushdy S et al (2016) Antimicrobial activity and green synthesis of silver nanoparticles using *Trichoderma viride*. Biotechnol Biotechnol Equip 30:299–304. https://doi.org/10.1080/13102818.2015.1133255

Firdhouse MJ, Lalitha P (2015) Biosynthesis of silver nanoparticles and its applications. J Nanotechnol

Foo ME, Anbu P, Gopinath SCB et al (2018) Antimicrobial activity of functionalized single-walled carbon nanotube with herbal extract of *Hempedu bumi*. Surf Interface Anal 50:354–361. https://doi.org/10.1002/sia.6375

Fujimaki M, Nomura K, Sato K et al (2010) Detection of colored nanomaterials using evanescent field-based waveguide sensors. Opt Express 18:15732–15740. https://doi.org/10.1364/OE.18.015732

Ge L et al (2014) Nanosilver particles in medical applications: synthesis, performance and toxicity. Int J Nanomed 9:2399–2407. https://doi.org/10.2147/IJN.S55015

Ghorbani HR, Safekordi AA, Attar H, Sorkhabadi SMR (2011) Biological and non-biological methods for silver nanoparticles synthesis. Res Pharm Sci 25:317–326

Gopinath SCB, Hilda A, Priya TL, Annadurai G (2002) Purification of lipase from *Cunninghamella verticillata* and optimization of enzyme activity using response surface methodology. World J Microbiol Biotechnol 18:449–458. https://doi.org/10.1023/A:1015579121800

Gopinath SCB, Hilda A, Lakshmi Priya T et al (2003) Purification of lipase from *Geotrichum candidum*: conditions optimized for enzyme production using Box-Behnken design. World J Microbiol Biotechnol 19:681–689. https://doi.org/10.1023/A:1025119222925

Gopinath SCB, Awazu K, Fujimaki M et al (2012) Surface functionalization chemistries on highly sensitive silica-based sensor chips. Analyst 137:3520. https://doi.org/10.1039/c2an35159e

Gopinath SCB, Anbu P, Lakshmipriya T, Hilda A (2013a) Strategies to characterize fungal lipases for applications in medicine and dairy industry. Biomed Res Int 2013:31–34. https://doi.org/10.1155/2013/154549

Gopinath SCB, Awazu K, Fujimaki M et al (2013b) Observations of immuno-gold conjugates on influenza viruses using waveguide-mode sensors. PLoS ONE 8:1–10. https://doi.org/10.1371/journal.pone.0069121

Gopinath SCB, Awazu K, Fujimaki M, Shimizu K (2013c) Evaluation of anti-A/Udorn/307/1972 antibody specificity to influenza A/H3N2 viruses using an evanescent-field coupled waveguide-mode sensor. PLoS ONE 8:e81396. https://doi.org/10.1371/journal.pone.0081396

Gopinath SCB, Tang TH, Chen Y et al (2014) Bacterial detection: from microscope to smartphone. Biosens Bioelectron 60:332–342. https://doi.org/10.1016/j.bios.2014.04.014

Gopinath SCB, Anbu P, Arshad MKM et al (2017) Biotechnological processes in microbial amylase production. Biomed Res Int 2017:1–9. https://doi.org/10.1155/2017/1272193

Gopinath SCB, Ramanathan S, Hann Suk K et al (2018) Engineered nanostructures to carry the biological ligands. In: MATEC web of conferences 150:06002. https://doi.org/10.1051/matecconf/201815006002

Gurunathan S, Han JW, Kwon D, Kim J (2014) Enhanced antibacterial and anti-biofilm activities of silver nanoparticles against gram-negative and gram-positive bacteria. Nanoscale Res Lett 1–17

Ibrahim HMM (2015) Green synthesis and characterization of silver nanoparticles using banana peel extract and their antimicrobial activity against representative microorganisms. J Radiat Res Appl Sci 8:265–275. https://doi.org/10.1016/j.jrras.2015.01.007

Iravani S, Korbekandi H, Mirmohammadi SV, Zolfaghari B (2014) Synthesis of silver nanoparticles: chemical, physical and biological methods. Res Pharm Sci 9:385–406

Jalal M, Ansari M, Alzohairy M et al (2018) Biosynthesis of silver nanoparticles from oropharyngeal *Candida glabrata* isolates and their antimicrobial activity against clinical strains of bacteria and fungi. Nanomaterials 8:586. https://doi.org/10.3390/nano8080586

Lakshmipriya T, Fujimaki M, Gopinath SCB et al (2013a) A high-performance waveguide-mode biosensor for detection of factor IX using PEG-based blocking agents to suppress non-specific binding and improve sensitivity. Analyst 138:2863–2870. https://doi.org/10.1039/c3an00298e

Lakshmipriya T, Fujimaki M, Gopinath SCB, Awazu K (2013b) Generation of anti-influenza aptamers using the systematic evolution of ligands by exponential enrichment for sensing applications. Langmuir 29:15107–15115. https://doi.org/10.1021/la4027283

Lakshmipriya T, Horiguchi Y, Nagasaki Y (2014) Co-immobilized poly(ethylene glycol)-block-polyamines promote sensitivity and restrict biofouling on gold sensor surface for detecting factor IX in human plasma. Analyst 139:3977–3985. https://doi.org/10.1039/c4an00168k

Lakshmipriya T, Gopinath SCB, Citartan M et al (2016a) Gold nanoparticle-mediated high-performance enzyme-linked immunosorbent assay for detection of tuberculosis ESAT-6 protein. Micro Nanosyst 8:92–98. https://doi.org/10.2174/1876402908666161026154223

Lakshmipriya T, Gopinath SCB, Tang TH (2016b) Biotin-streptavidin competition mediates sensitive detection of biomolecules in enzyme linked immunosorbent assay. PLoS ONE 11:16–20. https://doi.org/10.1371/journal.pone.0151153

Latha PS, Kannabiran K (2006) Antimicrobial activity and phytochemicals of *Solanum trilobatum* Linn. Afr J Biotech 5:2402–2404

Letchumanan I, Md Arshad MK, Balakrishnan SR, Gopinath SCB (2019) Gold-nanorod enhances dielectric voltammetry detection of c-reactive protein: a predictive strategy for cardiac failure. Biosens Bioelectron 130:40–47. https://doi.org/10.1016/j.bios.2019.01.042

Logaranjan K, Raiza AJ, Gopinath SCB et al (2016) Shape- and size-controlled synthesis of silver nanoparticles using aloe vera plant extract and their antimicrobial activity. Nanoscale Res Lett 11:520. https://doi.org/10.1186/s11671-016-1725-x

Małgorzata MP, Depciuch KJ (2016) Green synthesis and antibacterial effects of aqueous colloidal solutions of silver nanoparticles using camomile terpenoids as a combined reducing and capping agent. Bioprocess Biosyst Eng 39:1213–1223. https://doi.org/10.1007/s00449-016-1599-4

Musa H, Hafiz Kasim F, Nagoor Gunny AA et al (2018a) Enhanced halophilic lipase secretion by *Marinobacter litoralis* SW-45 and its potential fatty acid esters release. J Basic Microbiol 2018:1–14. https://doi.org/10.1002/jobm.201800382

Musa H, Kasim FH, Arbain D (2018b) Isolation, molecular identification and screening of halophilic and thermophiliclipase producing bacterial strains from extreme environmental conditions. Malays J Microbiol 14:413–423. http://dx.doi.org/10.21161/mjm.95716

Musa H, Kasim FH, Nagoor Gunny AA, Gopinath SCB (2018b) Salt-adapted moulds and yeasts: potentials in industrial and environmental biotechnology. Process Biochem 69:33–44

Nagarajan SM (2009) Comparative antimicrobial activity of callus and natural plant extracts of *Solanum trilobatum* L. Ancient Sci Life 28:3–5

Natheer SE, Thenmozhi A, Syed Abdul Rahman M (2015) Studies on antibacterial activity and phytochemical analysis of *Solanum trilobatum* against some human pathogens. Int J Curr Innov Res 1

Neena D, Devraj KM, Bhagat AP (2012) Role of capping agent in the synthesis of silver nanoparticles. J Pharm Res 5:4710–4712. https://doi.org/10.1021/acs.jpca.7b02186

Pattanayak S, Mollick MMR, Maity D et al (2017) Butea monosperma bark extract mediated green synthesis of silver nanoparticles: characterization and biomedical applications *Butea monosperma* bark extract mediated green synthesis of silver nanoparticles. J Saudi Chem Soc 21:673–684. https://doi.org/10.1016/j.jscs.2015.11.004

Pereira AC, Oliveira DF, Silva GH et al (2008) Identification of the antimicrobial substances produced by *Solanum palinacanthum* (Solanaceae). Anais da Academia Brasileira de Ciencias 80:427–432. https://doi.org/10.1590/S0001-37652008000300004

Perumal V, Saheed MSM, Mohamed NM et al (2018) Gold nanorod embedded novel 3D graphene nanocomposite for selective bio-capture in rapid detection of *Mycobacterium tuberculosis*. Biosens Bioelectron 116:116–122. https://doi.org/10.1016/j.bios.2018.05.042

Prabhu S, Poulose EK (2012) Silver nanoparticles: mechanism of antimicrobial action, synthesis, medical applications, and toxicity effects. Int Nano Lett 2:1–10

Pratheeba M, Rani KU, Ramesh B (2014) Studies on antimicrobial and anticancer activity of *Solanum trilobatum*. Asian J Pharm Clin Res 7:231–239

Ramanathan S, Gopinath SCB (2017) Potentials in synthesizing nanostructured silver particles. Microsyst Technol 23:4345–4357. https://doi.org/10.1007/s00542-017-3382-0

Ramanathan S, Gopinath SCB, Anbu P et al (2018) Eco-friendly synthesis of *Solanum trilobatum* extract-capped silver nanoparticles is compatible with good antimicrobial activities. J Mol Struct 1160:80–91. https://doi.org/10.1016/j.molstruc.2018.01.056

Reidy B, Haase A, Luch A et al (2013) Mechanisms of silver nanoparticle release, transformation and toxicity: a critical review of current knowledge and recommendations for future studies and applications. Materials 6:2295–2350. https://doi.org/10.3390/ma6062295

Rónavári A, Igaz N, Gopisetty MK et al (2018) Biosynthesized silver and gold nanoparticles are potent antimycotics against opportunistic pathogenic yeasts and dermatophytes. Int J Nanomed 13:695–703. https://doi.org/10.2147/IJN.S152010

Srikar SK, Giri DD, Pal DB et al (2016) Green synthesis of silver nanoparticles: a review. Sci Res 6:34–56

Suk KH, Gopinath SCB (2017) Drug encapsulated nanoparticles for treating targeted cells. Curr Med Chem 24:1–12. https://doi.org/10.2174/0929867324666170502122444

Suk KH, Gopinath SCB, Anbu P, Lakshmipriya T (2017) Cellulose nanoparticles encapsulated cow urine for effective inhibition of pathogens. Powder Technol 328:140–147. https://doi.org/10.1016/j.powtec.2018.01.010

Sulaiman GM, Mohammed WH, Marzoog TR et al (2013) Green synthesis, antimicrobial and cytotoxic effects of silver nanoparticles using *Eucalyptus chapmaniana* leaves extract. Asian Pac J Trop Biomed 3:58–63. https://doi.org/10.1016/S2221-1691(13)60024-6

Sumitha S, Vasanthi S, Shalini S et al (2018) Phyto-mediated photo catalysed green synthesis of silver nanoparticles using *Durio zibethinus* seed extract: antimicrobial and cytotoxic activity and photocatalytic applications. Molecules 23:3311. https://doi.org/10.3390/molecules23123311

Suominen A, Li Y, Youtie J, Shapira P (2016) A bibliometric analysis of the development of next generation active nanotechnologies. J Nanopart Res 18:270. https://doi.org/10.1007/s11051-016-3578-8

Tanner EEL, Tschulik K, Tahany R et al (2015) Nanoparticle capping agent dynamics and electron transfer: polymer-gated oxidation of silver nanoparticles. J Phys Chem C 119:18808–18815. https://doi.org/10.1021/acs.jpcc.5b05789

Theivasanthi T, Anne Christma FL, Toyin AJ et al (2018) Synthesis and characterization of cotton fiber-based nanocellulose. Int J Biol Macromol 109:832–836. https://doi.org/10.1016/j.ijbiomac.2017.11.054

Velusamy P, Su CH, Kumar GV et al (2016) Biopolymers regulate silver nanoparticle under microwave irradiation for effective antibacterial and antibiofilm activities. PLoS ONE 11:1–14. https://doi.org/10.1371/journal.pone.0157612

Velusamy P, Srinivasa CM, Kumar GV et al (2018) A pH stimuli thiol modified mesoporous silica nanoparticles: doxorubicin carrier for cancer therapy. J Taiwan Inst Chem Eng 87:264–271. https://doi.org/10.1016/j.jtice.2018.03.048

Wang F, Lakshmipriya T, Gopinath SCB (2018) Red spectral shift in sensitive colorimetric detection of tuberculosis by ESAT-6 antigen-antibody complex: a new strategy with gold nanoparticle. Nanoscale Res Lett 13:1–8

Wang L, Gopinath SCB, Anbu P et al (2019) Photovoltaic and antimicrobial potentials of electrodeposited copper nanoparticle. Biochem Eng J 142:97–104. https://doi.org/10.1016/j.bej.2018.10.009

Zhang S, Tang Y, Vlahovic B (2016) A review on preparation and applications of silver-containing nanofibers. Nanoscale Res Lett 11:80. https://doi.org/10.1186/s11671-016-1286-z

Use of Nanomaterials in the Pretreatment of Water Samples for Environmental Analysis

Sandra C. Bernardo, Ana C. A. Sousa, Márcia C. Neves and Mara G. Freire

Abstract The challenge of providing clean drinking water is of enormous relevance in today's human civilization, being essential for human consumption, but also for agriculture, livestock and several industrial applications. In addition to remediation strategies, the accurate monitoring of pollutants in water supplies, which most of the times are present at low concentrations, is a critical challenge. The usual low concentration of target analytes, the presence of interferents and the incompatibility of the sample matrix with instrumental techniques and detectors are the main reasons that render sample preparation a relevant part of environmental monitoring strategies. The discovery and application of new nanomaterials allowed improvements on the pretreatment of water samples, with benefits in terms of speed, reliability and sensitivity in analysis. In this chapter, the use of nanomaterials in solid-phase extraction (SPE) protocols for water samples pretreatment for environmental monitoring is addressed. The most used nanomaterials, including metallic nanoparticles, metal organic frameworks, molecularly imprinted polymers, carbon-based nanomaterials, silica-based nanoparticles and nanocomposites are described, and their applications and advantages overviewed. Main gaps are identified and new directions on the field are suggested.

1 Introduction

Since the beginning of industrialization and due to the continuous population growth, anthropogenic activities have been constantly rising, allowing many toxic compounds to reach the aquatic environment (Hussain and Kharisov 2016). According to the United Nations Environmental Program (UNEP) and the Food and Agriculture Organization of the United Nations (FAO), the contamination of the aquatic environment and consequently water pollution is a global challenge, both in developed and developing countries (UNEP/UNECE 2016; FAO 2017; WWAP 2017). The contamination

S. C. Bernardo · A. C. A. Sousa · M. C. Neves · M. G. Freire (✉)
Chemistry Department, CICECO—Aveiro Institute of Materials,
University of Aveiro, 3810-193 Aveiro, Portugal
e-mail: maragfreire@ua.pt

A. H. Bhat et al. (eds.), *Nanomaterials for Healthcare, Energy and Environment*,
Advanced Structured Materials 118, https://doi.org/10.1007/978-981-13-9833-9_6

of the aquatic environment has negative impacts on ecosystems and human health and, consequently, undermines economic growth. To address this important problem it is necessary to understand the extension of water contamination, where environmental monitoring is essential. However, some of the compounds that are being accumulated in the environment are sometimes close or under the detection limits of the currently available analytical techniques, despite being toxic at very low concentrations. In order to assess the current levels of these pollutants in the environment it is crucial to modernize and improve the performance of analytical techniques, so that pollutants that are present at ultra-low concentrations in environmental samples could be detected and quantified with accuracy and consistency (Scragg 2005; Hussain and Kharisov 2016).

Pollutants are occasionally ubiquitous contaminants; most of them can migrate from the point sources and contaminate other areas where they are not used (Scragg 2005). According to the United Nations Environmental Program (UNEP, UNEP b) and the Food and Agriculture Organization of the United Nations (FAO), the sources of contamination are diverse and include, for example, agriculture and food (e.g. land based farming, agroindustry, fisheries and aquaculture), energy (e.g. fossil fuels, biomass), industry (e.g. chemical, mining, and petrochemical), manufacturing (e.g. pharmaceuticals, home appliances, electronics), services (e.g. health care, tourism), transports (e.g. fuel use, engine emissions) and waste (e.g. municipal and industrial solid waste, hazardous waste, wastewater effluents, leachates) (Fig. 1).

The most common pollutants found in the environment (water, soil and air) include metals, polycyclic aromatic hydrocarbons (PAHs), persistent organic pollutants (POPs), pesticides, and pharmaceutical drugs (Liang and Guo 2009; Hussain and Kharisov 2016).

Metals are naturally occurring elements that due to human activities are being re-introduced in the environment, e.g. from agriculture activities and wastes from industrial processes (Scragg 2005; UNEP 2013). Once released, some metals such as cadmium, lead, copper, chromium, mercury, uranium, nickel, cobalt, zinc, arsenic, selenium, thallium, and antimony, tend to bioaccumulate (Akpor 2014). Some of these metals, such as mercury, are also able to bioconcentrate along the food chain and they can even cross the blood brain barrier, being considered as extremely toxic (Bjørklund et al. 2017). These pollutants have been extensively studied since water contamination with metals poses serious health risks due to their high toxicity (most are carcinogenic and teratogenic substances) (Scragg 2005; Vunain et al. 2016; Carolin et al. 2017). Hence, these toxic metals have been prioritized as the major inorganic contaminants in the environment; furthermore, they are non-biodegradable, highly persistent and present high mobility between ecosystems (Vunain et al. 2016).

Polycyclic aromatic hydrocarbons (PAHs) are hydrocarbons with two or more fused aromatic rings. PAHs are produced from the incomplete combustion or pyrolysis of carbonaceous and organic matter (coal, diesel, gas, wood, etc.). Their major anthropogenic sources into the atmosphere and environment include industrial processes, vehicle exhausts, energy producing plants, waste incineration, and domestic heating emissions (Pandey et al. 2011; Sajid et al. 2018). PAHs are highly toxic and some of them are mutagenic and carcinogenic (Scragg 2005; Sajid et al. 2018). PAHs

Fig. 1 Schematic representation of examples of human activities and pollutant's sources of water contamination

are ubiquitous in the environment being detected in all compartments (air, soils and water). In what concerns aquatic systems, PAHs have been detected in river, lakes, oceans and also in groundwater and drinking water (Menzie et al. 1992; Burgess et al. 2003).

Persistent organic pollutants (POPs) are a group of manmade chemicals that are persistent, toxic and globally distributed. These chemicals are even found at high levels in regions where they were not used, including the Arctic. POPs include different classes of chemicals, as for example, polychlorinated biphenyls (PCBs), dioxins and furans, organochlorine pesticides [e.g. dichlorodiphenyl trichloroethane (DDT), polybrominated diphenyl ethers (PBDEs), short-chained chlorinated paraffins or perfluorooctane sulfonic acids (PFOS)]. Their deleterious effects include cancer, neurotoxicity, immunotoxicity and endocrine disruption. They are regulated under the Stockholm convention and their use is mostly banned (UNEP a).

Another commonly found type of pollutants, especially due to agriculture activities, are **pesticides**, which include herbicides, fungicides and insecticides (UNEP c). Pesticides are a wide-ranging group of chemical compounds, including organochlorines, organophosphates, carbamates, chloroacetanilide, pyrethroids and neonicotinoids (Kaur et al. 2014; Wang et al. 2017). They are widely used to prevent crops destruction by any plague and to increase the agriculture productivity, leading to their accumulation in the food chain and in the contamination of air, water and soil (Kaur et al. 2014; Wang et al. 2017).

Despite the important role of **pharmaceuticals** for assuring public health safety, their excessive and prolonged use poses serious environmental concerns. Pharmaceuticals are not completely metabolized by the human body, being mainly discharged by urine and reaching wastewater treatment plants (WWTPs) (Lolic et al. 2015). Furthermore, daily tons of these compounds are directly loaded in the sewage systems. However, most of the pharmaceutical drugs and their bioactive metabolites are not completely removed since WWTPs are mainly designed to remove suspended solids and organic loadings, leading to a continuous discharge of these compounds into the environment (Lolic et al. 2015). Antibiotics, for instance, are amongst the pharmaceutical drugs presenting a higher threat; they induce toxic effects at very low concentrations and can in a long-term contribute to the growth of antibiotic-resistant bacteria (Wang et al. 2018). But other drugs are also of high concern due to their large consumption and release into the environment, such as non-steroidal anti-inflammatory drugs (NSAIDs), analgesics, β-blockers, lipid regulators and psychiatric drugs. The European Union established regulatory guidance to assess the presence of pharmaceuticals in the aquatic environment, in which several drugs are included in the watch list, namely the hormones 17-alpha-ethinylestradiol and 17-beta-estradiol, and the NSAID diclofenac, as they may pose a significant risk to the aquatic environment and for which monitoring data are insufficient (European Commission 2013; Lolic et al. 2015).

Due to the serious concern associated to these pollutants, the European union implemented the Water Framework Directive (WFD) (European Commission 2013), requiring their determination down to the ng L^{-1} range (Liang and Guo 2009). Accordingly, the improvement of the sensitivity of analytical techniques currently used to identify and quantify the several pollutants present in aqueous samples is in high demand (Liang and Guo 2009; Hussain and Kharisov 2016). However, the key difficulty in environmental analysis is connected to the low sensitivity and selectivity of the used analytical techniques. Most of the times these are insufficient for determining the low concentration levels and to eliminate interferences arising from the complex environmental matrices, turning mandatory a sample treatment step prior to analysis (Liang and Guo 2009). However, sample pretreatment is still one of the most time- and labor-consuming parts in environmental analysis, and has been subject of intensive research over the past 20 years (Hussain and Kharisov 2016).

2 Improvement of Environmental Analysis Using Nanomaterials

Environmental analysis can be defined as the use of analytical chemistry techniques to detect and quantify pollutants in the environment (Nilsson et al. 2007). The high demand for environmental monitoring is confirmed by the increasing number of

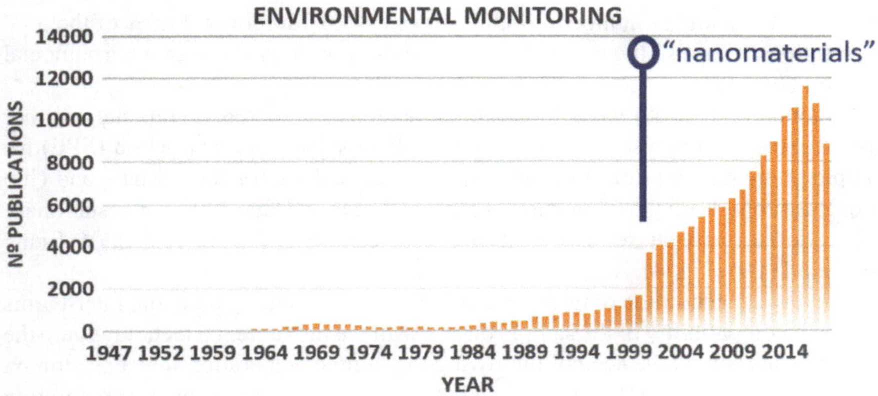

Fig. 2 Number of publications *per* year (from 1947 to 2018) focusing on environmental monitoring. The search was carried out in the PUBMED database in December 2018 using the keywords "environmental monitoring". The search was refined using the additional keyword "nanomaterials". Publications dealing with nanomaterials in environmental monitoring appeared after 2001

reports along the years, as shown in Fig. 2. Around 2001, a boom of publications on this matter arouse, especially due to the introduction of nanomaterials in environmental monitoring. Accordingly, more than 10% of the total of publications concerning the analysis of environmental samples are related with the use of nanomaterials, as summarized in Fig. 3. From these publications it is possible to ascertain the types of samples used, namely "water", "soil" and "air", being the water samples those that

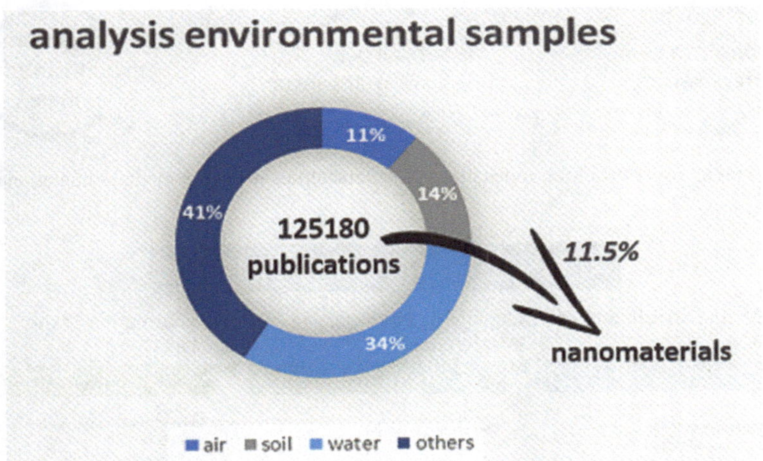

Fig. 3 Overview on publications (from 1947 to 2018) focusing on the analysis of environmental samples. The search was carried out in the PUBMED database using the keywords "analysis", "environmental" and "samples"

correspond to a higher number of publications (34%). Moreover, 11.5% of the total publications comprise the use of "nanomaterials" for the analysis of environmental samples (Fig. 3).

Nanomaterials could have different applications in environmental analysis. For instance, they can be used as sorbent materials in solid-phase extraction (SPE) for sample pretreatment prior to the analysis (Savage and Diallo 2005; Liang and Guo 2009), as supports for the fabrication of sensors (Liang and Guo 2009), or as stationary or pseudostationary phases in several analytical techniques (Zhang et al. 2006; Liang and Guo 2009) (Fig. 4).

The low concentration of target analytes, the high content of potential interferents and the incompatibility of the sample matrix with the instrumental techniques are the main reasons behind the need of improving the sample preparation step, envisioning the enhancement of speed, reliability and sensitivity. The goals of sample preparation methods are summarized in Fig. 5 (Xu et al. 2016). **Solid-phase extraction (SPE)** is one of the most attractive procedures for sample preparation since it can be easily implemented, can be combined with different detection techniques and it holds high ability for preconcentration (Hussain and Kharisov 2016). In SPE, nanomaterials, which will be discussed below, usually display a high performance for the adsorption

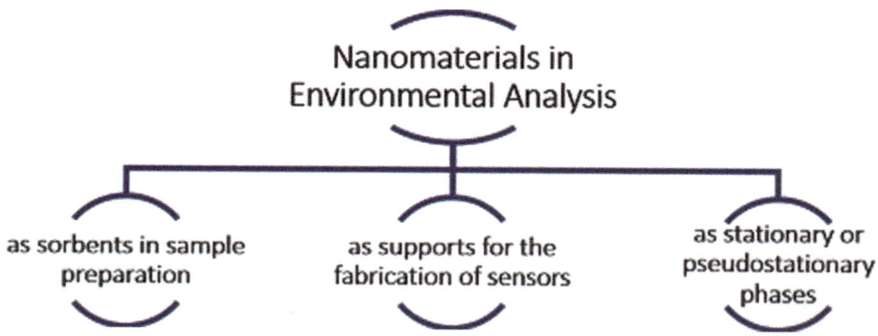

Fig. 4 General overview of the applications of nanomaterials in environmental samples analysis

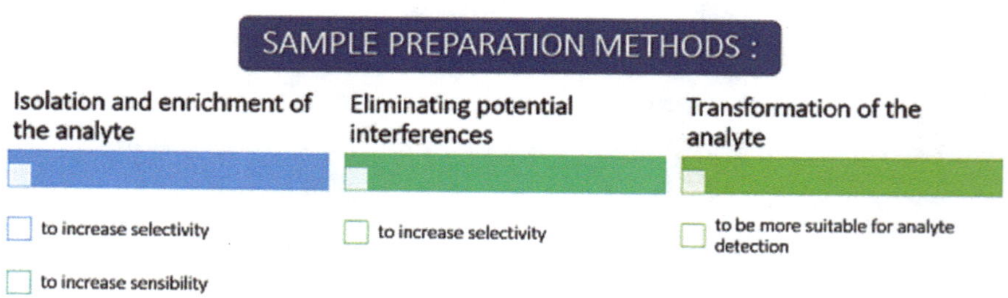

Fig. 5 Schematic representation of the main goals in the sample preparation step

of a wide range of analytes due to their high surface area and possibility of modifying their chemical nature and chemical groups at the surface.

3 Nanomaterials in Solid-Phase Extraction

As highlighted before, sample preparation is a crucial step in the analysis of aquatic environmental samples since these matrices tend to be highly complex, presenting most of the times trace level analytes (Ribeiro et al. 2014; Wierucka and Biziuk 2014). Sample preparation is essential to improve the analytes concentration and to remove interfering species (Tian et al. 2013), contributing to the enhancement of both accuracy and sensitivity (Xu et al. 2016). In sample preparation procedures, the most common difficulties are related with the use of time-consuming techniques, complete recovery of the target analytes, removal or isolation of impurities, and high associated costs (Wierucka and Biziuk 2014). It is highly desirable to improve the adsorption efficiency and selectivity, as well as the miniaturization and automation of various stages of sample preparation (Ribeiro et al. 2014; Wen et al. 2014; Wierucka and Biziuk 2014).

Amongst the different preconcentration techniques for sample preparation, SPE is advantageous over other pretreatment techniques, like liquid-liquid extraction, precipitation, co-precipitation, etc. (Das et al. 2012). SPE is a simple and fast operation, it allows high enrichment factors and recovery, may be of low cost, and can avoid the consumption of organic solvents (Das et al. 2012). Some classic materials, such as C8 and C18, mixed-mode/cationic-exchange (MCX), mixed-mode/anion-exchange (MAX) and weak anion-exchange (WAX), are commonly used in SPE (Wen et al. 2014). More recently, novel solid-phase sorbents based mainly on nanomaterials have been studied and applied for water samples pretreatment in environmental analysis. Although several sample preparation methods using nanomaterials have been proposed, in this chapter the use of nanomaterials in SPE for monitoring of environmental aqueous samples and drinking water samples is reviewed.

The term "nanomaterials" (NMs) usually refers to materials with nanometric size (typically 1–100 nm), at least in one dimension (length, width or thickness) (Tian et al. 2013; Xu et al. 2016). Therefore, nanomaterials may include nanometer sized particles and materials with porous size in the nanometer range or even nanometer sized particles dispersed within a porous matrix (Tian et al. 2013) (Fig. 6). These NMs present exceptional physical and chemical properties, such as high specific areas, and the possibility of displaying unique optical, electrical, thermal and magnetic characteristics. They can also be functionalized with various chemical groups to increase their affinity toward target compounds and present various morphologies, being tuneable according to the type of analyte and analytical technique (Tian et al. 2013; Xu et al. 2016).

During the past two decades, nanomaterials have been extensively employed to design novel pretreatment techniques focused on isolation and/or preconcentration of target analytes from different types of samples (Hussain and Kharisov 2016).

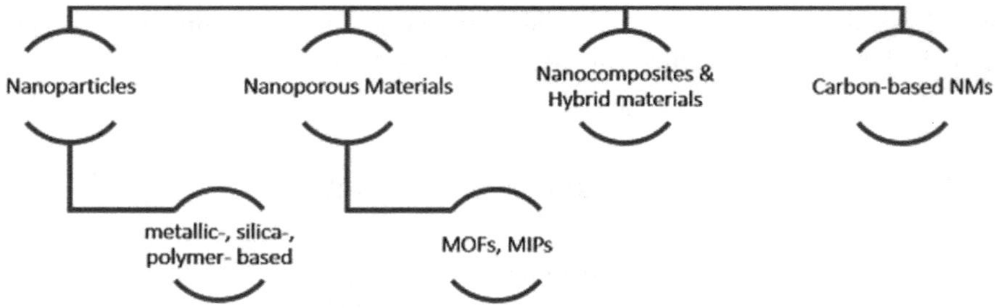

Fig. 6 Schematic representation of the different types of nanomaterials used in SPE for the analysis of environmental aqueous samples. *MOFs* metal organic frame-work materials; *MIPs* molecularly imprinted polymers

Herein, we summarize and discuss some of the latest (from 2009 to 2018) and most significant applications of nanomaterials in SPE for sample preparation, namely for environmental aqueous sample analysis. These NMs include metallic nanoparticles (MNPs), metal organic frame-work (MOF) materials, molecularly imprinted polymers (MIPs), carbonaceous nanomaterials and silica-based nanoparticles, which are described below in detail.

3.1 *Metallic Nanoparticles (MNPs) in SPE*

Metallic nanoparticles involve a wide range of different inorganic and inorganic/organic hybrid nanoparticles, such as **metal nanoparticles, metal oxide nanoparticles** and **magnetic nanoparticles** (Lucena et al. 2011; Tian et al. 2013).

Among **metal nanoparticles**, gold nanoparticles (Au NPs) and silver nanoparticles (Ag NPs) are the most well-known nanosorbents for metal ions and organic pollutants, such as PAHs (Wang and Campiglia 2010; Lo et al. 2012; Wu et al. 2012; Tian et al. 2013; Hussain and Kharisov 2016). In the past decade, Liu (2009) provided an overview on the characterization and applications of Au NPs, highlighting the incorporation of Au NPs in analytical systems, namely in the preparation and concentration of samples.

Au NPs are highly relevant in the field of environmental monitoring and analysis since organic molecules containing thiol (–SH) or amino ($-NH_2$) groups can adsorb spontaneously onto Au surfaces to form well-organized self-assembled monolayers (Liu 2009). Accordingly, Wang and Campiglia (2010) described the use of a gold nanoparticle-based SPE method for benzo[a]pyrene (PAH) determination in spiked tap water samples. The methodology was combined to laser-excited time-resolved Shpol'skii spectrometry (LETRSS), being able to improve the limit of detection (LOD) of the technique by an order of magnitude (LOD = 0.001 ng/mL) (Wang and Campiglia 2010). In addition to organic compounds, Karimipour et al. (2012) showed the applicability of Au NPs for the enrichment and preconcentration of trace amounts of metals, namely Co^{2+}, Cu^{2+}, Ni^{2+}, Fe^{2+}, Pb^{2+}, and Zn^{2+}, from real samples. By using a new sorbent based on Au NPs loaded in activated carbon (Au–NP–AC) and modified by bis(4-methoxy salicylaldehyde)-1,2-phenylenediamine (BMSAPD), a preconcentration factor of 200 was reported for all the metal ions under study, with detection limits ranging from 1.4 to 2.6 ng/mL (Karimipour et al. 2012). Also, Lo et al. (2012) tested the affinity of mercury species for Au NPs, namely by mixing Au NPs with aluminium oxide (Al_2O_3) particles. The authors reported removal efficiencies higher than 97% for the four mercury species, from spiked lake water, groundwater, and seawater samples (Lo et al. 2012). Moreover, by using this Au NP-Al_2O_3 adsorbent it was found that the LODs for the mercury species were in the sub-ppq level, which is about 24-fold lower than those obtained without preconcentration (LOD = 0.028 μg/mL) (Lo et al. 2012).

Ag NPs have also been described as adsorbents with high affinity towards mercury. In this line, Wu et al. (2012) increased the selectivity (more than 1000-fold) of the analytical system for Hg^{2+} over other metal ions, by combining nanoparticle extraction, fluorescent dye labelling, and flow injection analysis (FIA) detection. Silver nanoparticles were functionalized with an oligonucleotide composed of repeated 33 thymine nucleotides (T_{33}–AgNPs) that specifically captures Hg^{2+} from tap water samples (LOD = 0.006 ng/mL) (Wu et al. 2012). Not only surface modification with functional groups in Ag NPs provided extra selectivity, but also it prevents the prepared Ag NPs from aggregating (Tian et al. 2013).

Metal oxide nanoparticles (MONPs), such as TiO_2, Al_2O_3, ZnO, ZrO_2, MnO and CeO_2, have been reported in the literature as major inorganic nanoadsorbents for the enrichment and separation of metal trace elements (Lucena et al. 2011; Wen et al. 2014; Hussain and Kharisov 2016). TiO_2 NPs were used for Cd determination from groundwater and wastewater samples at the nanogram *per* kilogram level with high preconcentration factors (>10) (García-ruiz et al. 2011). As summarized in Table 1, several MONPs have been used to analyse metal ions in environmental aqueous samples (Afkhami et al. 2011; Baytak et al. 2011; Erdogan et al. 2011; García-ruiz et al. 2011; Yalçinkaya et al. 2012). Most of the times these MONPs are modified with functional coatings to increase their selectivity (Lucena et al. 2011); for instance, Baytak et al. (2011) described the use of a minicolumn of a yeast (*Yamadazyma spartinae*) immobilized on TiO_2 nanoparticles for the determination of Cr, Cu, Fe, Mn, Ni and Zn in environmental water samples by ICP-MS. Alumina (Al_2O_3) nanoparticles were also successfully used for the determination of trace amounts of Pb(II) and Cr(III) in environmental water samples (Afkhami et al. 2011). Afkhami et al. (2011) prepared modified MONPs by the immobilization of 2,4-dinitrophenylhydrazine on nano-alumina coated with sodium dodecyl sulphate, with recoveries of the metals ranging from 98 to 104% and a maximum preconcentration factor of 267 (Afkhami et al. 2011).

Magnetic nanoparticles (MNPs) are constituted by a core and a supported layer, in which the magnetic core is mainly composed of iron, nickel, cobalt, and their corresponding oxides (Wierucka and Biziuk 2014; Hussain and Kharisov 2016). The most common magnetic cores are composed of iron oxides, including magnetite (Fe_3O_4) and maghemite (Fe_2O_3) (Hussain and Kharisov 2016). Since pure iron-

Table 1 Summary on the use of metal oxide nanoparticles in SPE for metal ions detection (in the last 10 years)

Metal oxide NPs	Modification	Sample matrix	Analyte	Analytical technique	LOD (ng/mL)	References
TiO_2	–	Ground and waste waters	Cd(II)	ID-ICPMS	–	García-ruiz et al. (2011)
TiO_2	Coated with yeast (*Yamadazyma spartinae*)	Tap and lake waters	Cr(III), Cu(II), Fe(III), Mn(II), Ni(II) and Zn(II)	ICP-AES	0.1–0.45	Baytak et al. (2011)
Al_2O_3	Coated with 2,4-dinitrophenylhydrazine	Tap, bottled mineral, waste waters, industrial effluents	Pb(II) and Cr(III)	FAAS	0.43–0.55	Afkhami et al. (2011)
ZrO_2–B_2O_3	–	Tap and hot spring waters	As(III) and As(V)	HGAAS	0.009	Erdogan et al. (2011)
ZrO_2–B_2O_3	–	Waste water	Al(III)	FAAS	7.71	Yalçinkaya et al. (2012)

LOD limit of detection; *ID-ICPMS* inductively coupled plasma mass spectrometry; *ICP-AES* inductively coupled plasma atomic emission spectrometry; *FAAS* flame atomic absorption spectrometry; *HGAAS* hydride generation atomic absorption spectrometry oxide NPs easily form aggregates, which may alter their magnetic properties, the surface of the magnetic core is usually modified with a suitable coating (Wen et al. 2014; Wierucka and Biziuk 2014). The modifying layer can be either constituted by inorganic substances (e.g. silica, alumina, manganese oxide (IV), or graphene), organic substances (e.g. molecularly imprinted polymers, chitosan, divinylbenzene, polyamidoamines), or surfactants (Wierucka and Biziuk 2014; Hussain and Kharisov 2016). This coating layer protects not only the magnetic core from aggregation but also from oxidation and, if properly selected, may enhance the interaction between analytes and the adsorbent material.

Due to its several advantages, silica is the most commonly used material for coating MNPs. Silica carries silane groups that can be easily modified (Wierucka and Biziuk 2014). Figure 7 shows a scheme of the sol-gel process usually applied to synthetize silica-coated MNPs. Zhang et al. (2011) described the use of a core-shell magnetic (Fe_3O_4/SiO_2) mesoporous silica sorbent functionalized with C_{18} (octadecyltriethoxysilane) groups for the removal of PAHs from water samples. Also, surfactants such as sodium dodecyl sulphate (SDS) can be functionalized on the surface of MNPs, being suitable either for the adsorption of metal ions (mercury) (Faraji et al. 2010) or antidepressant drugs (fluoxetine) (Bagheri et al. 2011b). Moreover, some polymers and copolymers were described as coating agents of MNPs for the removal of toxic metal ions and organic compounds from environmental water samples and their pretreatment, such as polyaniline (Mehdinia et al. 2011), a copolymer of methyl methacrylate (MMA) and maleic anhydride (MA) (Masoumi et al. 2014), polydopamine (Iqbal et al. 2011) and polypyrrole (Iqbal et al. 2011; Meng et al. 2011).

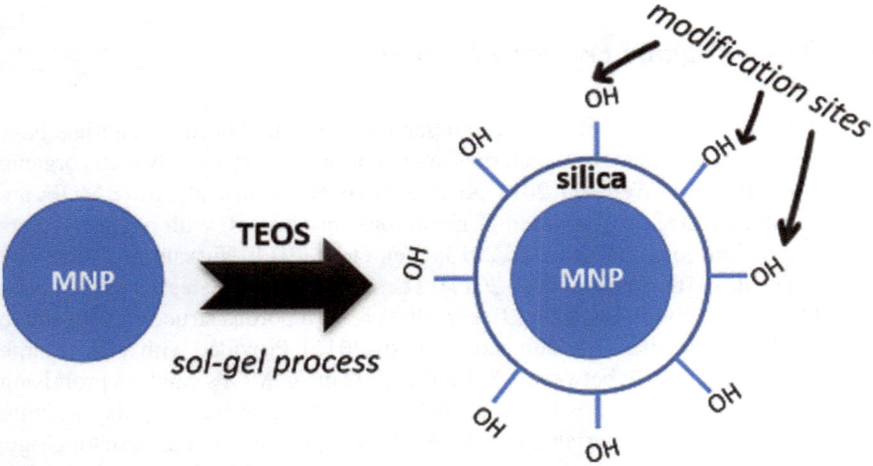

Fig. 7 Synthesis of silica-coated magnetic nanoparticles (MNPs), where hydroxyl groups could be then modified. *TEOS* tetraethyl orthosilicate

More recently, Tolmacheva et al. (2016) suggested the use of a new generation of highly porous polymers, the hypercrosslinked polystyrene (HCP), for the modification of Fe_3O_4 NPs. This absorbent was used for the preconcentration of four sulphonamide antibiotics from river water samples. The compounds were quantified by high-performance liquid chromatography with amperometric detection, with recoveries varying between 84 and 97% and the detection limit ranging between 0.6 and 1.0 ng/mL (Tolmacheva et al. 2016). Ionic liquids (ILs) such as 3-chloropropyl-N-methylimidazolium (Yang et al. 2011) and 1-hexadecyl-3-methylimidazolium bromide have also been successfully used to functionalize MNPs (Cheng et al. 2012), both works envisaging the preconcentration of chlorophenols in environmental water samples prior to HPLC-UV analysis. Table 2 compiles the latest uses of MNPs with different surface modifications in the preconcentration of several analytes in environmental water samples.

The latest described works correspond to magnetic solid-phase extraction (MSPE). This method presents a clear advantage over simple SPE, explaining the increased interest of researchers in such an approach. With MNPs, the sorbent with the adsorbed analytes can be easily separated from the aqueous solution using an external magnetic field (magnet) without the need for sample centrifugation or filtration, thus contributing to a faster separation process (Fig. 8). Due to these benefits, other types of magnetic absorbents have been reported for sample preparation in environmental analysis of water samples, such as the magnetic metal-organic frameworks (MMOFs), magnetic molecularly imprinted polymers (MMIP) and carbon-based magnetic hybrid nanocomposites, which will be discussed further ahead in this chapter.

3.2 *Metal Organic Frameworks (MOFs)*

In the last decade, a class of organic-inorganic hybrid microporous materials have emerged as improved absorbent materials in separation science, namely metal organic frameworks (MOFs) (Wen et al. 2014; Xu et al. 2016; Hashemi et al. 2017). MOFs are self-assembled materials composed of metal ions (or clusters) with organic linkers via coordination bonds (Tian et al. 2013; Wen et al. 2014; Hussain and Kharisov 2016; Xu et al. 2016; Hashemi et al. 2017). Their inorganic moieties act as the nodes of 3-D frameworks with ligands as linkers that create a porous structure with stable and well-defined cavities (Hussain and Kharisov 2016). Provided with a super large surface area, in the range between 1000 and 10,000 m^2/g, MOFs can be a promising material in sample preparation (Xu et al. 2016). Moreover, by changing the structure or size of guests, these materials can exhibit different pore size and diverse topology, and their surfaces can be also easily functionalized (Tian et al. 2013). The diversity of either metal ions or organic linkers allows the design and preparation of new and different types of MOFs in terms of topology, structure, and porosity (Hashemi et al. 2017). In summary, MOFs are remarkable porous materials as they possess the robustness of their inorganic metal structure and the tuneability of polymer materials

Table 2 Summary on the use of MSPE in the pretreatment of aqueous sample (in the last 10 years)

Magnetic core	Covering layer	Sample matrix	Analyte	Analytical technique	LOD (ng/mL)	References
Fe_3O_4	SDS	Tap, well and mineral water	Hg(II)	ICP-OES	0.04	Faraji et al. (2010)
	SDS	Tap and river water	Fluoxetine	LC-FD	20	Bagheri et al. (2011)
	Alkyl (C10-C18) carboxylates	River, reservoir and sea water	Benzo(a)pyrene, benzo(b)-fluoranthene, benzo(k)fluoranthene, benzo(ghi)perylene and indene(1,2,3-cd)pyrene (PAHs)	LC-FD	0.0002–0.0005	Ballesteros-Gomez and Rubio (2009)
	Polyaniline	Seawater	Methylmercury	GC–MS	0.1	Mehdinia et al. (2011)
	Polydopamine or Polypyrrole	–	Bisphenol A, proflavine, naphthalene acetic acid	CE-UV	–	Iqbal et al. (2011)
	Polypyrrole	Spiked water	Dimethyl phthalate, diethylphthalate, di-iso-butyl phthalate, di-n-butyl phthalate, benzylbutyl phthalate, di-(2-ethylhexyl) phthalate and di-n-octyl phthalate	GC–MS	0.006–0.068	Meng et al. (2011)
	APTMS and poly(MMA-co-MA)	–	Co^{2+}, Cr^{3+}, Zn^{2+}, and Cd^{2+}	–	–	Masoumi et al. (2014)
	HCP	River water	Sulfamethoxypyridazine, sulfamethazine, sulfamethoxazole and sulfachloropyridazine	HPLC-PDA	0.6–1.0	Tolmacheva et al. (2016)

(continued)

Table 2 (continued)

Magnetic core	Covering layer	Sample matrix	Analyte	Analytical technique	LOD (ng/mL)	References
	Silica	Lake and waste waters	Maphthalene, acenaphthene, anthracene, fluoranthene, and pyrene, benz[a]anthracene (PAHs)	HPLC-PDA	0.0028–0.0272	Huo and Yan (2012)
	Silica and 3-chloropropyl-N-methylimidazolium	River, spring and tap waters	3-chlorophenol, 2,4-dichlorophenol, pentachlorophenol	HPLC	0.20–0.35	Yang et al. (2011)
	1-hexadecyl-3-methylimidazolium bromide	River and tap water	2,4-dichlorophenol and 2,4,6-trichlorophenol	HPLC	0.12–0.13	Cheng et al. (2012)
	C18-functionalized mesoporous silica	Spiked ultra-pure water	Fluoranthene, pyrene, benzo(a)anthracene, benzo[b]fluoranthene, benzo(a)pyrene, benzo[g,h,i]perylene (PAHs)	HPLC-FD	–	Zhang et al. (2011)
	Mesoporous silica (MCM-41) and functionalization with pyridine	Tap, river and sea water	Cd^{2+}	FAAS	0.04	Omidi et al. (2015)
	$Cu_3(BTC)_2$ MOF and functionalization with dithizone	Tap and lake water	Pb^{2+}	FAAS	0.0046	Wang et al. (2013)

LOD limits of detection; *SDS* sodium dodecyl sulphate; *APTMS* 3-aminopropyltrimethoxysilane; *HCP* hypercrosslinked polystyrene; *ICP-OES* inductively coupled plasma optical emission spectrometry; *LC-FD* liquid chromatography with fluorescence detection; *GC-MS* gas chromatography-mass spectrometry; *CE-UV* capillary electrophoresis with UV-detection; *HPLC* high-performance liquid chromatography; *HPLC-PAD* HPLC with pulsed amperometric detection; *HPLC-FD* HPLC with fluorescence detection; *FAAS* flame atomic absorption spectrophotometry

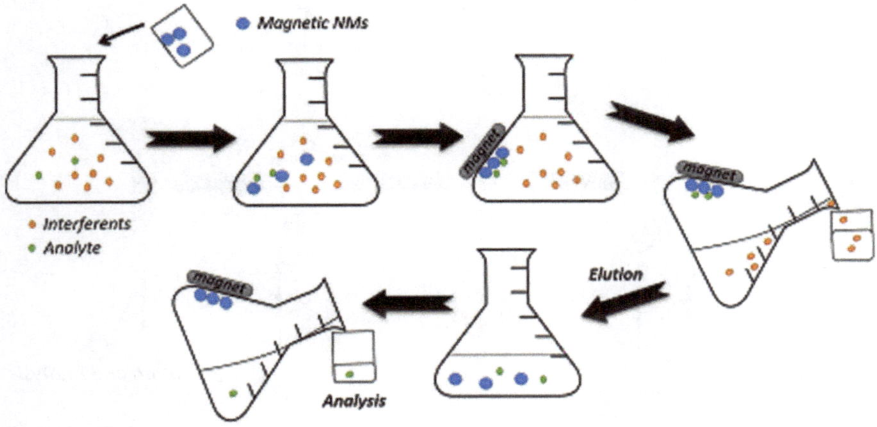

Fig. 8 Schematic representation of the MSPE procedure

(Tian et al. 2013; Wen et al. 2014; Hussain and Kharisov 2016; Xu et al. 2016; Hashemi et al. 2017)

Hashemi et al. (2017) reviewed the available data on the use of MOFs for the separation of diverse organic and inorganic pollutants from several types of samples. The authors also described the chemical and physical features responsible for the high performance of this class of materials; metal ions are linked to organic ligands or organic secondary building units (SBU) by strong bonds, whereas the common inorganic secondary building units include metal clusters (e.g. metal-carboxylate clusters and metal-azolate clusters), metal atoms, or rod-shaped clusters, and the common used organic linkers contains oxygen or nitrogen donors (e.g. carboxylates, azoles, nitriles, etc.) to offer linkage sites with inorganic SBU (Hashemi et al. 2017).

When dealing with water-based matrices, water-stable MOFs have to be considered, such as the "Material Institute Lavoisier" (MIL)-series MOFs (e.g. MIL-101(Cr), MIL-100(Fe), and MIL-53(M) (M: Fe, Al, and Cr), Zr-MOFs (e.g. UiO-66 and UiO-66-NH_2), and zeolite imidazolate frameworks (ZIFs: e.g. ZIF-8) (Hashemi et al. 2017). Due to its excellent thermal and chemical stability, MIL-101(Cr) was widely used in diverse fields, such as in chromatography, catalysis, gas adsorption and in SPE (Li et al. 2014). Li et al. (2014) used MIL-101(Cr) as a SPE sorbent material combined with fast detection analysis in real time mass spectrometry (DART-MS) for the analysis of triazine herbicides from lake water samples. LODs for the five studied herbicides were below the ng/mL level, meeting the regulatory requirements of the European Union (Li et al. 2014). Moreover, the recoveries were in the range of 85–110%. The obtained high adsorption capacity is probably due to the π-π interactions between the organic ligand network of the MOF and the phenyl ring of the target molecules (Li et al. 2014). Another commonly used MOF, especially for metals recovery, is $Cu_3(BTC)_2$ (BTC = benzene-1,3,5-tricarboxylate). Sohrabi (2014) reported a novel thiol-functionalized MOF, where thiol-functionalized silica NPs were incorporated in the $Cu_3(BTC)_2$ MOF, for the removal and preconcentra-

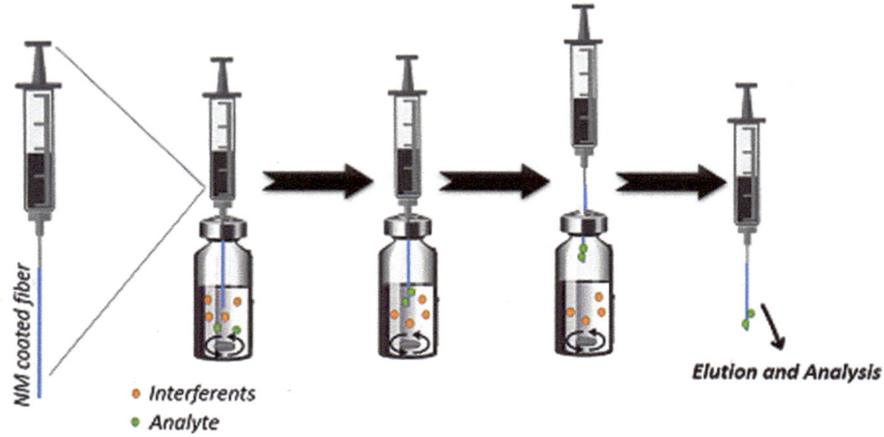

Fig. 9 Schematic representation of SPME by the direct immersion extraction mode

tion of Hg(II) ions from environmental water samples. When compared with other solid-phases, this MOF sorbent allowed higher enrichment factors and lower LODs, reaching 20 pg/mL (Sohrabi 2014).

MOFs can be used not only in SPE but also in the coating of fibers for solid-phase microextraction (SPME), as summarized in Fig. 9. Solid-phase microextraction (SPME) was proposed in 1990 (Arthur and Pawliszyn 1990), integrating sampling and sample preparation in one step (Lucena et al. 2011; Ribeiro et al. 2014). SPME consists in the exposure of a small fused-silica fiber coated with an appropriate sorbent layer to a gas or liquid sample that can be easily coupled with chromatographic and electrophoretic techniques (Lucena et al. 2011; Ribeiro et al. 2014). Chen et al. (2012) explored the application of MIL-53(Al, Cr, Fe) as fiber coatings for the SPME-GC-MS/MS of 16 PAHs in environmental water samples. It was found that MIL-53(Al) showed the highest adsorption efficiency towards PAHs, where very low detection limits (0.10 pg/mL) were achieved. These materials were compared with commercial fibers, showing that the MIL-53(Al) coatings provide better results in terms of adsorption efficiency of PAHs, while being of lower cost and higher robustness (Chen et al. 2012).

Similar to SPME, stir bar sorptive extraction (SBSE) is an additional equilibrium technique, possessing significant advantages, such as high sensitivity and good reproducibility. Contrastingly, the coating amount on the stir bar is 50–250 times higher than that on SPME fibers, resulting in an increase in the adsorption and recovery values (Frank and Pat 2007). Hu et al. (2013) reported the synthesis of three novel PDMS/MOFs (MOF-5, MOF-199 and IRMOF-3) coatings by the sol-gel technique for the extraction of 7 estrogens from environmental water samples by SBSE. Under the optimal experimental conditions, LODs were found to be in the range of 0.15–0.35 ng/mL using the best-found material (PDMS/IRMOF-3), with enrichment factors ranging from 30.3 to 55.6-fold (Hu et al. 2013) (Fig. 10).

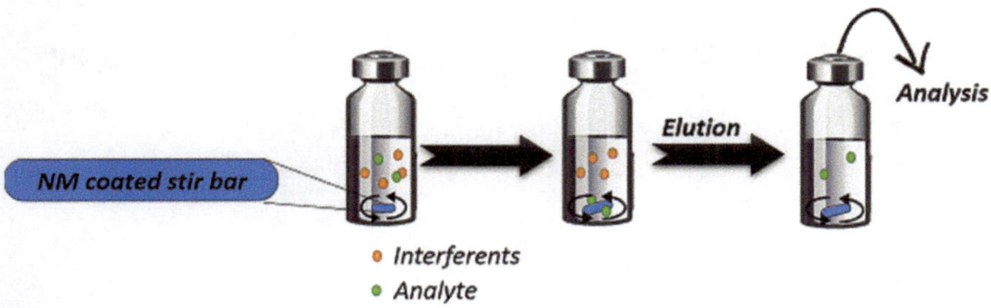

Fig. 10 Schematic representation of the SBSE process

Recently, composites of MOFs with magnetic NPs have also been developed for the preconcentration of different analytes using the MSPE technique. Wang et al. (2013) used dithizone (H2Dz, 1,5-diphenylcarbazone), a commonly used chelating reagent that can form a stable complex with lead, to functionalize a magnetic MOF ($Fe_3O_4/Cu_3(BTC)_2$). The developed method was applied for the preconcentration and determination of lead in environmental water samples, with recoveries in the range of 97–102% from real samples, and a detection limit as low as 0.0046 ng/mL (Wang et al. 2013). Ma et al. (2016) developed different kinds of magnetic MOFs for MSPE; for instance, they have reported a SPE procedure based on a magnetic porous carbon derived from an MOF for the preconcentration of steroid hormones from wastewater samples. In this approach, MIL-53-C was prepared through the direct carbonization of MIL-53, and the recoveries of the analytes for this novel methodology ranged from 96.2 to 98.7%. This method was also compared with other reported ones for the determination of steroid hormones, being found that it had a similar or even better performance in terms of LOD (0.005–0.01 ng/mL) and recovery. In Table 3 the use of MOFs for the preconcentration of different analytes in environmental aqueous samples are summarized (in the last decade).

3.3 *Molecularly Imprinted Polymers (MIPs)*

Molecularly imprinted polymers (MIPs) are a specific class of synthetic polymers that have been used as selective sorbents for SPE since 1994 (Sellergren 1994), which can be tailor-made for target analytes (Lucena et al. 2011; Hussain and Kharisov 2016; Xu et al. 2016). These selective sorbent materials are built by the introduction of templates in order to form specific nanosized cavities or surface binding sites (Lucena et al. 2011; Xu et al. 2016). The main components for MIPs synthesis include a functional monomer, a cross-linking agent, a template, a reaction initiator and an appropriate solvent (Hussain and Kharisov 2016). A reaction of copolymerization occurs, where the functional monomer forms a stable complex with the template, resulting in a three-dimensional crosslinked polymer (Hussain and Kharisov 2016). After the template elution, imprinted pores/cavities are left in the polymer network

Table 3 Application of MOFs in SPE, SPME, SBSE and MSPE in environmental monitoring (in the last 10 years)

MOFs	Metal	Modification	Preconcentration technique	Sample matrix	Analyte	Analytical technique	LOD (ng/mL)	References
MOF-5	Zn	–	SPE	Tap, spring and river water	PAHs	HPLC-PDA	0.0004–0.004	Yang et al. (2013)
MIL-101	Cr		SPE	Lake water	Triazine herbicides	DART-MS	0.1–0.2	Li et al. (2014)
ZIF-8 (Zn(2-methylimidazole)2)	Zn	–	SPE	River water	1H-benzotriazole, 5-tolyltriazole	–	–	Jiang et al. (2013)
Cu3(BTC)2	Cu	Thiol-modified silica NPs	SPE	River, tap, waste water and sea water	Hg(II)	CV-AAS	0.02	Sohrabi (2014)
[Ag12 (MA)8(mal)6.18 H2O]n MOF	Ag	–	SPE	Distilled, tap, river, lake, waste and sea water	Pb(II)	FAAS	0.5	Salarian et al. (2014)
MIL-53	Al	–	SPME	Waste water	PAHs	GC-MS/MS	0.0001–0.00073	Chen et al. (2012)
MOF-5	Zn	Polydimethylsiloxane	SBSE	Lake, river and fishpond waters	Estrogens	HPLC	–	Hu et al. (2013)
MOF-199	Zn						–	
IRMOF-3	Cu						0.15–0.35	
MIL-101	Cr	Addition of $Fe_3O_4@SiO_2$	MSPE	Lake and waste waters	PAHs	HPLC-PDA	0.0028–0.0272	Huo and Yan (2012)
MIL-101	Cr	Addition of $Fe_3O_4@SiO_2$	MSPE	Reservoir, river, and seawater	Pyrazole/pyrrole pesticides	HPLC	0.3–1.5	Ma et al. (2016)

(continued)

Table 3 (continued)

MOFs	Metal	Modification	Preconcentration technique	Sample matrix	Analyte	Analytical technique	LOD (ng/mL)	References
MOF-177	Zn	Addition of $Fe_3O_4@SiO_2$	MSPE	River, lake and waste waters	Phenols	GC-MS	0.0168–0.2083	Wang et al. (2014)
Magnetic MOF ($Fe_3O4/Cu_3(BTC)_2$)	Cu	Dithizone	MSPE	Tap and lake water	Pb(II)	ETAAS	0.0046	Wang et al. (2013)
Magnetic MOF ($Fe_3O4/Cu_3(BTC)_2$)	Cu	Fe_3O_4 NPs with Dithizone	MSPE	Distilled, tap, river, sea and mineral water	Cd(II), Pb(II), Ni(II) and Zn(II)	FAAS	0.12–1.2	Taghizadeh et al. (2013)
Magnetic MOF ($Fe_3O4/Cu_3(BTC)_2$)	Cu	Fe_3O_4 NPs with pyrine	MSPE	Tap, river and sea water	Cd(II) and Pb(II)	FAAS	0.2–1.1	Sohrabi et al. (2013)
Magnetic MOF ($Fe_3O4/Cu_3(BTC)_2$)	Cu	Fe_3O_4 NPs with pyrine	MSPE	Distilled, tap, River, and mineral water	Pd(II)	FAAS	0.37	Bagheri et al. (2012)
MIL-53-C	Fe	Carbonization of the MIL-53	MSPE	Waste water	Hormones	HPLC	0.005–0.01	Ma et al. (2016)

LOD limit of detection; *HPLC* high-performance liquid chromatography; *HPLC-PAD* HPLC with pulsed amperometric detection; *DART-MS* direct analysis in real time-mass spectrometry; *CV-AAS* cold vapor atomic absorption spectroscopy; *FAAS* flame atomic absorption spectrophotometry; *GC-MS* gas chromatography-mass spectrometry; *ETAAS* electrothermal atomic absorption spectrometry

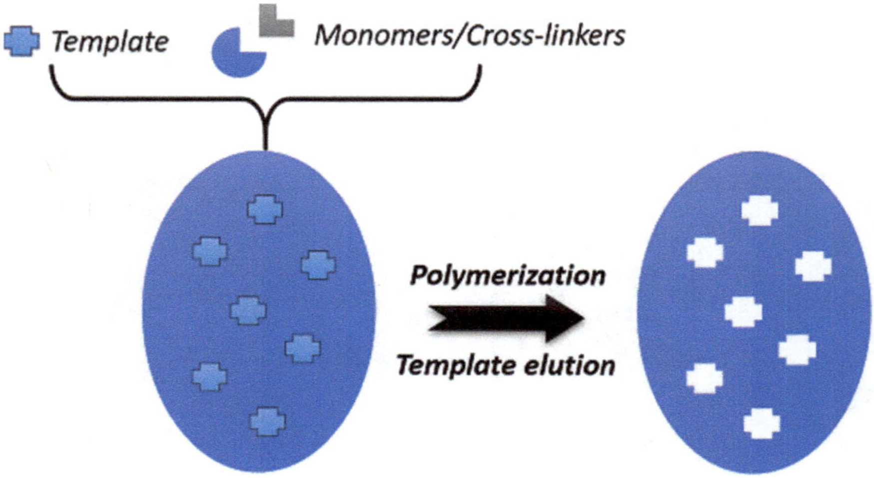

Fig. 11 Schematic representation for MIPs synthesis

that are sterically complementary to the target compounds (Lucena et al. 2011; Hussain and Kharisov 2016). Figure 11 shows a schematic representation of the MIPs synthesis. Regarding the templates, neutral organic molecules, ionic compounds and epitopes have been employed (Lucena et al. 2011).

Due to the high specificity and selectivity of the molecular recognition mechanism of molecular imprinting, several studies regarding the applicability of MIPs for sample preparation have been carried out. Lian and Wang (2016) reported a molecularly imprinted solid-phase extraction step followed by high-performance liquid chromatography with fluorescence detection for the clean-up, preconcentration and further analysis of ciprofloxacin, a fluoroquinolone antibiotic, in natural seawater samples. Ciprofloxacin was used as a template molecule and methacrylic acid was selected as the functional monomer because it was favourable for hydrogen bond recognition between the monomer and the target pollutant (Lian and Wang 2016). The recoveries of spiked seawater on the prepared MIP-based cartridges ranged from 75.2 to 112.4% and the obtained LOD was 0.2 ng/mL (Lian and Wang 2016). Canale et al. (2010) reported a method to analyse bisphenol A (BPA), 17-b-estradiol (17β-E2,) and ethynylestradiol (EE2) at ultratrace levels, with the goal of developing a screening procedure to monitor these endocrine-disrupting chemicals (EDCs) in environmental water samples. In their study, the fluorinated BPA derivative (4,40-(hexafluoroisopropylidene)-diphenol) was used as a mimic template, where the synthetized imprinted polymer (poly-4-VP-co-TRIM) showed high affinity and selectivity for BPA, 17β-E2 and EE2, providing recoveries close to 100% and low LOD (0.010 pg/mL) for BPA in real drinking water samples (Canale et al. 2010).

Traditionally, the use of MIPs involves the loading of MIPs into SPE cartridges; however, more recently, MSPE became a popular extraction mode based on the use of MIPs that are grafted onto magnetic NPs. Ji et al. (2009) and Chen et al.

(2010) prepared magnetic MIPs for the preconcentration and detection of BPA and fluoroquinolone antibiotics in environmental water samples. Additionally, both silica particles (Zhu et al. 2011; Song et al. 2012; Yin et al. 2012) and carbon nanotubes (Tan et al. 2011; Yang et al. 2013) can be an excellent support material of MIPs, especially MWCNTs due to their extremely large surface area (Wen et al. 2014). Accordingly, Tan et al. (2011) and Yang et al. (2013) described the preparation of MIPs on the surface of MWCNTs for the determination of several herbicides in environmental water samples.

Table 4 lists some applications of MIPs in sample preparation and extraction of toxic analytes from environmental water samples.

3.4 Carbon-Based Nanomaterials

Carbon-based nanomaterials have been described in the literature as one of the most important trends in SPE technology (Wen et al. 2014). Carbon exists in various allotropes, such as fullerenes, carbon nanocones, carbon nanodisks, carbon nanofibers, carbon nanohorns, diamond, **graphene (G) and graphene oxide (GO)**, and **carbon nanotubes (CNTs)**, including single-walled CNTs (SWCNTs) and multi-wall CNTs (MWCNTs) (Tian et al. 2013; Wen et al. 2014; Xu et al. 2016). However, applications in environmental analysis, namely regarding sample preparation, have been mainly focused on the use of fullerenes, CNTs and G/GO (Tian et al. 2013; Wen et al. 2014).

The earliest report on carbon-based nanomaterials dates from 1985, when fullerene C_{60} was firstly prepared by Smalley et al. (Kroto et al. 1985). Nevertheless, only in 1994 Gallego et al. (1994) fabricated a continuous-flow system using a C_{60} fullerene mini-column for the preconcentration of lead for environmental analysis (Wen et al. 2014; Xu et al. 2016). Fullerenes are polyhedral nanostructures, in a spherical arrangement, in which carbons are bonded to form five-to-six-membered rings (Lucena et al. 2011). They present a low tendency to form aggregates, and thus fullerenes are promising alternatives to be used as sorbents in SPE procedures based on continuous flow devices. In this perspective, Jurado-Sanchez et al. (2009) described the use of fullerenes for aromatic and non-aromatic N-nitrosamines separation. The authors compared the performance of C_{60} and C_{70} fullerenes and nanotubes, reporting that C_{60} fullerene is the best choice to selectively retain the aromatic fraction (sorption efficiency >99%). Fullerenes act as filters retaining only the aromatic fraction via π-π interactions (Jurado-sánchez et al. 2009). Thus, by using an automatic system containing two sequential sorbent columns coupled to GC-MS, the quantification at the low nanogram per litter level of non-aromatic amines in drinking water and wastewater samples was possible, in which LODs of 0.004–0.012 ng/mL were reached (Jurado-sánchez et al. 2009).

Graphene is a single-atom-thick honeycomb two-dimensional lattice carbon nano-scaled material (Tian et al. 2013; Wen et al. 2014). Graphene exhibits a nanosheet structure with a large specific theoretical surface area and possesses a

Table 4 Application of MIPs in SPE, SPME, and MSPE for environmental monitoring (in the last 10 years)

Template	Monomer/cross-linker	Modification	Sample matrix	Analyte	Preconcentration technique	Analytical technique	LOD (ng/mL)	References
Ciprofloxacin	MAA/EGDMA		Seawater	Ciprofloxacin	SPE	HPLC-FD	0.2	Lian and Wang (2016)
BPA	4-VP/TRIM		Tap, waste and river waters	BPA	SPE	CE–UV	3.0–6.9	Mei et al. (2011)
TMB	γ-MPTMS/TMOS	Grafted on the silica gel surface	Tap, lake and sea water	Biphenyls	SPE	GC-ECD	1	Zhu et al. (2011)
Fluorinated BPA derivative	4-VP/TRIM		Potable water	BPA, 17β-E2, EE2	SPE	GC-MS	0.00001	Canale et al. (2010)
PAHs	PTMS/TEOS	Grafted on the silica gel surface	Seawater	PAHs	SPE	GC-MS	0.0052–0.0126	Song et al. (2012)
DPA or BPA	APTES/TEOS	Grafted on the silica gel surface	Tap, river and lake water	TBBPA	SPE	RRLC-DAD	2	Yin et al. (2012)
Ciprofloxacin	MAA/EGDMA	Incorporated with Fe_3O_4 NPs	Lake, river, waste waters	Fluoroquinolone antibiotics	MSPE	HPLC-MS/MS	0.0032–0.0062	Chen et al. (2010)
BPA	2-VP/EGDMA	Incorporated with Fe3O4 NPs	River and waste water	BPA	MSPE	HPLC	0.014	Ji et al. (2009)

(continued)

Table 4 (continued)

Template	Monomer/cross-linker	Modification	Sample matrix	Analyte	Preconcentration technique	Analytical technique	LOD (ng/mL)	References
Prometryn	MAA/TRIM	Polymerized on MWCNTs	River and waste water	Triazine herbicides	SPME	HPLC	0.08–0.38	Tan et al. (2011)
2,4-D	Acrylamide and styrene/EGDMA	Polymerized on SiO_2-coated MWCNTs	River water	2,4-D	SPE	HPLC	100	Yang et al. (2013)

LOD limit of detection; *BPA* bisphenol A; *TMB* 3,3′,5,5′-tetramethylbenzidine; *DPA* diphenolic Acid; *TBBPA* tetrabromobisphenol A; *2,4-D* 2,4-dichlorophenoxyacetic acid; *17β-E2* 17-b-estradiol; *EE2* ethynylestradiol; *MMA* Methacrylic acid; *EGDMA* ethylene glycol dimethacrylate; *4-VP* 4-vinylpyridine; *TRIM* trimethylolpropane trimethacrylate; *γ-MPTMS* γ-mercaptopropyltrimethoxysilane; *TMOS* tetramethoxysilane; *PTMS* propyltrimethoxysilane; *TEOS* tetraethyl orthosilicate; *APTES* (3-aminopropyl)triethoxysilane; *HPLC* high-performance liquid chromatography; *HPLC-FD* HPLC with fluorescence detection; *CE-UV* capillary electrophoresis with UV-detection; *GC-ECD* gas chromatography with electron capture detection; *GC-MS* gas chromatography-mass spectrometry; *RRLC-DAD* reversed-phase rapid-resolution liquid-chromatography with photodiode array detection; *HPLC-MS/MS* HPLC coupled with tandem mass spectrometry large delocalized π-electron system, offering a strong affinity for π-π interactions with aromatic analytes. Due to these characteristics, it is a very popular option for use as an adsorbent material in SPE (Tian et al. 2013; Wen et al. 2014). Since many types of organic pollutants, such as PAHs, organophosphorus pesticides (OPPs), and phthalic acid esters (PAEs) hold aromatic residues, there are several available studies reported in the literature describing graphene as a good sorbent in SPE. For instance, Luo et al. (2014) and Wu et al. (2013) reported the use of graphene as a SPE sorbent for the determination of PAEs in environmental water samples, while Wang et al. (2013) used graphene in a SPE disks-based column for a fast separation and preconcentration of trace PAHs from environmental water samples. Different types of SPE approaches and analytical techniques were used in these studies. In the first study (Luo et al. 2014), graphene nanoplatelets were used as a novel sorbent material for a SPE cartridge to analyse PAEs by HPLC, showing that this method was highly effective for the extraction of target analytes with recoveries ranging from 87.7 to 100.9%, being superior to other sorbents (C18 silica, HLB Oasis, activated carbon, and MWCNTs) (Luo et al. 2014). A simple and inexpensive method of dispersive solid-phase extraction (DSPE) combined GC–MS was used by Wu et al. (2013), where graphene was dispersed into water samples in a centrifuge tube to adsorb PAEs; graphene was then isolated from the solution by centrifugation (recoveries between 71 and 117%). Sun et al. (2014) proposed a miniaturized self-assembly pipette tip graphene SPE cartridge (PT-G-SPE) coupled with liquid chromatography with fluorescence detection (LC-FD) for the preconcentration and determination of three sulphonamides in environmental water samples, allowing to reach detection limits in the range of pg/mL.

Graphene oxide (GO) is a monolayer of two-dimensional carbon-based materials, containing multi-functional groups, such as carboxyl, epoxy, ketone and hydroxyl groups, which can easily offer functionalized sites to enhance the selectivity for certain analytes (Tian et al. 2013; Wen et al. 2014). However, the direct use of graphene or GO as SPE adsorbents may rise several concerns, regarding aggregation or the leakage of the G or GO sheets from the SPE cartridges/columns (Liu et al. 2011). Therefore, new types of SPE sorbents produced by graphene bounded silica and graphene oxide bounded silica were introduced by Liu et al. (2011), with this new material yielding excellent performance with recoveries of chlorophenols from water samples approaching 100%. Recent applications of graphene in environmental sample preparation are summarized in Table 5.

Carbon nanotubes (CNTs) are amongst the most currently used nanomaterials for sample preparation in environmental analysis. CNTs can be described as a graphene sheet derived from the scrolling up into a nanoscale-tube (Tian et al. 2013; Wen et al. 2014). CNTs can be divided into single-walled carbon nanotubes (SWCNTs) and multi-walled carbon nanotubes (MWCNs). Like other carbon-based compounds, CNTs also exhibit good performance for the adsorption of several organic compounds (Yu et al. 2010; Hadjmohammadi et al. 2010; Katsumata et al. 2010; Ma et al. 2010; See et al. 2010; Wu et al. 2010; Guo et al. 2011; Herrera-herrera et al. 2011; Gao et al. 2012; Gonzalez-Curbelo et al. 2013; Lasarte-aragonés et al. 2013) and metal ions with or without chelating agents (Chen et al. 2009; Ozcan et al.

Table 5 Application of graphene nanomaterials in SPE for environmental monitoring (in the last 10 years)

Carbon material	Modification	Sample	Analyte	Operation mode	Analytical technique	LOD (ng/mL)	References
Graphene nanoplatelets (GN)		Tap and commercial bottled water	PAEs	SPE cartridge	HPLC	0.09–0.33	Luo et al. (2014)
Graphene		Tap, sea, river, and waste waters	Pb(II)	SPE column	FAAS	0.61	Wang et al. (2012)
Graphene		Ultrapure, river and sea water	PAEs	DSPE	GC–MS	2.0–6.0	Wu et al. (2013)
Graphene		Lake, river, waste water	Sulfonamide antibiotics	SPE cartridge	HPLC-FD	0.0005–0.0017	Sun et al. (2014)
Graphene		Tap and river water	PAHs	SPE column	GC–MS	0.00084–0.013	Wang et al. (2013)
Graphene/Graphene oxide	Bound to silica		Chlorophenols	SPE cartridge	MS		Liu et al. (2011)
Graphene oxide		River and sea water	Chlorophenoxy acid herbicides	SPE cartridge	CE	0.3–0.5	Tabani et al. (2013)
Graphene oxide	Oxidized with H_2SO_4	River, lake and sea water	Pb(II), Cd(II), Bi(III) and Sb(III)	SPE	AFS	0.002–0.01	Deng et al. (2013)

LOD limit of detection; *PAEs* phthalate esters; *HPLC* high-performance liquid chromatography; *FAAS* flame atomic absorption spectrometry; *GC-MS* gas chromatography-mass spectrometry; *HPLC-FD* HPLC with fluorescence detection; *CE* capillary electrophoresis; *AFS* atomic fluorescence spectrometry

2010; El-sheikh et al. 2011; López-garcía et al. 2011; Soylak and Unsal 2012; Cui et al. 2012; Nabid et al. 2012; Behbahani et al. 2013; Soliman et al. 2013; Kocot et al. 2013; Madadizadeh et al. 2013; Soylak and Topalak 2014; Sahmetlioglu et al. 2014). Chen et al. (2009) described the use of a microcolumn packed with SWCNTs as a new adsorption material for the preconcentration of trace metal ions in environmental samples. SWCNTs were first oxidized with nitric acid and filled into a PTFE microcolumn for the sample pretreatment, in which lake water samples were used. The levels of Cu, Co and Pb ions present in the samples were determined by ICP-MS, reaching LOD values as low as 1.2 pg/mL (Chen et al. 2009).

The modification of CNTs with specific functional groups has been also adopted for sample preparation (Tian et al. 2013; Xu et al. 2016). Gao et al. (2012) shown the applicability of cyano-functionalized MWCNTs for the preconcentration of phenolic compounds from river water and seawater samples. By using SPE cartridges filled with this cyano-functionalized MWCNTs adsorbent material, it was found that the limits of detection of the method were 0.45, 0.09, 0.08, and 3.00 ng/mL for *p*-chlorophenol, 1-naphthol, 2-naphthol, and 2,4-dichlorophenol, respectively (Gao et al. 2012). Furthermore, the functionalization of CNTs is also very valuable for metal ion analysis since several chelating reagents can be coupled to the surfaces of CNTs (Cui et al. 2012; Nabid et al. 2012; Behbahani et al. 2013; Madadizadeh et al. 2013; Soliman et al. 2013; Sahmetlioglu et al. 2014; Soylak and Topalak 2014). Cui et al. (2012) covalently grafted phenyl-iminodiacetic acid groups onto MWCNTs for the preconcentration of trace quantities of several metal ions (Fe (III), Cu(II) and Pb(II) ions) from tap and river water samples prior to their determination by ICP-OES. When compared with pristine (unmodified) MWCNTs, the functionalized PIAD-MWCNTs show better dispersibility in water and also higher sorption affinities for the tested metals (Cui et al. 2012). The recent applications of CNTs in SPE for the analysis of environmental water samples are summarized in Table 6.

3.5 Silica-Based Nanomaterials

Silica-based nanomaterials for sample preparation include silica nanoparticles (SiO_2 NPs), and mesoporous silica materials (e.g. MCM-41 and SBA-15). These silica-based materials are characterized by high surface areas and high adsorption capacity (Lucena et al. 2011; Tian et al. 2013). Moreover, compared with other materials, silica-based NMs provide good biocompatibility (Xu et al. 2016) and are also easy to synthesize. Because they display intrinsic surface reactivity, it is possible to easily introduce functional groups on the surface (Lucena et al. 2011; Tian et al. 2013). Most of the silica-based materials are chemically modified when used as SPE sorbents or used as a support in other nanomaterials to form composites/hybrid materials (Gao et al. 2010; Bagheri et al. 2011a; Zhang et al. 2011; Zhu et al. 2011; Liu et al. 2011; Yang et al. 2011; Song et al. 2012; Yin et al. 2012; Wang et al. 2013; Mehdinia et al. 2014; Sohrabi 2014; Omidi et al. 2015).

Table 6 Application of CNTs in SPE for environmental monitoring (in the last 10 years)

CNTs	Modification	Sample	Analyte	Analytical technique	LOD (ng/mL)	References
SWNTs	Oxidized with nitric acid	Lake water	Cu(II), Co(II) and Pb(II)	ICP-MS	0.0012–0.039	Chen et al. (2009)
MWCNTs	–	Fortified lake, tap, and bottled mineral water	Fe(III), Cu(II), Mn(II) and Pb(II)	FAAS	3.5–8.0	Ozcan et al. (2010)
MWCNTs	–	River water	Triazine herbicides	RRLC-MS/MS	0.0001	Yu et al. (2010)
MWCNTs	–	Sea, river and waste water	Cd(II) and Pb(II)	TXRF	1.0–2.1	Kocot et al. (2013)
MWCNTs	–	Run-off, mineral and tap water	Organophosphorus pesticides	GC-NPD	0.00116–0.0936	Gonzalez-Curbelo et al. (2013)
MWCNTs	–	Tap, river and sea water	Triazine herbicides	GC-MS	0.0025–0.005	Katsumata et al. (2010)
MWCNTs	–	River, tap and sea waters	PAHs	GC-MS	0.002–0.0085	Ma et al. (2010)
MWCNTs	–	River water	Triazine herbicides	Micro-LC	0.2–0.5	See et al. (2010)
MWCNTs	–	River water	PAHs	GC-MS	0.001–0.15	Wu et al. (2010)
MWCNTs	–	Fortified lake, well and waste water	Fe(III) and Pb(II)	FAAS	1.3–2.9	Soylak and Unsal (2012)

(continued)

Table 6 (continued)

CNTs	Modification	Sample	Analyte	Analytical technique	LOD (ng/mL)	References
MWCNTs	–	Tap, well and river water	Chlorpyrifos and phosalone pesticides	HPLC	0.403–3.03	Hadjmohammadi et al. (2010)
MWCNTs	–	Bottled mineral, tap, river and sea water	As (III, V) and Sb (III, V)	ETAAS	0.02–0.05	López-garcía et al. (2011)
MWCNTs	–	Tap and river water	Linear alkylbenzene sulfonates	HPLC	0.02–0.03	Guo et al. (2011)
MWCNTs		River, well and tap water	Triazine herbicides	UPLC	0.15–0.40	Lasarte-aragonés et al. (2013)
MWCNTs	Oxidized with nitric acid	Commercial mineral, tap and waste water	Quinolone antibiotics	CE-DAD	0.028–0.094	Herrera-herrera et al. (2011)
MWCNTs	Oxidized with nitric acid	Tap, reservoir and mineral waters	Hg(II) ions	–	–	El-sheikh et al. (2011)
MWCNTs	Oxidized with nitric acid and 2-(5-bromo-2-pyridylazo)-5-diethylaminophenol modification	River, waste and sea waters	Cd(II) ions	ETAAS	0.00014	Madadizadeh et al. (2013)
MWCNTs	Immobilization of 5-aminsalicylic acid	Distilled, tap, sea and waste water	Pb(II) ions	ICP-OES	0.25	Soliman et al. (2013)

(continued)

Table 6 (continued)

CNTs	Modification	Sample	Analyte	Analytical technique	LOD (ng/mL)	References
MWCNTs	Tartrazine impregnated	Tap, mineral and lake water	Cd(II) and Pb(II) ions	FAAS	0.8-6.6	Soylak and Topalak (2014)
MWCNTs	Polypyrrole coating	Well and wastewater	Pb(II) ions	FAAS	1.1	Sahmetlioglu et al. (2014)
MWCNTs	Poly(2-amino thiophenol) coating	Distilled, tap, river, lake and waste waters	Cd(II) and Pb(II) ions	FAAS	0.3–10	Nabid et al. (2012)
MWCNTs	Cyano-functionalized	River and sea waters	Phenolic compounds	HPLC	0.08–3.00	Gao et al. (2012)
MWCNTs	Diphenylcarbazide functionalized	Distilled, tap, river, and waste waters	Cd(II) ions	FAAS	0.05	Behbahani et al. (2013)
MWCNTs	Phenyl-iminodiacetic-acid functionalized	Tap and river water	Fe (III), Cu(II) and Pb(II) ions	ICP-OES	0.15–0.26	Cui et al. (2012)

LOD limit of detection; *ICP-MS* inductively coupled plasma-mass spectrometry; *FAAS* flame atomic absorption spectrometry; *RRLC-DAD: RRLC-MS/MS* reversed-phase rapid-resolution liquid-chromatography coupled with tandem mass spectrometry; *TRXF* total reflection X-ray fluorescence; *CG-NPD* gas chromatography with nitrogen phosphorous detection; *GC-MS* gas chromatography-mass spectrometry; *Micro-LC* micro liquid-chromatography; *HPLC* high-performance liquid chromatography; *ETAAS* electrothermal atomic absorption spectrometry; *UPLC* ultra-performance liquid chromatography; *CE-DAD* capillary electrophoresis with photodiode array detection; *ICP-OES* inductively coupled plasma optical emission spectrometry

Rao et al. (2016) tested the adsorption capacity of silica NPs in the preconcentration of residues of pyrethroid pesticides in water samples. Spiked water samples were passed through silica NPs-filled cartridges for the simultaneous determination of the pesticide residues using GC-EI-MS (Rao et al. 2016). It was then found that silica NPs-based cartridges exhibit higher capacity to trap pesticides when compared with conventional C18 cartridges, leading also to a low LOD value of 0.05 ng/mL (Rao et al. 2016). Additionally, silica gel was used as a support material in SPE to extract and preconcentrate metal ions from environmental water samples. In particular, Leopold et al. (2009) reported the use of gold-coated silica as a preconcentration method for the determination of mercury in river water samples. Due to the high potential of nanogold surfaces for the adsorption of mercury species from aqueous solutions, the gold-coated silica nanomaterials proved to be highly efficient for the adsorption of different mercury species from acidified waters without using chelating agents, with recoveries higher than 90% (Leopold et al. 2009). The microcolumn filled with gold-coated silica was then coupled to atomic fluorescence spectrometry (AFS), resulting in a low detection limit, of 0.18 pg/mL (Leopold et al. 2009). The application of silica-supported ionic liquids (ILs) in SPE has also been reported, namely in the preconcentration of trace metal ions from water samples. Liang and Peng (2010) and Ayata et al. (2011) described the preparation of a microcolumn packed with IL-modified silica combined with flame atomic absorption spectrometry (FAAS) for the determination of trace amounts of cadmium and lead in environmental water samples. The 1-butyl-3-methylimidazolium hexafluorophosphate (Liang and Peng 2010) and the 1-methyl-3-butylimidazolium bromide (Ayata et al. 2011) ILs were used to modify the silica surface for the determination of trace amount of metal ions in lake, tap water and river water samples with satisfactory results (LOD = 0.6 ng/mL (Liang and Peng 2010) and 0.7 ng/mL (Ayata et al. 2011)). A summary of the materials and applications of silica-based nanomaterials in SPE in the last decade are listed in Table 7.

4 Conclusions

The contamination of the aquatic environment is a serious concern, which has negative impacts on the ecosystem and human health. To better address this concern, environmental monitoring is mandatory. However, most of the ubiquitous pollutants are close or under the detection limits of the currently available analytical techniques. There is thus a crucial need on improving the sample preparation step, envisioning the enhancement of speed, reliability and sensitivity of analysis. In this chapter, the most recent applications of nanomaterials in SPE for environmental water samples pretreatment were described and discussed. These NMs include metallic and magnetic nanoparticles, MOFs, MIPs, and carbonaceous- and silica-based nanomaterials.

In general, nanomaterials exhibit relevant physical and chemical properties, making of them superior candidates as sorbents in sample preparation techniques. Furthermore, the ability of nanomaterials for surface modification with a wide variety

Table 7 Application of silica-based nanomaterials in SPE for environmental monitoring (in the last 10 years)

Type of material	Functionalization	Sample matrix	Analyte	Analytical technique	LOD (ng/mL)	References
Silica NPs	–	Water samples	Pyrethroid pesticides	GC-MS	0.05	Rao et al. (2016)
Silica NPs	Resacetophenone	River water	Pb(II)	UV	0.58	Kaur and Gupta (2009)
Silica gel	Gold deposition	Estuary and river water	Hg(II)	AFS	0.00018	Leopold et al. (2009)
Silica gel	Hydrophobic IL (SG-1,10-PhenanNTf2)	Ground, lake and waste water	Coumarin	HPLC		Marwani and Bakhsh (2013)
Silica gel	1-butyl-3-methylimidazolium hexafluorophosphate	Lake and tap water	Cd(II)	FAAS	0.6	Liang and Peng (2010)
Silica gel	1-methyl-3-butylimidazolium bromide	River water	Pb(II)	FAAS	0.7	Ayata et al. (2011)
Mesoporous silica particles (UVM-7)	Triethanolamine	Tap, river and waste water	Cr(VI)	FAAS	1.2	Shirkhanloo et al. (2015)

LOD limit of detection; *GC-MS* gas chromatography-mass spectrometry; *AFS* atomic fluorescence spectrometry; *UPLC* ultra-performance liquid chromatography; *FAAS* flame atomic absorption spectrometry of functional groups promotes higher adsorption efficiencies and selectivity, while decreasing the limits of detection of the analytical techniques. All these features turn possible the analysis of several toxic compounds that are found in ultratrace levels in water samples.

In addition to the described NMs, in the future it is expected that new nanomaterials emerge in sample preparation techniques. It is also expected that an increasing number of novel nanocomposites and/or hybrid nanomaterials will appear by combining advantageous properties of the different types of nanomaterials. For instance, the use of composite/hybrid materials containing MNPs combined with MIPs, MOFs, CNTs, graphene, among others, are an example of such new solutions, further allowing an easy recovery by an external magnetic field without the need of centrifugation or filtration steps.

Among the most recent functional groups in NMs, ionic liquids (ILs) have emerged as a new class of sorbents coatings in SPE, named as supported ionic-liquid phases (SILPs). ILs are described as designer solvents, and can be widely modulated by altering their cation/anion chemical structures, thus allowing the establishment of specific interactions with the target analyte and the improvement on analysis. Although their numerous advantages, the use of SILPs in the pretreatment of samples for environmental monitoring is still in its infancy.

A new trend in analytical chemistry and environmental monitoring is focused on the development of more sustainable pretreatment methods and analysis. To this end, it is necessary to eliminate, or at least to reduce, the amounts of organic solvents used in sample preparation techniques. Moreover, the recovery and reuse of solvents and/or nanomaterials, and the techniques automatization and miniaturization, are also major issues to improve.

Presently, most of the applications of nanomaterials for sample preparation in water environmental analysis are focused on metal ions, toxic organic compounds and pesticides. Nonetheless, due to their current serious concerns, pharmaceutical drugs and endocrine disrupting chemicals need to be more deeply addressed in the future.

Acknowledgements This work was developed within the scope of the project CICECO-Aveiro Institute of Materials, FCT Ref. UID/CTM/50011/2019, financed by national funds through the FCT/MCTES. This work was financially supported by the project POCI-01-0145-FEDER-031106 (IonCytDevice) funded by FEDER, through COMPETE2020—Programa Operacional Competitividade e Internacionalização (POCI), and by national funds (OE), through FCT/MCTES, and developed in the scope of the "Smart Green Homes" Project [POCI-01-0247-FEDER-007678], a co-promotion between Bosch Termotecnologia S.A. and the University of Aveiro. It is financed by Portugal 2020 under the Competitiveness an Internationalization Operational Program, and by the European Regional Development Fund. A.C.A.S. acknowledges University of Aveiro, for funding in the scope of the framework contract foreseen in the numbers 4, 5 and 6 of the article 23, of the Decree-Law 57/2016, of August 29, changed by Law 57/2017, of July 19. M. C. N. acknowledges FCT, I.P. for the research contract CEECIND/00383/2017 under the CEEC Individual 2017.

References

Afkhami A et al (2011) Flame atomic absorption spectrometric determination of trace amounts of Pb (II) and Cr (III) in biological, food and environmental samples after preconcentration by modified nano-alumina. Microchim Acta 172:125–136. https://doi.org/10.1007/s00604-010-0478-y

Akpor OB (2014) Heavy metal pollutants in wastewater effluents: sources, effects and remediation. Adv Biosci Bioeng 2:37

Arthur CL, Pawliszyn J (1990) Solid phase microextraction with thermal desorption using fused silica optical fibers. Anal Chem 62(19):2145–2148

Ayata S, Bozkurt SS, Ocakoglu K (2011) Separation and preconcentration of Pb (II) using ionic liquid-modified silica and its determination by flame atomic absorption spectrometry. Talanta. 84:212–215. https://doi.org/10.1016/j.talanta.2011.01.006

Ballesteros-Gómez A, Rubio S (2009) Hemimicelles of alkyl carboxylates chemisorbed onto magnetic nanoparticles: study and application to the extraction of carcinogenic polycyclic aromatic hydrocarbons in environmental water samples. Anal Chem 81(21):9012–9020

Bagheri H, Ayazi Z, Aghakhani A (2011a) A novel needle trap sorbent based on carbon nanotube-sol-gel for microextraction of polycyclic aromatic hydrocarbons from aquatic media. Anal Chim Acta 683(2):212–220. https://doi.org/10.1016/j.aca.2010.10.026

Bagheri H, Zandi O, Aghakhani A (2011b) Extraction of fluoxetine from aquatic and urine samples using sodium dodecyl sulfate-coated iron oxide magnetic nanoparticles followed by spectrofluorimetric determination. Anal Chim Acta 692(1–2):80–84. https://doi.org/10.1016/j.aca.2011.02.060

Bagheri A et al (2012) Synthesis and characterization of magnetic metal-organic framework (MOF) as a novel sorbent, and its optimization by experimental design methodology for determination of palladium in environmental samples. Talanta 99:132–139. https://doi.org/10.1016/j.talanta.2012.05.030

Baytak S, Zereen F, Arslan Z (2011) Preconcentration of trace elements from water samples on a minicolumn of yeast (*Yamadazyma spartinae*) immobilized TiO_2 nanoparticles for determination by ICP-AES. Talanta 84:319–323. https://doi.org/10.1016/j.talanta.2011.01.020

Behbahani M et al (2013) Application of multiwalled carbon nanotubes modified by diphenylcarbazide for selective solid phase extraction of ultra traces Cd (II) in water samples and food products. Food Chem 141:48–53. https://doi.org/10.1016/j.foodchem.2013.03.011

Bjørklund G et al (2017) The toxicology of mercury: current research and emerging trends. Environ Res 159(Supplement C):545–554

Burgess RM, Ahrens MJ, Hickey CW (2003) Geochemistry of PAHs in aquatic environments: source, persistence and distribution. In: PAHs: an ecotoxicological perspective

Canale F et al (2010) Development of a molecularly imprinted polymer for selective extraction of bisphenol A in water samples. J Sep Sci 33:1644–1651. https://doi.org/10.1002/jssc.201000013

Carolin CF et al (2017) Efficient techniques for the removal of toxic heavy metals from aquatic environment: a review. J Environ Chem Eng 5:2782–2799. https://doi.org/10.1016/j.jece.2017.05.029

Chen S et al (2009) Solid-phase extraction of Cu, Co and Pb on oxidized single-walled carbon nanotubes and their determination by inductively coupled plasma mass spectrometry. J Hazard Mater 170:247–251. https://doi.org/10.1016/j.jhazmat.2009.04.104

Chen L et al (2010) Determination of fluoroquinolone antibiotics in environmental water samples based on magnetic molecularly imprinted polymer extraction followed by liquid chromatography-tandem mass spectrometry. Anal Chim Acta 662(1):31–38. https://doi.org/10.1016/j.aca.2010.01.001

Chen X et al (2012) Metal-organic framework MIL-53 (Al) as a solid-phase microextraction water samples by gas chromatography-tandem mass spectrometry. Analyst 137:5411–5419. https://doi.org/10.1039/c2an35806a

Cheng Q et al (2012) Mixed hemimicelles solid-phase extraction of chlorophenols in environmental water samples with 1-hexadecyl-3-methylimidazolium bromide-coated Fe_3O_4 magnetic nanopar-

ticles with high-performance liquid chromatographic analysis. Anal Chim Acta 715:113–119. https://doi.org/10.1016/j.aca.2011.12.004
Cui Y et al (2012) Novel phenyl-iminodiacetic acid grafted multiwalled carbon nanotubes for solid phase extraction of iron, copper and lead ions from aqueous medium. Microchim Acta 176:359–366. https://doi.org/10.1007/s00604-011-0725-x
Das D, Gupta U, Das AK (2012) Recent developments in solid phase extraction in elemental speciation of environmental samples with special reference to aqueous solutions. Trends Anal Chem 38:163–171. https://doi.org/10.1016/j.trac.2011.01.020
Deng D et al (2013) Organic solvent-free cloud point extraction-like methodology using aggregation of graphene oxide. Anal Chem 86(1):758–765. https://doi.org/10.1021/ac403345s
El-sheikh AH et al (2011) Effect of oxidation and geometrical dimensions of carbon nanotubes on Hg (II) sorption and preconcentration from real waters. Desalination 270:214–220. https://doi.org/10.1016/j.desal.2010.11.048
Erdogan H, Yalçinkaya Ö, Türker AR (2011) Determination of inorganic arsenic species by hydride generation atomic absorption spectrometry in water samples after preconcentration/separation on nano ZrO_2/B_2O_3 by solid phase extraction. Desalination 280:391–396. https://doi.org/10.1016/j.desal.2011.07.029
European Commission (2013) Directive 2013/39/EU of the European Parliament and of the Council amending Directives 2000/60/EC and 2008/105/EC as regards priority substances in the field of water policy
FAO (Food and Agriculture Organization) (2017) Water pollution from agriculture: a global review. Available at: http://www.fao.org/3/a-i7754e.pdf
Faraji M, Yamini Y, Rezaee M (2010) Extraction of trace amounts of mercury with sodium dodecyle sulphate-coated magnetite nanoparticles and its determination by flow injection inductively coupled plasma-optical emission spectrometry. Talanta. 81:831–836. https://doi.org/10.1016/j.talanta.2010.01.023
Frank D, Pat S (2007) Stir bar sorptive extraction for trace analysis. J Chromatogr A 1152(1–2):54–69
Gallego M, de Pena YP, Valcarcel M (1994) Fullerenes as Sorbent materials for metal preconcentration. Anal Chem 66(22):4074–4078
Gao R et al (2010) Synthesis and evaluation of molecularly imprinted core-shell carbon nanotubes for the determination of triclosan in environmental water samples. J Chromatogr A 1217(52):8095–8102. https://doi.org/10.1016/j.chroma.2010.10.121
Gao W et al (2012) Preparation of cyano-functionalized multiwalled carbon nanotubes as solid-phase extraction sorbent for preconcentration of phenolic compounds in environmetal water. J Sep Sci 35:1967–1976. https://doi.org/10.1002/jssc.201200045
García-ruiz S et al (2011) Cadmium determination in natural waters at the limit imposed by European legislation by isotope dilution and TiO_2 solid-phase extraction. Anal Bioanal Chem 401:2785–2792. https://doi.org/10.1007/s00216-011-5256-0
Gonzalez-Curbelo MA et al (2013) Analysis of pesticides residues in environmental water samples using multiwalled carbon nanotubes dispersive solid-phase extraction. J Sep Sci 36:556–563. https://doi.org/10.1002/jssc.201200782
Guo P et al (2011) Determination of linear alkylbenzene sulfonates by ion-pair solid-phase extraction and high-performance liquid chromatography. Talanta 84:587–592. https://doi.org/10.1016/j.talanta.2011.01.014
Hadjmohammadi MR, Peyrovi M, Biparva P (2010) Comparison of C 18 silica and multi-walled carbon nanotubes as the adsorbents for the solid-phase extraction of Chlorpyrifos and Phosalone in water samples using HPLC. J Sep Sci 33:1044–1051. https://doi.org/10.1002/jssc.200900494
Hashemi B et al (2017) Metal-organic frameworks as advanced sorbents for the extraction and determination of pollutants from environmental, biological, and food media. Trends Anal Chem 97:65–82. https://doi.org/10.1016/j.trac.2017.08.015
Herrera-herrera AV et al (2011) Oxidized multi-walled carbon nanotubes for the dispersive solid-phase extraction of quinolone antibiotics from water samples using capillary electrophoresis and

large volume sample stacking with polarity switching. J Chromatogr A.1218(31):5352–5361. https://doi.org/10.1016/j.chroma.2011.06.031

Hu C et al (2013) Polydimethylsiloxane/metal-organic frameworks coated stir bar sorptive extraction coupled to high performance liquid chromatography-ultraviolet detector for the determination of estrogens in environmental water samples. J Chromatogr A 1310:21–30. https://doi.org/10.1016/j.chroma.2013.08.047

Huo S, Yan X (2012) Facile magnetization of metal–organic framework MIL-101 for magnetic solid-phase extraction of polycyclic aromatic hydrocarbons in environmental water samples. Analyst 137(15):34–45. https://doi.org/10.1039/c2an35429b.

Hussain CM, Kharisov B (2016) Advanced environmental analysis, applications of nanomaterials. R Soc Chem

Iqbal Z et al (2011) Rapid CE-UV binding tests of environmentally hazardous compounds with polymer-modified magnetic nanoparticles. Electrophoresis 32:2181–2187. https://doi.org/10.1002/elps.201100106

Ji Y et al (2009) Preparation of magnetic molecularly imprinted polymer for rapid determination of bisphenol A in environmental water and milk samples. Anal Bioanal Chem 395:1125–1133. https://doi.org/10.1007/s00216-009-3020-5

Jiang J, Yang C, Yan X (2013) Zeolitic imidazolate framework-8 for fast adsorption and removal of benzotriazoles from aqueous solution. ACS Appl Mater Interf 5:9837−9842

Jurado-sánchez B, Ballesteros E, Gallego M (2009) Fullerenes for aromatic and non-aromatic N-nitrosamines discrimination. J Chromatogr A 1216:1200–1205. https://doi.org/10.1016/j.chroma.2008.12.033

Karimipour G et al (2012) Modification of gold nanoparticle loaded on activated carbon with Bis(4-methoxysalicylaldehyde)-1,2-Phenylenediamine as new sorbent for enrichment of some metal ions. Biol Trace Elem Res 145:109–117. https://doi.org/10.1007/s12011-011-9153-3

Katsumata H et al (2010) Preconcentration of atrazine and simazine with multiwalled carbon nanotubes as solid-phase extraction disk. Microchem J 96(2):348–351. https://doi.org/10.1016/j.microc.2010.06.005

Kaur A, Gupta U (2009) Sorption and preconcentration of lead on silica nanoparticles modified with resacetophenone. E-J Chem 6(3):633–638

Kaur R et al (2014) 'Synthesis and surface engineering of magnetic nanoparticles for environmental cleanup and pesticide residue analysis: a review. J Sep Sci 37:1805–1825. https://doi.org/10.1002/jssc.201400256

Kocot K et al (2013) Dispersive micro solid-phase extraction using multiwalled carbon nanotubes combined with portable total-reflection X-ray fluorescence spectrometry for the determination of trace amounts of Pb and Cd in water samples. J Anal At Spectrom 28:736–742. https://doi.org/10.1039/c3ja50047k

Kroto HW et al (1985) C60: Buckminsterfullerene. Nature 318:162–163

Lasarte-aragonés G et al (2013) Effervescence-assisted carbon nanotubes dispersion for the micro-solid-phase extraction of triazine herbicides from environmental waters. Anal Bioanal Chem 405:3269–3277. https://doi.org/10.1007/s00216-013-6718-3

Leopold K, Foulkes M, Worsfold PJ (2009) Gold-coated silica as a preconcentration phase for the determination of total dissolved mercury in natural waters using atomic fluorescence spectrometry. Anal Chem 81:3421–3428. https://doi.org/10.1039/B820701A

Li X et al (2014) Solid-phase extraction with the metal-organic framework MIL-101(Cr) combined with direct analysis in real time mass spectrometry for the fast analysis of triazine herbicides. J Sep Sci 37:1489–1495. https://doi.org/10.1002/jssc.201400151

Lian Z, Wang J (2016) Determination of ciprofloxacin in Jiaozhou Bay using molecularly imprinted solid-phase extraction followed by high-performance liquid chromatography with fluorescence detection. Mar Pollut Bull 111:411–417. https://doi.org/10.1016/j.marpolbul.2016.07.034

Liang M, Guo L (2009) Application of nanomaterials in environmental analysis and monitoring. J Nanosci Nanotechnol 9(4):2283–2289. https://doi.org/10.1166/jnn.2009.SE22

Liang P, Peng L (2010) Ionic liquid-modified silica as sorbent for preconcentration of cadmium prior to its determination by flame atomic absorption spectrometry in water samples. Talanta 81:673–677. https://doi.org/10.1016/j.talanta.2009.12.056

Liu F (2009) Analysis and applications of nanoparticles in the separation sciences: a case of gold nanoparticles. J Chromatogr A 1216:9034–9047. https://doi.org/10.1016/j.chroma.2009.07.026

Liu Q et al (2011) Graphene and graphene oxide sheets supported on silica as versatile and high-performance adsorbents for solid-phase extraction. Angew Chem Int Ed 50:5913–5917. https://doi.org/10.1002/anie.201007138

Lo S et al (2012) Gold nanoparticle-aluminum oxide adsorbent for efficient removal of mercury species from natural waters. Environ Sci Technol 46:2724–2730. https://doi.org/10.1021/es203678v

Lolic A et al (2015) Assessment of non-steroidal anti-inflammatory and analgesic pharmaceuticals in seawaters of North of Portugal: occurrence and environmental risk. Sci Total Environ 508:240–250. https://doi.org/10.1016/j.scitotenv.2014.11.097

López-garcía I, Rivas RE, Hernández-córdoba M (2011) Use of carbon nanotubes and electrothermal atomic absorption spectrometry for the speciation of very low amounts of arsenic and antimony in waters. Talanta 86:52–57. https://doi.org/10.1016/j.talanta.2011.07.105

Lucena R et al (2011) Potential of nanoparticles in sample preparation. J Chromatogr A 1218:620–637. https://doi.org/10.1016/j.chroma.2010.10.069

Luo X et al (2014) Graphene nanoplatelets as a highly efficient solid-phase extraction sorbent for determination of phthalate esters in aqueous solution. Talanta 120:71–75. https://doi.org/10.1016/j.talanta.2013.11.079

Ma J et al (2010) Determination of 16 polycyclic aromatic hydrocarbons in environmental water samples by solid-phase extraction using multi-walled carbon nanotubes as adsorbent coupled with gas chromatography-mass spectrometry. J Chromatogr A 1217(34):5462–5469. https://doi.org/10.1016/j.chroma.2010.06.060

Ma R et al (2016) Magnetic porous carbon derived from a metal-organic framework as a magnetic solid-phase extraction adsorbent for the extraction of sex hormones from water and human urine. J Sep Sci 39:3571–3577. https://doi.org/10.1002/jssc.201600347

Madadizadeh M, Taher MA, Ashkenani H (2013) Determination of ultra-trace amounts of cadmium by ET-AAS after column preconcentration with a new sorbent of modified MWCNTs. Environ Monit Assess 185:4097–4105. https://doi.org/10.1007/s10661-012-2852-1

Marwani HM, Bakhsh EM (2013) Silica gel supported hydrophobic ionic liquid for selective extraction and determination of coumarin. Am J Anal Chem 04(01):8–16

Masoumi A, Ghaemy M, Bakht AN (2014) Removal of metal ions from water using poly (MMA-co-MA)/Modified-Fe_3O_4 magnetic nanocomposite: isotherm and kinetic study. Ind Eng Chem Res 53:8188–8197. https://doi.org/10.1021/ie5000906

Mehdinia A, Roohi F, Jabbari A (2011) Rapid magnetic solid phase extraction with in situ derivatization of methylmercury in seawater by Fe_3O_4/polyaniline nanoparticle. J Chromatogr A 1218(28):4269–4274. https://doi.org/10.1016/j.chroma.2011.04.070

Mehdinia A et al (2014) Magnetic solid phase extraction using gold immobilized magnetic mesoporous silica nanoparticles coupled with dispersive liquid-liquid microextraction for determination of polycyclic aromatic hydrocarbons. J Chromatogr A 1364:20–27. https://doi.org/10.1016/j.chroma.2014.08.063

Meng J et al (2011) Preparation of polypyrrole-coated magnetic particles for micro solid-phase extraction of phthalates in water by gas chromatography-mass spectrometry analysis. J Chromatogr A 1218:1585–1591. https://doi.org/10.1016/j.chroma.2011.01.057

Menzie CA, Potocki BB, Santodonato J (1992) Exposure to carcinogenic PAHs in the environment. Environ Sci Technol 26(7):1278–1284

Mei S et al (2011) Determination of trace bisphenol A in complex samples using selective molecularly imprinted solid-phase extraction coupled with capillary electrophoresis. Microchem J 98(1):150–155. https://doi.org/10.1016/j.microc.2011.01.003

Nabid MR et al (2012) Preparation and application of poly (2-amino thiophenol)/MWCNTs nanocomposite for adsorption and separation of cadmium and lead ions via solid phase extraction. J Hazard Mater 203–204:93–100. https://doi.org/10.1016/j.jhazmat.2011.11.096

Nilsson C, Birnbaum S, Nilsson S (2007) Use of nanoparticles in capillary and microchip electrochromatography. J Chromatogr A 1168:212–224. https://doi.org/10.1016/j.chroma.2007.07.018

Omidi F et al (2015) Solid phase extraction and trace monitoring of cadmium ions in environmental water and food samples based on modified magnetic nanoporous silica. J Mag Magn Mater 395:213–220. https://doi.org/10.1016/j.jmmm.2015.07.093

Ozcan SG, Satiroglu N, Soylak M (2010) Column solid phase extraction of iron (III), copper (II), manganese (II) and lead (II) ions food and water samples on multi-walled carbon nanotubes. Food Chem Toxicol 48(8–9):2401–2406. https://doi.org/10.1016/j.fct.2010.05.078

Pandey SK, Kim K, Brown RJC (2011) A review of techniques for the determination of polycyclic aromatic hydrocarbons in air. Trends Anal Chem 30(11):1716–1739. https://doi.org/10.1016/j.trac.2011.06.017

Rao TN et al (2016) Pre-concentration of pesticide residues in environmental water samples using silica nanoparticles and identification of residues By GC-MS Method. Orient J Chem 32(4):2221–2230

Ribeiro C et al (2014) New trends in sample preparation techniques for environmental analysis. Crit Rev Anal Chem 44(2):142–185. https://doi.org/10.1080/10408347.2013.833850

Sahmetlioglu E et al (2014) Polypyrrole/multi-walled carbon nanotube composite for the solid phase extraction of lead (II) in water samples. Talanta 119:447–451. https://doi.org/10.1016/j.talanta.2013.11.044

Sajid M et al (2018) Removal of heavy metals and organic pollutants from water using dendritic polymers based adsorbents: a critical review. Sep Purif Technol 191:400–423. https://doi.org/10.1016/j.seppur.2017.09.011

Salarian M et al (2014) A metal-organic framework sustained by a nanosized Ag12 cuboctahedral node for solid-phase extraction of ultra traces of lead(II) ions. Microchim Acta 181(9–10):999–1007. https://doi.org/10.1007/s00604-014-1200-2

Savage N, Diallo MS (2005) Nanomaterials and water purification: opportunities and challenges. J Nanopart Res 7:331–342. https://doi.org/10.1007/s11051-005-7523-5

Scragg A (2005) Environmental monitoring. In: Environmental biotechnology, 2nd edn. Oxford University Press

See HH et al (2010) Determination of triazine herbicides using membrane-protected carbon nanotubes solid phase membrane tip extraction prior to micro-liquid chromatography. J Chromatogr A 1217:1767–1772. https://doi.org/10.1016/j.chroma.2010.01.053

Sellergren B (1994) Direct drug determination by selective sample enrichment on an imprinted polymer. Anal Chem 66(9):1578–1582

Shirkhanloo H et al (2015) On-line micro column preconcentration system based on amino bimodal mesoporous silica nanoparticles as a novel adsorbent for removal and speciation of chromium (III , VI) in environmental samples. J Environ Health Sci Eng 13(47):1–12. https://doi.org/10.1186/s40201-015-0205-z

Sohrabi MR (2014) Preconcentration of mercury (II) using a thiol-functionalized metal-organic framework nanocomposite as a sorbent. Microchim Acta 181:435–444. https://doi.org/10.1007/s00604-013-1133-1

Sohrabi MR et al (2013) Solid phase extraction of Cd(II) and Pb(II) using a magnetic metal-organic framework, and their determination by FAAS. Microchim Acta 180(7–8):589–597. https://doi.org/10.1007/s00604-013-0952-4

Soliman EM, Marwani HM, Albishri HM (2013) Novel solid-phase extractor based on functionalization of multi-walled carbon nano tubes with 5-aminosalicylic acid for preconcentration of Pb (II) in water samples prior to determination by ICP-OES. Environ Monit Assess 185:10269–10280. https://doi.org/10.1007/s10661-013-3331-z

Song X et al (2012) Determination of 16 polycyclic aromatic hydrocarbons in seawater using molecularly imprinted solid-phase extraction coupled with gas chromatography-mass spectrometry. Talanta 99:75–82. https://doi.org/10.1016/j.talanta.2012.04.065

Soylak M, Unsal YE (2012) Determination of traces of iron and lead in food and water samples after preconcentration on multiwalled carbon nanotubes. J AOAC Int 95(4):1183–1189. https://doi.org/10.5740/jaoacint.11-168

Soylak M, Topalak Z (2014) Multiwalled carbon nanotube impregnated with tartrazine: solid phase extractant for Cd (II) and Pb (II). J Ind Eng Chem Korean Soc Ind Eng Chem 20(2):581–585. https://doi.org/10.1016/j.jiec.2013.05.017

Sun N et al (2014) A self-assembly pipette tip graphene solid-phase extraction coupled with liquid chromatography for the determination of three sulfonamides in environmental water. Anal Chim Acta 810:25–31. https://doi.org/10.1016/j.aca.2013.12.013

Tabani H et al (2013) Combination of graphene oxide-based solid phase extraction and electro membrane extraction for the preconcentration of chlorophenoxy acid herbicides in environmental samples. J Chromatogr A 1300:227–235. https://doi.org/10.1016/j.chroma.2013.04.026

Taghizadeh M et al (2013) A novel magnetic metal organic framework nanocomposite for extraction and preconcentration of heavy metal ions, and its optimization via experimental design methodology. Microchim Acta 180(11–12):1073–1084. https://doi.org/10.1007/s00604-013-1010-y

Tan F et al (2011) Evaluation of a novel microextraction technique for aqueous samples: porous membrane envelope filled with multiwalled carbon nanotubes coated with molecularly imprinted polymer. J Sep Sci 34:707–715. https://doi.org/10.1002/jssc.201000791

Tian J et al (2013) Application of nanomaterials in sample preparation. J Chromatogr A 1300:2–16. https://doi.org/10.1016/j.chroma.2013.04.010

Tolmacheva VV et al (2016) Facile synthesis of magnetic hypercrosslinked polystyrene and its application in the magnetic solid-phase extraction of sulfonamides from water and milk samples before their HPLC determination. Talanta 152:203–210. https://doi.org/10.1016/j.talanta.2016.02.010

UNEP (2019 a) All POPs listed in the stockholm convention. Available at: http://chm.pops.int/TheConvention/ThePOPs/ListingofPOPs/tabid/2509/Default.aspx. Accessed 25 Feb 2019

UNEP (2019 b) Major sources of today's pollution. Available at: https://environmentlive.unep.org/media/imgs/pollution/infographics/sources_of_pollution.pdf. Accessed 25 Feb 2019

UNEP (2019 c) Persistent organic pollutants (POPs). Available at: https://www.unenvironment.org/explore-topics/chemicals-waste/what-we-do/persistent-organic-pollutants-pops. Accessed 25 Feb 2019

UNEP (2013) Working group on the global metal flows to the international resource panel. In: van der Voet E, Salminen R, Eckelman M, Mudd G, Norgate T, Hischier R (eds) Environmental risks and challenges of anthropogenic metals flows and cycles. Available at: http://orbit.dtu.dk/files/54666484/Environmental_Challenges_Metals_Full_Report.pdf

UNEP/UNECE (2016) GEO-6 assessment for the pan-European region. Nairobi, Kenya. Available at: http://uneplive.unep.org/media/docs/assessments/GEO_6_Assessment_pan_European_region.pdf

Vunain E, Mishra AK, Mamba BB (2016) Dendrimers, mesoporous silicas and chitosan-based nanosorbents for the removal of heavy-metal ions: a review. Int J Biol Macromolecules 86:570–586. https://doi.org/10.1016/j.ijbiomac.2016.02.005

Wang G, Lei Y, Song H (2014) Evaluation of Fe3O4@SiO2–MOF-177 as an advantageous adsorbent for magnetic solid-phase extraction of phenols in environmental water samples. Anal Methods 6(19):7842–7847. https://doi.org/10.1039/C4AY00822G

Wang H, Campiglia AD (2010) Direct determination of benzo[a]pyrene in water samples by a gold nanoparticle-based solid phase extraction method and laser-excited time-resolved Shpol'skii spectrometry. Talanta 83(1):233–240. https://doi.org/10.1016/j.talanta.2010.09.013

Wang W et al (2013a) Magnetic microsphere-confined graphene for the extraction of polycyclic aromatic hydrocarbons from environmental water samples coupled with high performance liquid

chromatography-fluorescence analysis. J Chromatogr A 1293:20–27. https://doi.org/10.1016/j.chroma.2013.03.071
Wang X et al (2017) Facile synthesis of phenyl-modified magnetic graphene/mesoporous silica with hierarchical bridge-pore structure for efficient adsorption of pesticides. Mater Chem Phys 198:393–400. https://doi.org/10.1016/j.matchemphys.2016.12.017
Wang Y et al (2012) Graphene-based solid-phase extraction combined with flame atomic absorption spectrometry for a sensitive determination of trace amounts of lead in environmental water and vegetable samples. Anal Chim Acta 716:112–118
Wang Y et al (2013a) Preparation of a functionalized magnetic metal-organic framework sorbent for the extraction of lead prior to electrothermal atomic absorption spectrometer analysis. J Mater Chem A 1:8782–8789. https://doi.org/10.1039/c3ta11178d
Wang Z et al (2013b) Graphene-based solid-phase extraction disk for fast separation and preconcentration of trace polycyclic aromatic hydrocarbons from environmental water samples. J Sep Sci 36:1834–1842. https://doi.org/10.1002/jssc.201300186
Wang T et al (2018) Adsorption of agricultural wastewater contaminated with antibiotics, pesticides and toxic metals by functionalized magnetic nanoparticles. J Environ Chem Eng 6:6468–6478. https://doi.org/10.1016/j.jece.2018.10.014
Wen Y et al (2014) Recent advances in solid-phase sorbents for sample preparation prior to chromatographic analysis. Trends Anal Chem 59:26–41. https://doi.org/10.1016/j.trac.2014.03.011
Wierucka M, Biziuk M (2014) Application of magnetic nanoparticles for magnetic solid-phase extraction in preparing biological, environmental and food samples. Trends Anal Chem 59:50–58. https://doi.org/10.1016/j.trac.2014.04.007
Wu H et al (2010) Flow injection solid-phase extraction using multi-walled carbon nanotubes packed micro-column for the determination of polycyclic aromatic hydrocarbons in water by gas chromatography-mass spectrometry. J Chromatogr A 1217(17):2911–2917. https://doi.org/10.1016/j.chroma.2010.02.040
Wu Z, Lin J, Tseng W (2012) Oligonucleotide-functionalized silver nanoparticle extraction and laser-induced fluorescence for ultrasensitive detection of mercury (II) ion. Biosens Bioelectron 34(1):185–190. https://doi.org/10.1016/j.bios.2012.01.041
Wu X et al (2013) Graphene-dispersive solid-phase extraction of phthalate acid esters from environmental water. Sci Total Environ 444:224–230. https://doi.org/10.1016/j.scitotenv.2012.11.060
WWAP (United Nations World Water Assessment Programme) (2017) The United Nations world water development report, 2017: wastewater: the untapped resource
Xu L et al (2016) Recent advances in applications of nanomaterials for sample preparation. Talanta 146:714–726. https://doi.org/10.1016/j.talanta.2015.06.036
Yalçinkaya Ö et al (2012) Preconcentration of aluminum on nano ZrO_2/B_2O_3 and its determination by flame atomic absorption spectrometry. Spectrosc Lett 45:344–351. https://doi.org/10.1080/00387010.2012.666700
Yang F et al (2011) Magnetic microsphere confined ionic liquid as a novel sorbent for the determination of chlorophenols in environmental water samples by liquid chromatography. J Environ Monit 13:440–445. https://doi.org/10.1039/c0em00389a
Yang W et al (2013) Molecularly imprinted polymers coated on multi-walled carbon nanotubes through a simple indirect method for the determination of 2,4-dichlorophenoxyacetic acid in environmental water. Appl Surf Sci 284:692–699. https://doi.org/10.1016/j.apsusc.2013.07.157
Yin Y et al (2012) Dummy molecularly imprinted polymers on silica particles for selective solid-phase extraction of tetrabromobisphenol A from water samples. J Chromatogr A 1220:7–13. https://doi.org/10.1016/j.chroma.2011.11.065
Yu Z et al (2010) Application of SPE using multi-walled carbon nanotubes as adsorbent and rapid resolution LC-MS-MS for the simultaneous determination of 11 triazine herbicides residues in river water. Chromatographia 72(11):1073–1081. https://doi.org/10.1365/s10337-010-1769-5
Zhang Z et al (2006) Applications of nanomaterials in liquid chromatography: opportunities for separation with high efficiency and selectivity. J Sep Sci 29:1872–1878. https://doi.org/10.1002/jssc.200600154

Zhang X et al (2011) A core-shell magnetic mesoporous silica sorbent for organic targets with high extraction performance and anti-interference ability. Chem Commun 47:4454–4456. https://doi.org/10.1039/c1cc10300h

Zhu H et al (2011) Preparation of a molecularly imprinted polymer using TMB as a dummy template and its application as SPE sorbent for determination of six PBBs in water and fish samples. Anal Methods 3:393–399. https://doi.org/10.1039/c0ay00479k

Nano Ceramics and Their Applications

Khalid Mujasam Batoo

Abstract In this article, we present an overview with the emphasis to highlight the basics of the origin of magnetism, size and shape effects on magnetic properties of materials and in particular the origin of nano magnetism which is also known as superparamagnetism when the size of the material particles lies between 1 and 30 nm. Besides this, a brief review of, various applications of the superparamagnetism in the areas like biosensors, data storage, drug delivery, and hyperthermia are discussed.

Keywords Nano ceramics · Nano materials · Drug delivery · Biosensors

1 Introduction

Materials whose length scale lies within 1–100 nm are termed as nanomaterials. On nanoscale, the properties of material are found to change remarkably from their bulk counter parts. The physiochemical, properties of these materials are found very sensitive to the size and shape and hence exhibit properties which are entirely different when studied in bulk materials. This uniqueness in the properties nanomaterials is pronounced due to their large surface to volume ratio. Nano ceramic materials show great refractory properties, chemical resistance, mechanical resistance and hardness both at normal and high temperatures; they are especially amenable to sintering and reactions with different oxides (Khalil 2012). The research in the field of nanomaterials have attracted lot of attention of scientific and research community over last two decades because of their unique and fascinating properties in the different fields of modern sciences like materials science biotechnology and genetics (Šajgalík et al. 2013).

The interest to study the nano ceramic materials developed through various means has been encouraged due to the large number of applications they offer to modernize our tools and modifications in the existing materials and also offer of new materials like fabrication of capacitors, corrosion-resistant coatings, dense ceramics, sensors, solid electrolytes for fuel cells, batteries, cosmetic health, thermal barrier coatings,

K. M. Batoo (✉)
King Abdullah Institute for Nanotechnology, King Saud University, Riyadh 11451, Saudi Arabia
e-mail: kbatoo@ksu.edu.sa

A. H. Bhat et al. (eds.), *Nanomaterials for Healthcare, Energy and Environment*, Advanced Structured Materials 118, https://doi.org/10.1007/978-981-13-9833-9_7

catalysts, bioengineering, automotive, optoelectronics, computers, and electronics etc. The nanomaterials have shown their importance also in the field of luminescence, especially, as they show unique structural and optoelectronic properties. A lot of focus is given in the design and development of new ways and techniques to synthesize nanomaterials (Virk and Poonam 2010). So far ceramic materials having different shapes and sizes have been produced such as nanoparticles, nanorods, nanoribbins, nanotubes, nanosheets, nanofluids. Research has shown that the modifications can be done in optical, luminescence and other physiochemical properties of nanomaterials by modifying their shape, size or by doping different impurity elements at different crystallographic sites or by creation of defects or voids such as point defects, cluster defects or columnar defects using ion implantation technique or high energy beam line (Miyake et al. 1988).

The nanoceramic materials promise unique properties which are size and molecular structure dependent. These properties are often shown in terms of various electrical and magnetic physics phenomenon which include (Darweesh 2018): (a) Dielectric; (b) Ferroelectric; (c) Piezo-electric; (d) Pyro-electric; (e) Ferromagnetic; (f) Magnetoresistive; (g) Superconductive; (h) Electro-optical.

2 Ceramic Oxides

Among the ceramic materials, oxides are one such category of materials which have been investigated over the last 6 decades and still they fascinate us by offering new and unique properties when tried in various forms such as multilayered structures as GMR and CMR, etc. The different oxides have been prepared by various chemical and solid state reaction methods and studied for various structural, chemical and electromagnetic properties. It is worth to mention that oxides have always fascinated researchers with their remarkable properties when studied in different combinations, such as by doping with different valence state elements, as thin films, multilayer structures. As the research in the area of nanotechnology is growing everyday researchers started investigating the effect of size and shape on the different electromagnetic properties to see how new wonders can be made in ceramic oxides (Heiligtag and Niederberger 2013) etc.

3 Magnetism in Oxides

The interplay between magnetic properties and structure is a very interesting phenomenon in materials science, which has been explored by many researchers over the last century. The magnetic properties of materials are mainly determined by their structure and the various magnetic interactions between neighouring atoms.

It is quite interesting to understand what happens when the physical dimension of a bulk magnetic material are reduced comparably to or smaller than a magnetic domain wall width in nanoscopic regime? This would lead the magnetic particle to possess only a single-domain and show superparamagnetic behaviors. The process of magnetization reversal can be achieved by introducing thermal fluctuations which are sensitive to the size and shape of the magnetic particles. The possibility of using these unique size and shape dependent magnetic characteristics in modern technologically has been proved in many new and traditional fields, such as ultrahigh density magnetic data storage (Weller and Moser 1999), giant magnetoresistance (GMR) sensor (Weller and Moser 2000), magnetocaloric refrigerator (Sun et al. 2000), magneto-electrics, magnetic resonance imaging (MRI) contrast enhancement agents (Weller and Doerner 2000), magnetically directed target-specific drug delivery systems, cells, DNAs and genes sorting and delivery (Herzer et al. 2005), and ferrofluids (Knobel et al. 1999).

Magnetic properties of a matter are fundamentally due to the presence of electrons in an atom. An electron in an atom has got two kinds of motions; spin and orbital motion and each has a magnetic moment associated with it. When a magnetic field is applied to material, the response to this field (**H**) by the material is the characteristic of the magnetic induction or the flux density (**B**) and the response that a material shows through magnetic induction in a magnetic field is represented by the magnetization (**M**). The equation relating these three magnetic quantities is:

$$\mathbf{B} = \mu_0(\mathbf{H} + \mathbf{M}) \tag{1}$$

Also

$$\mathbf{B} = \mu\mathbf{H} \tag{2}$$

where μ_0 is a universal constant which represents permeability in free space and μ is permeability of a material known as magnetic permeability and is a characteristic of a material. From Eq. (1), it can be seen that $\mu_0\mathbf{H}$ represents the term for magnetic induction generated by the applied field while the term $\mu_0\mathbf{M}$ represents the additional magnetic induction contributed by a material. The term magnetization of a material '**M**' can be described as the processes of turning a non-magnetic bar into a magnetic bar and is measured as magnetic moment per unit volume. The magnetic susceptibility 'χ' is defined as:

$$\chi = \frac{M}{H} \tag{3}$$

The relative permeability is given by the ratio of the magnetic permeability to that of free space:

$$\mu_r = \frac{\mu}{\mu_0} \tag{4}$$

Based on this response, magnetic materials are broadly divided into six classes: (a) Diamagnetism; (b) Paramagnetism; (c) Ferromagnetism; (d) Ferrimagnetism; (e) Antiferromagnetism; (f) Superparamagnetism

4 Nanomagnetism

A nanometer length scale is comparable to an electron's mean free path. When the particle size is less than a characteristic length such as the mean free path, it is possible to see different physical or chemical characteristics than those exhibited by their larger counterparts. When the particle size approaches to the nano scale, properties like electrical resistivity, reactivity, melting temperature and optical absorption have all been found to change drastically (Lubitz and Rachford 2002). With this in mind, one can immediately deduce that magnetic properties will be affected by particle size, especially when particle size is of the same scale as the magnetic domains of the material. At the nanometer scale, magnetic particles have been found to behave differently from that of bulk magnetic materials. As particle size decreases, remanent magnetization and saturation magnetization increases. This implies that scaling particles down to the nanometer scale can greatly improve the quality of magnets fabricated from them (Moskowitz 1991). When the size of a single domain particle reduces below a critical diameter, a new type of magnetism is generated, known as *superparamagnetism*, where the coercivity and remanent magnetization goes zero and the net magnetic moment in zero field at T > 0 K averages to zero for superparamagnetic particles, average to zero. On the other hand, there will be a net statistical alignment of magnetic moment when field is applied, which is analogous to paramagnetism, except now the magnetic moment is not that of a single atom, but of a single domain particle containing 10^5 atoms. Therefore, superparamagnetism denotes a much higher susceptibility value than that for paramagnetism (Kittle 1996). The efficiency of systems can be improved which are subjected to rapidly alternating ac magnetic fields like transformers and rotating electrical machinery through superparamagnetism. A traditional magnet when exposed to an alternating magnetic field, the magnetic field cycles through its hysteresis loop often causing a loss of efficiency and a rise in temperature. This rise in temperature is attributed to the frictional heating which occurs due to the magnetic domains having different orientations and tries to align along the field direction of applied field. The amount of energy loss in each cycle is proportional to the area enclosed by the loop, so a small or non-existent coercivity is desirable (Ruan et al. 2000). It has also been shown that particle size has a large effect on microwave absorption. Particles of nano meter size greatly improve the absorptive efficiency and broaden the bandwidth (Kodama et al. 1996).

To accomplish better magnetic coupling, the high concentration of interfaces in nanocomposites enables one to take benefits of different magnetic phases like ferromagnetic–nonmagnetic, and ferrimagnetic–antiferromagnetic. However, applications of magnetic nanomaterials are limited due to the cap put on by operating temperature, which causes loss in magnetic properties as the temperature rises.

Permanent magnets are used in number of applications which range from power sources to actuators and switches (Kodama et al. 1997). They can be subjected or tested to temperatures higher than or near the Curie temperature and consequently, thermal management of nanocomposite magnets has become an important criteria. At Curie temperature magnetization of magnetic materials reduces considerably and a conventional approach to avoid this to happen is to increase the Curie temperature through material engineering and by using nanomaterials (Bean and Jacobs 1956). A barrier in order to enhance heat conduction properties of nanomagnets is a possibility of degrading the magnetic characteristics. Thus, one must miniaturize to increase the thermal conductivity without substantial loss in magnetization.

5 Superparamagnetism

Among different oxide materials studied, iron oxide is one such material which has been synthesized and studied extensively over the last 6 decades. The interest in this kind of oxides keeps on growing especially when it is about superparamagnetic iron oxide nanoparticles (SPIONs) or ultra-small superparamagnetic iron oxide nanoparticles (USPIONs), with hydrodynamic diameter smaller than 30 nm, such as magnetite Fe_3O_4 and maghemite γ-Fe_2O_3 with sizes between 1 and 20 nm. Magnetite one of the main iron ores and one of the oxides of iron exhibits ferrimagnetism is attracted to a magnet and can be converted into permanent magnet itself through magnetization. Among all the naturally-occurring minerals on Earth it is the most magnetic. The naturally magnetized pieces of magnetite, called lodestone, will attract small pieces of iron, which is how ancient peoples first discovered the property of magnetism and today it is mined as iron ore.

A bulk magnetic material is comprised of magnetic domains and the magnetization inside each domain is uniform, but varies from domain to domain as they are separated by an interface layer known as the domain wall. By reducing the dimension of a magnetic material, the sizes of the domains are accordingly decreased and their structures may change in terms of domain wall width and wall structures. As far as the energy is concerned, when the size reaches a critical size, the magnetic material possesses only a *single domain*, since the energy cost for the formation of domain walls becomes energetically unfavorable, the energy gain from the formation of domain walls is higher than the energy reduction by dividing the *single domain* into even smaller domains. The critical size (radius $\boldsymbol{R_C}$) for domain formation has been estimated by Kittle (1996), which depends on spontaneous magnetization ($\mathbf{M_s}$), the anisotropy constant (**K**), and the exchange energy density or constant (A) as given in Eq. (5) (Cullity 1972):

$$\mathrm{R_C} = 36(\mathrm{KA})^{1/2}/\mu_0\mathrm{M_S^2} \tag{5}$$

The critical size for typical magnetic materials is in the range of 10–800 nm. A small magnetic particle less than critical size ($\leq$40 nm) prefers to be uniformly magnetized along one of its anisotropy easy axes, and is accompanied by a strong enhancement in coercivity. If the size of magnetic material is below the critical size (up to 20 nm) magnetic materials can only acquire a single domain. On still reduction in size (<20 nm), the magnetic anisotropy energy further deceases and may become so low as to be comparable to or even lower than the thermal energy ($\boldsymbol{K_B}$**T**, as $\boldsymbol{K_B}$ being Boltzmann constant). As a result, the energy barrier for magnetization reversal may thermally overcome, and the magnetic moment is thermally fluctuated, like a single spin in a paramagnetic material. As a result, the spins within particle remain magnetically coupled to each other leads to the formation of a '*superspin*' with a relative enormous moment per particle, and the phenomenon is known as *superparamagnetism* (Tagantsev et al. 2003).

In a simple model described by Stoner-Wohlfarth (1948) for a non-interacting single domain spherical particle with uniaxial anisotropy in zero magnetic field the magnetic anisotropy energy is given by an expression:

$$\mathrm{E_A} = \mathrm{KV}\ \sin^2\theta \tag{6}$$

where E_A represents the energy barrier, K the magnetocrystalline anisotropy constant, V the volume of the particle, and θ the angle between magnetization and easy axis of the particle. The dependency of anisotropy energy on θ is depicted schematically in Fig. 1. Clearly, either $\theta = 0$ or π is a direction of minimum energy and these directions are symmetrically separated by an energy barrier as high as KV. The magnetic field can also be a form of energy that can lower the magnetic anisotropy energy barrier for magnetization reversal. When a magnetic field **H** is applied along the easy axis, the anisotropy energy will be:

$$\mathrm{E_A} = \mathrm{KV}\ \sin^2\theta\ -\ \mathrm{HVM_{nr}}\cos\theta \tag{7}$$

where $\mathrm{M_{nr}}$ is the non-relaxing magnetization.

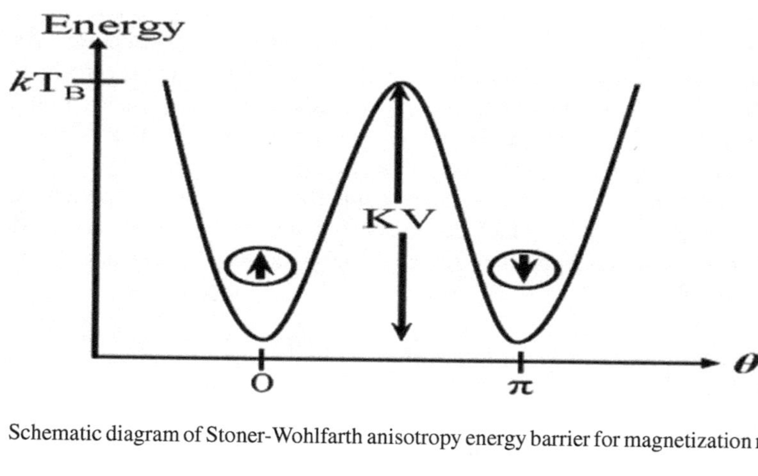

Fig. 1 Schematic diagram of Stoner-Wohlfarth anisotropy energy barrier for magnetization reversal

Superparamagnetism is also characterized by a relaxation time (τ) and thus the actual magnetic behaviour depends on the value of measuring time (t_m) in a particular experimental technique. Now, the magnetic anisotropy serves as an energy barrier to the total spin reorientation, while in a bistable system, the probability for such magnetic moment thermally overcoming energy barrier is proportional to Boltzmann factor as derived by Néel as:

$$\tau = \tau_0 \exp\left(\frac{-KV}{K_B T}\right) \tag{8}$$

where τ_0 is an attempt frequency factor equal to approximately 10^{-9} s, and depends on several factors such as temperature, gyromagnetic ratio, magnetic field, magnetization, particle size and damping constant, and yet is treated as a constant. At given temperature, if the measurement time is much longer than relaxation time, the moment is rapidly relaxed by thermal fluctuation, and consequently the entire system is in superparamagnetic state. On the contrary, as the measurement time is much shorter than relaxation time, the moment relaxes so slow that it seems to be blocked. The *blocking temperature* is thus defined as the temperature at which the magnetic moment relaxation time is equal to the measurement time, or the temperature where the moment is able to overcome the energy barrier into the superparamagnetic state at a certain measurement time. Some typical instrument measurement times are 100 s for DC SQUID, 10^{-7}–10^{-9} s for Mössbauer spectroscopy, and 10^{-8}–10^{-12} s for neutron diffraction.

Magnetite is an iron oxide commonly found in nature. Magnetite shows inverse spinel structure with a face centered cubic (fcc) close-packed oxygen array, and iron in both fourfold (tetrahedral) and six fold (octahedral) coordination. Fe^{2+} cations occupy octahedral sites and half of the Fe^{3+} cations are placed in the tetrahedral sites, whereas the other half occupies octahedral ones (shown in Fig. 2). The unit cell has Fd_3m space group with a lattice parameter a = 8.3941 Å. The chemical formula of the unit cell can be written as $(Fe^{3+}{}_8)^A[Fe^{2+}{}_8, Fe^{3+}{}_8]^B O_{32}$, where A and B represent tetrahedral and octahedral sites, respectively. The maghemite γ-Fe_2O_3 structure is similar to magnetite. However, maghemite only has Fe^{3+} cations distributed in a cubic unit cell with P213 spatial group with a = 8.33 Å. Similarly to magnetite, in the unit cell the Fe^{3+} cations occupy 1/8 of tetrahedral sites, whereas the rest of Fe^{3+} cations occupy the octahedral, thus leaving vacancies in the octahedral sublattice. The chemical formula by unit cell can be written as $(Fe^{3+}{}_8)^A[Fe^{3+}{}_{40/3}, Fe^{2+}{}_{8/3}]^B O_{32}$, where A and B mean tetrahedral and octahedral sites, respectively and □ means vacancies.

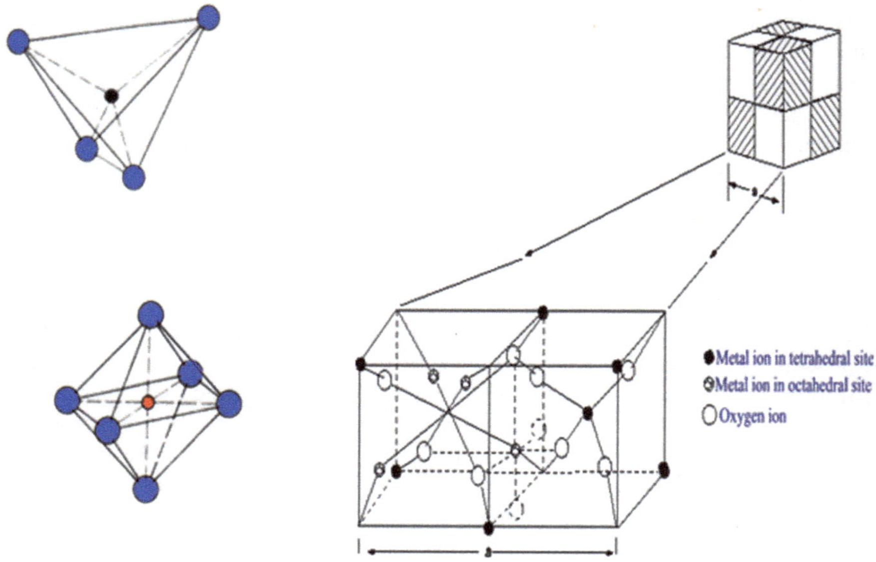

Fig. 2 Crystal structure of a cubic ferrite

6 Magnetic Interaction

The metallic ions occupy two different crystallographic sites in ferrites, i.e., octahedral (B) sites and tetrahedral (A) sites. Three kinds of magnetic interactions are possible (shown in Fig. 3) between the metallic ions, mediated by oxygen ions, by superexchange mechanism, namely, j_{AB}, j_{BB} and j_{AA} interactions. As the distance between the two ions is large; hence, direct interaction is negligible. The experimental results have shown that these interaction energies are negative, and hence induce an anti-parallel alignment.

In general, the magnitude of the interaction energy between two magnetic ions, Me^{I} and Me^{II} depends upon:

a. Bond length (represented by the quantities I^{I} and I^{II}).
b. Bond angle (represented by the term ø as shown in Fig. 3).

The exchange energy is having its maximum value at an angle of 180° and it decreases rapidly with increasing bond distance. Gorter investigated the various types of effective magnetic interactions between the various possible configuration of ion pairs with favourable bond lengths and bond angles (Gorter 1954). Experiments have shown that the values of bond length are smaller and the values of bond angle are fairly high (~180°) for A-B interaction, which makes it the strongest interaction among the three interactions. The B-B interaction (bond length is too large for the effective interaction) is weak interaction as compared to the A-B interaction, while the A-A interaction is the weakest of all the interactions for which ø ~ 80°.

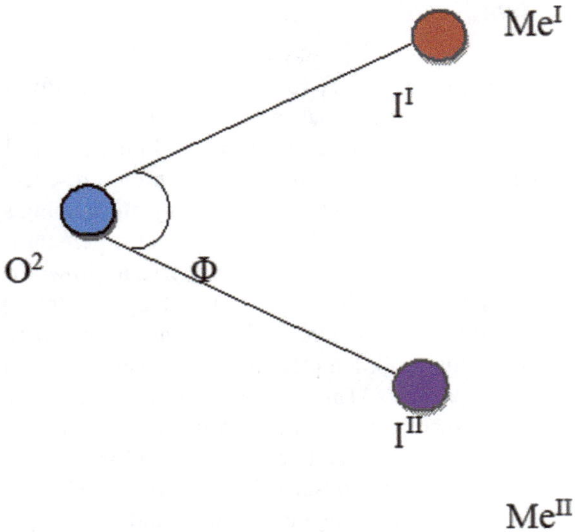

Fig. 3 Me^{1}-O^{2-}-Me^{II} angles in spinel ferrites

Since, it is the only A-B interaction which is predominating, the spins of the ions in A- and B-sites, in ferrites are anti-parallel which gives the resultant magnetic moment equal to the difference between those of the A- and B-site ions. In general, the value of saturation magnetic moment for the B-lattice ($\mathbf{M_B}$) is greater than that of the A-lattice ($\mathbf{M_A}$), such that the resultant saturation magnetization ($\mathbf{M_S}$) is written as:

$$Ms = M_B - M_A \tag{9}$$

The magnetic features of ferrites closely resemble with ferromagnets. Hysteresis phenomenon is observed in ferromagnetic materials during magnetization cycle. Ferrites have specific Curie temperature.

The mixed ferrites show most interesting characteristics while doped with different non-magnetic/magnetic elements (Guillaud 1951). The values of the magnetization and of Curie temperatures can be conveniently varied by suitable variation in composition. If a small concentration of non-magnetic material is increased in the ferrite material, the value of saturation magnetization increases (Gilllaud and Creveaux 1950). But at higher concentrations deviations occur due to the weakening of A-B interaction and consequent stronger B-B interaction. Although, an increase in the value of $\mathbf{M_S}$ is observed with the substitution of divalent non-magnetic ions, the Curie temperature decreases, as the reduced value of $\mathbf{M_A}$ weakens the A-B interaction, which is responsible for the observed ferrimagnetism.

7 Iron Oxide Nanoparticles

Nanoparticles of Fe_2O_3 (both SPIONs and USPIONs) have been synthesized by different techniques, such as controlled precipitation from soluble salts, aerosol assisted methods, co-precipitation etc. In the area of biomedical applications, most SPIONs and USPIONs are prepared in the presence of the coating agent through co-precipitation method. In this strategy, the coating agent is mandatory for colloid stability. Primarily, to avoid excessive enlargement of the nanoparticles, it acts as capping agent in the nucleation and crystal growth process. Secondly, the coating exerts a colloid stabilization function based on inter-particle repulsions of electrostatic and/or steric nature. The decrease in magnetic properties compared to the corresponding bulk counterpart is a consubstantial problem while preparing magnetic nanoparticles. This is due to the reason that coating agents introduce defects as well as surface effects (non-collinear spins, spin canting, spin-glass-like behavior, etc.) into the crystalline structure and increases further as the volume/surface ratio decreases. In this sense, typical values of saturation magnetization for magnetite nanoparticles range between 30 and 50 emu g^{-1}, significantly lower than the 90 emu g^{-1} reported for magnetite in bulk. In order to explore the full potential of these systems, more research focus is required specifically in target magnetite nanoparticles towards cell, organelles or even molecules and developing synthesis strategies to improve the magnetic properties of nanoparticles without compromising the colloidal stability.

8 Biomedical Applications

The versatility and advantages in choosing nanoparticles of iron oxide over other nanomaterials for biological applications is due to the fact that some of the crystalline phases of iron oxide can be modified when exposed to magnetic field gradient. Due to this uniqueness it can be sued in magnetic separation such as in cells, mechanical manipulation of cells or targeted drug delivery. Among the major biomedical applications of iron oxide, targeted drug delivery is at present largely being discussed investigated. With the help of externally applied magnetic field magnetic nanoparticles are used as drug carriers to deliver the drug at the specific sites to achieve controlled release of high, localized concentrations of drug by retention of the carriers in the region of interest. The aim of controlled release at the selected target can be manifested through a combination of a thermosensitive polymer and iron oxide magnetic nanoparticles which act as localized heat sources when exposed to an alternating magnetic field. Through a single particle by combining drug delivery and gene therapy, it promises the transfection efficiency or to achieve a synergic effect of drug and gene therapy.

8.1 MRI Contrast Agents

The use of nanoparticles as new therapeutic and diagnosis tools represents an outstanding advance for biomedical sciences. The application of magnetic nanoparticles (MNPs) for diagnosis was introduced in the 1990s, as contrast agents for magnetic resonance imaging (MRI) of liver. One of the powerful techniques nowadays used for imaging is magnetic resonance imaging (MRI) because of its noninvasive nature, tomographic capabilities and high spatial resolution, but major limitation of it is low signal sensitivity. The emergence of MRI as a clinical diagnostic modality has pushed for the search of new and better contrast agents. In the past, as MRI contrast agents, small paramagnetic metal chelates were most commonly, which act by shortening T_1 relaxation times (the time constant describing the return movement of a group of nuclei, protons usually, to the field direction) on the other hand, SPIONs act by mainly shortening T_2 relaxation times (the time constant describing the relaxation constant of usually protons interfering with each other) (Rummery et al. 2008). Basically, the presence of SPIONs significantly modifies the local magnetic fields around protons. A little progress has been made by signal enhancement caused by conventional iron oxide nanoparticles (the only FDA approved product); however, it is not that much satisfactory when compared to that obtained with other imaging modalities such as fluorescence and PET. Currently, there are several products in the market which are undergoing clinical tests that point toward better benefits for diagnosis (mainly in MRI) as well as for the treatment of tumors by hyperthermia, magnetofection and targeted drug delivery vehicles. The capability of MNPs for modifying the relaxation of the protons in living tissues, thus acting as contrast agents, is the most developed biomedical application of these nano devices. MNPs in the form of superparamagnetic iron oxide nanoparticles (SPIONs) have found applications as contrast agents in the bowel (Lumiren® and Gastromark®), as well as for the imaging of liver and spleen (Endorem® and Feridex IV®). For clinical applications, Magnetic nanoparticles having sixe between 10 and 200 nm are required because particles with sizes larger than 200 nm can be filtered by the human spleen while the MNPs smaller than 10 nm can be removed by renal clearance. The benefit or usage of grown MNPs in medical field is mainly associated with the coating materials which are biocompatible, nontoxic and biodegradable. Several contrast agents based on ultra-small superparamagnetic iron oxide nanoparticles (USPIONs) are currently under clinical research (for instance Combidex), due to their capability for reaching lymphatic nodules and contribution to the detection of metastases.

8.2 Drug Delivery

With every increasing demands and advancement in the field of nanotechnology and molecular biology, researchers are looking for new ways to develop nanoparticles with functional properties to address the shortcomings of traditional disease diagnosis and therapeutic agents. Advanced nanoparticles are produced for particular applications to address issues like brighter, tissue specific imaging probes in order to visualize and diagnose diseases at earliest stages, in some cases, even prior to disease manifestation. Researchers are working on development of nanoparticles which can be used as drug carriers with tailored drug release properties, low immunogenicity, etc. for the better yield and improved treatment efficacy with lesser unwanted side effects (Torchilin 2006). A significant development has been made where imaging and delivery facilities have been tuned together in unique nanoparticle formulations through smart combinations of nanoscale materials, which has enabled simultaneous in vivo diagnostic imaging and drug delivery for real-time treatment tracking (Sun et al. 2008).

Currently, numbers of SPOINS developed are going through clinical testing and trails and many formulations have been so far which are used in clinics in medical imaging and therapeutic applications (Lee et al. 2007). Without any doubt the physicochemical profiles of SPIONs have provided passive targeting, but not to that extent as being offered by bioligands. It has been seen that if bioactive molecules are added to the SPION surface, it enhances the targeting specificity of NPs (Goya et al. 2008), which produces contrast agents that specifically illuminate targeted tissue and drug carriers which don't interact with healthy tissues (Dufay et al. 2009). The advancement in the area of targeted drug delivery represents a majority of SPION research today. A very careful engineering of the size, shape, coating, and surface modifications is required to fabricate the next generation SPIONs which can specifically target and eliminate or illuminate damaged tissue. A detailed investigation is required to consider and to design parameters in order to produce NPs that can overcome biological obstacles and carry out their function. To achieve the goal, focus should be given on the physical properties of targeting molecules in addition to their binding characteristics, and joined with nanoparticle system in such a way that they remain functionally active. Iron oxide NPs have considerable capacity to be used in highly personalized medicine. Specialists build up a library of synthesis protocols, and discrete nanoscale modules with explicit jobs for malignant growth theranostics, individualized nanoparticle details displaying a full-suite of treatment and analytic capacities might be made in a proficient and powerful way.

8.3 Biosensors

Iron oxide attractive particles are reasonable for use in biosensors on the grounds that most biological species are not non-magnetic in nature, which implies that there is intrinsically low background noise. New techniques have been displayed so as to evaluate the amount of biomolecules appended to iron oxides in a fluid using magnetic sensors dependent on various advancements (magnetorelaxometry (Ennen et al. 2016), magnetoresistance (He et al. 1999), Hall impact (Tagantsev et al. 2003) or SQUID sensors (Ignatyeva et al. 2016) or utilizing their characteristic capacity to improve electrochemical signals or optical properties of the noble metal shaping a multifunctional segment with the nano magnets and the biomolecule (magneto plasmonic) (Gilchrist et al. 1957). The magnetic properties of a nanocomposite are being used for segregation purpose and optical properties of the noble metal for its recognition. In order to qualify for the next generation biosensors, the iron oxide nanoparticles are required to improve merits of high sensitivity, small size, speedy reaction, protection from forceful medium, and low cost (not so restrictive in nanomedicine). A multicomponent approach appears a decent pathway to follow so as to expand detecting capacities.

8.4 Hyperthermia

An iron oxide nanoparticle for magnetic hyperthermia treatment of cancers was first demonstrated by Gilchrist et al. (1957). Following this seminal work different research groups started exploring and investigating the important operational parameters to effectively carry out the use of magnetic hyperthermia in cancer therapy (Gordon et al. 1979). It was year 2004 when the first clinical magnetic hyperthermia treatment system was developed at Charité—Medical University of Berlin (Gneveckow et al. 2004) and a few years later, Magforce® (Magforce) obtained European regulatory approval to treat patients with brain tumor using magnetic hyperthermia. Though for brain tumor treatment in Europe the magnetic hyperthermia has been clinically approved, yet it is not fully practiced in the clinics for diagnosis purpose. It has not been approved as a treatment process in hospitals in the USA and other parts of the world. The absence of far reaching adaptation of this treatment methodology can be halfway ascribed to gaps in the improvement of upgraded magnetic nanoparticles as hyperthermia agents.

Magnetic nanoparticles of Iron oxide with low electrical conductivity when exposed to applied magnetic field might act as localized heat sources at specified target regions in side human body. The heat generation in magnetic nanoparticles happens due to frictional heat when particles try to rotate environment of appropriate viscosity or during reversal of coupled spin within nanoparticles. Since, the nanoparticles have very low electrical conductivity therefore, the inductive heating is negligible (i.e., through eddy currents) (Jordan et al. 1993) One of the major approaches or trends which are being adopted in hyperthermia treatment is utilization of superparamagnetic nanoparticles having size range in monodomain to multidomain transition. One more approach or trend which is quite used is above mentioned general one in nanomedicine where different therapy approaches with targeting and monitoring are combined together. For instance, results on the chemistry of photosensitizers and the particular take-up of porphyrin derivatives in tumors are in the premise of the interest in getting iron oxide nanoparticles doped with porphyrin derivatives (Gneveckow et al. 2004). Phototherapy which is a type of photocatalysis together with magnetoheating, magnetodiagnostics and magnetotargeting establish a typical example of the so-called theranostic agents (Abenojar et al. 2016).

8.5 Memory Devices

Magnetic nanoparticles are significant due to their size dependent properties. These materials carry nanoparticles which exhibit superparamagnetism, which is a key essential characteristic to use magnetic materials for storage applications. Generally, during magnetization, the magnetic moments are oriented along the field direction via two distinct processes. In the first process, each magnetic nanoparticle aligns itself in the field direction in such a way that the alignment of the magnetic moment gets closer to the applied field direction while the magnetic moments remain fixed against the single domain nanoparticles. On the other hand in second process the single domain nanoparticle remain fixed while the magnetic moments rotate against the nanoparticle. Under such conditions, a nanoparticle system maintains its thermodynamic balance is known as magnetic relaxation (Rosensweig 2002). Two types of relaxation processes can appear due these mechanisms: one is known as Brown relaxation attributed to the nanoparticles rotation, and the N´eel relaxation attributed to the rotation of the magnetic moment inside the nanoparticle (Asgari and Bowen 2017).

To determine the applications of a magnetic material, it is actually the time duration (relaxation time) for which the magnetic response of effect of applied magnetic field is observed in a material. In other words, it is magnetic time measurement of nanomaterials which determines their unique properties. We observe the transition from superparamagnetic state to a blocked state usually when the measuring time is equal to the Neel-Relaxation time.

For industrial applications demands, hence thrust is always given to keep the measurement time constant while temperature is varied in order to observe the transition between the superparamagnetic and the blocked state. Since magnetic materials offer critical and excellent prompt response, nanomaterials have been promoted for accomplishing distinguished out puts in various industrial scale applications, for example, magnetic seals in motors, as contrast agents in biomedical applications and data storage systems (Saeed et al. 2018). The general essence of the distinctive sort of magnetic reactions at the nanoscale is the fact that as the material dimensions are reduced from bulk to nanoscale, it remains at the edge of losing its ferromagnetic properties, as far as applications for magnetic storage are concerned, we require a material which is nanosized and also possesses ferromagnetism (Hansen and Morup 1998). The major challenge in information or data storage is to reduce the scale of the materials to nano and also keep ferromagnetic behavior intact. But this is not the general case where a material loses it ferromagnetic properties just by reducing its size to nanoscale. Materials show zero coercivities and remnant magnetization due to thermal annealing effects at nanoscale (Koo et al. 2019) which results in superparamagnetism, while on the contrary for data storage applications a material should have considerable values of coercivity and remnant magnetization values. Which in tune means every sort of demand or application is fundamentally based on particular properties of the material which it inherits. For example, in case of memory applications, material should possess particles which are stable and having switchable magnetic state and compatible to represent information stored in the form of bits and should not be perturbed by the changes occurring in temperature.

9 Conclusion

The book chapter presents brief overview about the nano ceramic materials especially class of magnetism known as superparamagnetism observed when the size of bulk material is reduced to nano scale. Superparamagnetic materials show quite interesting and unique applications in various fields like medical science as biosensors and as drug carriers in computer science as data storage, etc. Different aspects of these materials are tested and studied to wider the range of their applications by playing with the fundamental techniques of preparation using different parameters such as different solvents, sintering temperature and pressure but still many challenges are faced like loss of ferromagnetism when the size of material is reduced from bulk to nanoscale. Demand of new approach and new technique is high to overcome these shortcomings. Since nanoparticles of iron oxide are nonhazardous and non-toxic they have great potential to be used in clinic diagnosis and in green technology.

References

Abenojar EC, Wickramasinghe S, Bas-Concepcion J, Samia ACS (2016) Prog Nat Sci Mater Int 26:440–448
Asgari, Bowen (2017) Nat Prod Chem Res 5:6
Bean CP, Jacobs IS (1956) J Appl Phys 27:1448
Cullity BD (1972) Introduction to magnetic materials. Addison-Wesley, Inc. Philippines. ISBN 0-201-1218-9
Darweesh HHM (2018) Nanoscience 1:40–66
Dufay B, Saez S, Denoual M, Dolabdjian C, Ludwig F, Heim E, Schilling M, Melo L, Yelon A, Menard D (2009) Sens Lett 7:1–4
Ennen I, Kappe D, Rempel T, Glenske C, Hütten A (2016) Sensors 16:904
Gilchrist RK, Medal R, Shorey WD, Hanselman RC, Parrott JC, Bruce Taylor C (1957) Ann Surg 146:596–606
Gilllaud C, Creveaux H (1950) Compt Rend Acad Sci Paris 230:1256–1458
Gneveckow U, Jordan A, Scholz R, Brüss V, Waldöfner N, Ricke J, Feussner A, Hildebrandt B, Rau B, Wust P (2004) Med Phys 31:1444–1451
Gordon T, Hines JR, Gordon D (1979) Med Hypotheses 5:83–102
Gorter EW (1954) '*Philips Res*. Rept'. 9:321
Goya GF, Grazu V, Ibarra MR (2008) Current Nanosci 4:1–16
Guillaud C (1951) J Phys Radium 12:239
Hansen MF, Morup S (1998) J Magn Magn Mater 184:262
He DF, Krause H-J, Zhang Y, Bick M, Soltner H, Wolters N, Wolf W, Bousack H, Trans IEEE (1999) Appl Supercond 9:3684–36870
Heiligtag FJ, Niederberger M (2013) Mater Today 16:262–271
Herzer G, Vazquez M, Knobel M, Zhukov A, Reininger T, Davies HA, Groessinger R, Sanchez JL (2005) J Magn Magn Mater 294:252

Ignatyeva DO, Kapralov PO, Knyazev GA, Borovkova OV, Sekatskii SK, Nur-E-Alam M, Vasiliev M, Alameh K, Belotelov VI (2016) In: 10th international conference on advanced electromagnetic materials in microwaves and optics metamaterials, Crete, Greece
Jordan A, Wust P, Fahling H, John W, Hinz A, Feliz R (1993) Int J Hyperth 9:51–68
Khalil KA (2012) Advanced sintering of nano-ceramic materials. In: Ceramic materials—progress in modern ceramics
Kittle C (1996) Introduction to solid state physics, 7th edn. Wiley, New York. ISBN 9971-51-180-0
Knobel M, Ferrari EF, Silva FCS (1999) Mater Sci Forum 169:302–303
Kodama RH, Berkowitz AE, McNiff EJ Jr, Foner S (1996) Phys Rev Lett 77:394
Kodama RH, Berkowitz AE, McNiff EJ Jr, Foner S (1997) Mater Sci Forum 643:235–238
Koo KN, Ismail AF, Hafiz zarfan Othman M, Rahman MA, Sheng TZ (2019) Malays J Fund Appl Scien. 15:21–23
Lee JH, Huh YM, Jun Y, Seo J, Jang J, Song HT, Kim S, Cho EJ, Yoon HG, Suh JS, Cheon J (2007) Nat Med 13:95–99
Lubitz P, Rachford FJ (2002) J Appl Phys 91:7613
Magforce®: The nanomedicine company (n.d.) http://www.magforce.de/en/unternehmen/ueber-uns.html
Miyake H, Yuba Y, Gamo K, Namba S (1988) J Vac Sci Technol B Microelectron Process Phenom 6:1001
Moskowitz BM (1991) Hitchhiker's guide to magnetism. In: Environmental magnetism workshop, Institute for Rock Magnetism. www.magneticmicrosphere.com
Rosensweig RE (2002) J Magn Magn Mater 252:370–374
Ruan S, Xu B, Suo H, Wu F, Xiang S, Zhao M (2000) J Magn Magn Mater 212:175
Rummery EJ, Reimer P, Heindel W, Thieme (2008) MRI imaging of the body. Stuttgart, New York
Saeed M, Ren W, Wu A (2018) Biomater Sci 6:708
Šajgalík P, Dusza J, Lenčéš Z, Hnatko M, Galusek D, Ghillányová K (2013) Chapter 8 in Ceramic science and technology. Wiley Online Library. ISBN: 9783527631940
Stoner EC, Wohlfarth EP (1948) Trans R Soc A 240:599
Sun S, Murray CB, Weller D, Folks L, Moser A (2000) Science 287:1989
Sun C, Lee JSH, Zhang MQ (2008) Adv Drug Deliv Rev 60:1252–1265
Tagantsev AK, Sherman VO, Astafiev KF, Venkatesh J, Setter N (2003) J Electroceram 11:5
Torchilin VP (2006) Multifunctional nanocarriers. Adv Drug Deliv Rev 58:1532–1555
Virk HS, Poonam S (2010) Int J Adv Eng Technol 1:114–129
Weller D, Doerner MF (2000) Annu Rev Mater Sci 30:611
Weller D, Moser A (1999) IEEE Trans Magn 35:4423

Nanomaterials for Removal of Toxic Metals Ions from the Water

Meena Bisht

Abstract The rapid deterioration in water quality has become a global concern. Heavy metal ions are the most dangerous water pollutants for living organisms; hence, there is a necessity to remove these toxic pollutants from water. Traditional water purification methods are expensive and inefficient to provide adequate quality of water. In the past few decades, nanotechnology has gained remarkable attention in many areas including water purification processes. Nanomaterials have unique properties such as greater surface area, exceptional adsorption capability and high selectivity which make them more promising materials for removal of heavy metal ions, and other pollutants from water. Nanomaterials are capable of removing toxic metal ions with high efficiency and selectivity even at their lower concentration. This chapter gives an overview of various nanomaterials especially carbon nanomaterials (e.g., graphene and carbon nanotubes) for the removal of highly toxic metal ions such as arsenic (As^{5+}), lead (Pb^{2+}), cadmium (Cd^{2+}), and mercury (Hg, Hg^{2+}) from water. This chapter will also highlight the toxic effects and main barriers of nanomaterials in sustainable water treatment.

Keywords Nanotechnology · Nanomaterials · Water purification · Toxic metal ions · Carbon nanomaterials · Carbon nanotubes · Graphene

1 Introduction

Water is an important and essential component of life (John 1990). Globally, 844 million people do not have access to potable water. It is estimated that 10–20 million people die every year due to waterborne and nonfatal infections (Nemerow and Dasgupta 1991; Leonard et al. 2003; WHO 2017; Damia 2005). The quality of water resources is degrading due to population growth, depletion of water resources and the continuous addition of undesirable chemicals into them (Vörösmarty et al. 2000; Rockstrom 2003). As industrialization and civilization increase over the last century,

M. Bisht (✉)
CICECO—Aveiro Institute of Materials, Chemistry Department, University of Aveiro, 3810-193 Aveiro, Portugal
e-mail: meenabisht@ua.pt

A. H. Bhat et al. (eds.), *Nanomaterials for Healthcare, Energy and Environment*, Advanced Structured Materials 118, https://doi.org/10.1007/978-981-13-9833-9_8

the shortage of clean water supply is also increasing due to the contamination of various water sources by chemicals, fertilizers, pharmaceutical wastes, algal blooms, dyes, detergents, pesticides, heavy metals, etc. (Fawell and Nieuwenhuijsen 2003; Forgacs et al. 2004; Helmer 1997). Contamination of water by toxic heavy metal ions has become a serious issue for the entire biosphere (Ritter et al. 2002; Tchounwou et al. 2012). Heavy metal ions are able to induce toxicity even at a low level of exposure. Industries such as mining, paints, tannery, metallurgical, chemical manufacturing, fossil fuel, and battery manufacturing industries release heavy metals (toxic metal ions) into the water bodies at levels that far exceed the World Health Organization (WHO) limits (WHO 2017; Duffus 2002; Ali et al. 2011; Gupta and Nayak 2012). Heavy metal ions such as Arsenic (As), cadmium (Cd), lead (Pb), mercury (Hg), chromium (Cr), etc. are very dangerous to human health. The maximum contaminant level (MCL) for some toxic metals in drinking water is provided in Table 1 (Li et al. 2002; Ali 2012; Diniz et al. 2002; Arias et al. 2002; Barakat 2011). Some metal ions such as Pb^{2+} can substitute other important metals like Mg^{2+}, Ca^{2+}, and Fe^{2+} which ultimately interfere with metabolic processes. Long-term accumulated of heavy metals can damage the functioning of the brain, kidney, lungs, liver, and other important organs (Barakat 2011).

Therefore the presence of these toxic metal ions in water is a major threat to living organisms; hence, it is extremely important to develop a sensitive technique for the effective determination and removal of these pollutants from aqueous source (Diniz et al. 2002; Arias et al. 2002; Barakat 2011). Conventional water treatment techniques such as ion-exchange, filtration, adsorption via activated carbon, micro- and ultrafiltration, gravity separation, reverse osmosis (RO), coagulation, membrane separation, and electrochemical treatment have been used to eliminate the toxic metal ions from contaminated water (Binns 2010; Obare and Meyer 2004; Bottero et al. 2006; Upadhyayula et al. 2009; Monser and Adhoum 2002; Fahrner 2005; Stefaniuk et al. 2016; Ali et al. 2018; Bolisetty et al. 2019). However, due to the complicated

Table 1 The MCL standards for some toxic metal ions

Heavy metal	Toxicities	MCL (mg/L)
Cadmium (Cd)	Kidney damage, human carcinogen, renal disorder, Itai-Itai	0.01
Arsenic (As)	Skin manifestations, vascular disease, visceral cancers	0.01
Mercury (Hg)	Rheumatoid arthritis, kidney, circulatory system and nervous system diseases	0.00003
Lead (Pb)	Damage the fetal brain, diseases of the kidneys, circulatory system, and nervous system	0.006
Chromium (Cr)	Headache, nausea, vomiting, diarrhea, carcinogenic to human	0.05
Zinc (Zn)	Lethargy, depression, neurological signs and increased thirst	0.80
Nickel (Ni)	Dermatitis, nausea, chronic asthma, coughing, carcinogenic to human	0.20

Reprinted with permission from Barakat (2011). Copyright 2006 Elsevier operating mechanisms, the use of toxic chemicals, low efficiency, low selectivity, and high production cost, their employability is limited. Figure 1 represents some well-established techniques for the removal of toxic metal ions (Bolisetty et al. 2019). Among them some techniques are not able to detect the low concentration of pollutants; therefore, conventional methods remain ineffective to provide adequate quality of water (Bolisetty et al. 2019; Ali et al. 2017).

In this context, nanotechnology is the most promising technique to overcome all water-related issues (Obare and Meyer 2004; Bottero et al. 2006; Upadhyayula et al. 2009; Monser and Adhoum 2002; Fahrner 2005; Stefaniuk et al. 2016; Ali et al. 2017, 2018; Bolisetty et al. 2019). Nanomaterials offer promising potential for removing toxic metals from contaminated water more quickly and efficiently as compared to conventional materials (Upadhyayula et al. 2009; Monser and Adhoum 2002). Nanomaterials are non-toxic, selective, effective, stable and recyclable due to their unique physicochemical and surface properties (Fahrner 2005; Stefaniuk et al. 2016). In recent years, nanomaterials are the most extensively investigated materials that have many applications in various fields. Among them, carbon nanomaterials have high capacity/selectivity for toxic metal ions, viruses, organic, inorganic as well as for pollutants present in water (Ali et al. 2017, 2018; Bolisetty et al. 2019). The efficiency and selectivity of carbon nanomaterials can further be improved by functionalizing it with various chemical groups (Mishra 2014).

In this chapter, we will provide an overview of recent developments in the field of nanotechnology for water treatment. Here, we will focus on the application of various nanomaterials particularly carbon nanomaterials for the removal of highly toxic metal ions such as As(V), Cd(II), Pb(II), and Hg(II) from contaminant water. This chapter

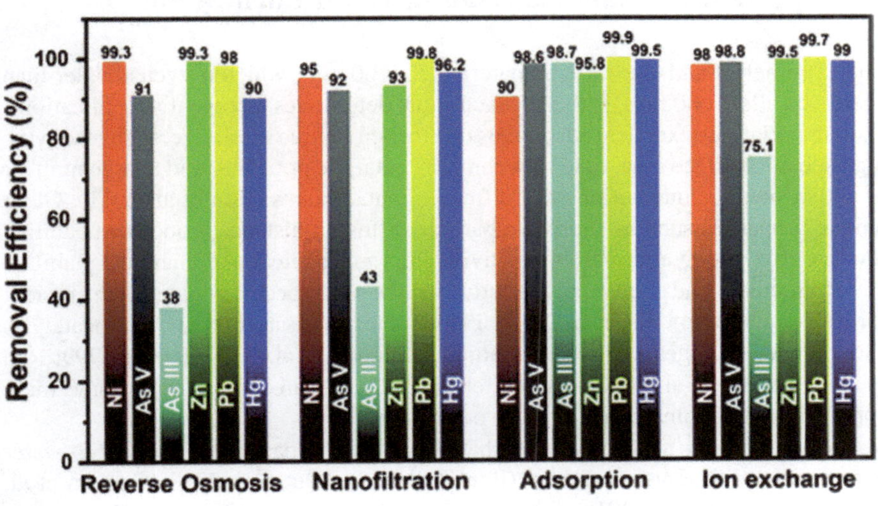

Fig. 1 Current technologies to remove the heavy metal ions. Reprinted with permission from Bolisetty et al. (2019). Copyright 2019 Royal Society of Chemistry

will also highlight the toxicological effects and main barriers of nanomaterials in sustainable water treatment.

2 What Are Nanomaterials

Nanotechnology uses materials of particle size between 1 and 100 nm (Amin et al. 2014). At the nanoscale, materials possess different novel properties as compare to the macro scale (Rickerby and Morrison 2007). Nanomaterials can be separated into different categories depending on their physicochemical and surface characteristics. There are different types of nanomaterials such as carbon-based nanomaterials, metal oxide nanoparticles (Fe_2O_3, ZnO, TiO_2, Al_2O_3, NiO), metallic nanoparticles (Au, Ag and Pt NPs), mixed oxide nanoparticles (Fe–Ti NPs), nanofibers, polymer-based nanoadsorbents, etc. (Iijima et al. 1999; Novoselov et al. 2004; Gehrke et al. 2015; Carpenter et al. 2015). These new generation nano-materials can be prepared using inert gas condensation, spray pyrolysis, pulsed laser ablation, laser pyrolysis, spark discharge generation, ion sputtering, photothermal synthesis, and flame spray pyrolysis, etc. (Ali 2012; Carpenter et al. 2015; Sharma et al. 2009; Palkar 1999). The nanomaterials are characterized by scanning electron microscopy (SEM), transmission electron microscopy (TEM) techniques and X-ray diffraction (XRD) (Palkar 1999).

3 Importance of Nanomaterials in Water Purification

The extremely small size of nanomaterials (1–100 nm), which is even smaller than bacterial cells (1000 nm), is the key factor that determines its potential applications. Nanomaterials are excellent adsorbents due to their unique small size, high reactivity, high specific surface area, ease of separation, catalytic potential, and a large number of active sites for interaction with different contaminants (Ren et al. 2011). Other unique properties such as short intraparticle diffusion distance, good permeability, low toxicity, strong antimicrobial activity, high mechanical and chemical stability, better selectivity and photocatalytic activity in the solar spectrum make them efficient for water purification (Qu et al. 2013). These can be use as adsorbents, photocatalysts and antibacterial agents in water treatment (Sharma et al. 2009; Palkar 1999; Ren et al. 2011; Qu et al. 2013; Gautam et al. 2013). Figure 2 represents some main application of nanomaterials in water purification.

There are many nanoadsorbents that have been prepared and applied in water treatment by different researchers (Ren et al. 2011; Qu et al. 2013; Gautam et al. 2013; Kalhapure et al. 2015; Pradeep 2009; Coston et al. 1995; Van Benschoten et al. 1994; Santhosh et al. 2019; Liu et al. 2016; Ravindranath et al. 2017; Cao et al. 2007). **Metal nanoparticles** (NPs) are nano-sized metals with dimensions within the size range of 1–100 nm. The most important feature of nanoparticles is their

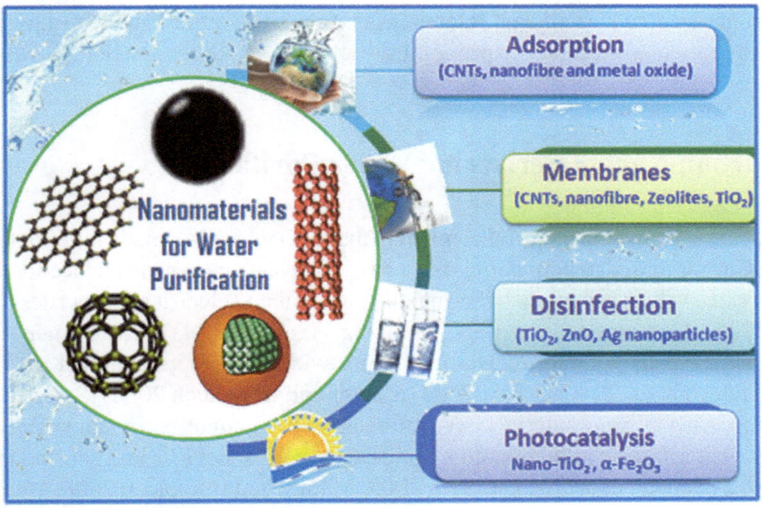

Fig. 2 Application of nanomaterials for water purification

high surface area to volume ratio and thus, it can absorb toxicants on its surface. The most studied metal NPs are nano silver (nAg), nano gold (nAu), and nano iron (nFe). These metal NPs show enhanced antibacterial effects against harmful pathogens (Kalhapure et al. 2015; Pradeep 2009). Therefore, metal NPs seem to be promising in solving the problem of contaminated water. **Metal oxide** based nanomaterials are composed of one or more metals and their oxides. These nanomaterials are good candidates for wastewater treatment and reuse processes (Pradeep 2009). Nanosized metal oxides are classified as the promising adsorbents for the removal of heavy metal ions. These include Fe_2O_3, NiO, MnO_2, TiO_2, ZnO, and Co_3O_4 (Coston et al. 1995; Van Benschoten et al. 1994). Among them, Fe_2O_3 has gain enormous attention due to its non-toxicity, eco-friendliness, low production cost, high photostability, easy separation ability and high efficiency (Santhosh et al. 2019; Liu et al. 2016). In addition, mixed metal oxides have also gained sufficient attention in removing the toxic metal ions from water. In this regard, Ravindranath et al. (2017) prepared cube-like heterostructures of Fe_2O_3/Al_2O_3 composites by the precipitation method. These Fe_2O_3/Al_2O_3 composite had a large surface area (208.3 m^2/g), which can effectively remove toxic Hg(II) from aqueous solution (Ravindranath et al. 2017). The metal oxides such as copper oxide and Iron(III)-Titanium(IV) binary mixed oxide have shown promising results for the removal of toxic metals from aqueous solution (Cao et al. 2007; Gupta and Ghosh 2009). **Nanofiltration (NF)** is a relatively new technique which is also designed to remove nano size pollutants from water (Peydayesh et al. 2017; Li et al. 2017). **Nanoscale biopolymers** offer an eco-friendly route for the remediation of heavy metals such as Hg and As from water (Savage and Diallo 2005). In the present scenario, **Carbon-based nanomaterials** such as Carbon nanotubes

(CNTs) have gained significant interest as compared to other nanomaterials due to their exceptional high catalytic properties, high surface adsorption and reactivity.

4 Carbon Nanomaterials for Water Purification

Carbon is the third most plentiful element in the earth which has much potential application. Carbon-based nanomaterials have an exclusive place in nanoscience for owing to their extraordinary chemical, thermal, mechanical and electrical properties (Kroto et al. 1985). Carbon nanomaterials, including graphene, CNTs, carbon nanofibers, fullerenes have shown enormous potential for widespread application (Kroto et al. 1985; Sadegh et al. 2017; Colvin 2003; Mauter and Elimelech 2008). Since the first invention, carbon nanomaterials have attracted special attention in nanoelectronics, catalyst, water purification, quantum computers, optical devices and sensors (Kroto et al. 1985; Sadegh et al. 2017; Colvin 2003; Mauter and Elimelech 2008; Wan et al. 2012). Carbon nanomaterials such as fullerenes, CNTs, and graphene are topologically confined objects in zero, one, or two dimensions, respectively are shown in Fig. 3 (Wan et al. 2012).

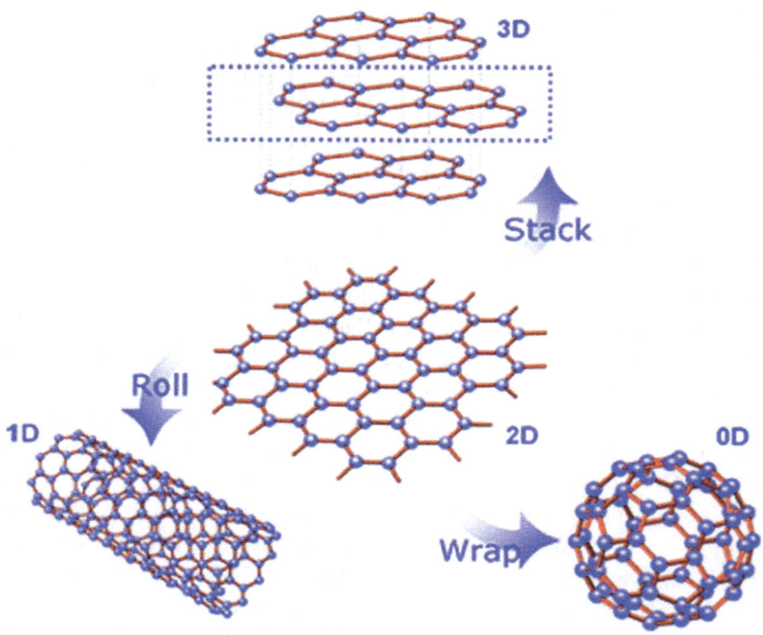

Fig. 3 Allotropes of carbon nanomaterials in different dimensions. Reprinted with permission from Wan et al. (2012). Copyright 2012 American Chemical Society

4.1 Removal of Toxic Metal Ions from Water by Fullerenes

Fullerenes are the third allotropic form of carbon after graphite and diamond. The chemical properties of fullerenes were first reported by Kroto's research group (Kroto et al. 1985). In recent years, fullerenes have received much attention because of their exceptional properties and cage-like structure (Savage and Diallo 2005; Kroto et al. 1985). Fullerenes are produced by chemical vapor deposition, arc-discharge vaporization of graphite, and combustion methods (Carpenter et al. 2015; Sharma et al. 2009; Palkar 1999; Ren et al. 2011; Qu et al. 2013). In the family of fullerenes, Buckminsterfullerene (C60) is the most studied structure. Nanoparticles of C_{60} have a wide range of antimicrobial activity (Scida et al. 2011). The low tendency for aggregation (due to a hydrophobic surface) and high surface/volume ratio make it good adsorbents for the removal of toxic heavy metal ions from water (Scida et al. 2011; Alekseeva et al. 2016). Fullerenes significantly enhance the oxidative processes that destroy a broad range of pollutants including carcinogenic pollutants (Ouyang et al. 2016; Maleki Dizaj et al. 2015). At present, fullerenes are generally used to remove organic impurities from aqueous media (Gupta and Saleh 2013). The adsorption behavior of fullerenes was found higher than that of activated carbon for organic pollutants.

4.2 Removal of Toxic Metal Ions from Water by Graphene

Graphene is a two-dimensional carbon allotrope in which carbon atoms are arranged in the form of a honeycomb lattice (Fig. 3). Graphene consists of a single or thin layer of graphite. It has extraordinary physicochemical properties in terms of tunable surface properties, controllable structure and easy modification (Sitko et al. 2013; Zhao et al. 2014). In recent years, there is enormous growth in the use of graphene-based materials such as graphene oxide (GO), graphene/semiconductor hybrids, composites graphene/metal nanoparticles for water purification (Sitko et al. 2013; Zhao et al. 2011, 2014). GO has the great theoretical specific surface area and high content of oxygen groups which make it efficient adsorbents for heavy metal ions from water (Zhao et al. 2011; Huang et al. 2011; Li et al. 2012; Mueller and Nowack 2010). A composite with GO and metal oxide can be used as effective adsorbents for the removal of Zn^{2+}, Pb^{2+}, and Cd^{2+} ions from water (Lee and Yang 2012; Anitha et al. 2015). Gollavelli et al. (2013) have shown high removal efficiencies ($\approx$99%) of Cr^{6+}, As^{5+}, and Pb^{2+} using magnetic graphene. Additionally, this graphene exhibited efficient antibacterial activity towards E. *coli* and low toxicity for zebrafish (Gollavelli et al. 2013). For effective metal ions adsorption, GO is further functionalized with ethylenediaminetetraacetic acid (EDTA) (Cui et al. 2015; Madadrang et al. 2012). Since, EDTA can form stable chelates with metal ions hence, it can improve the efficiency of GO to remove toxic metal ion. In this context, Madadrang et al. (2012) successfully linked EDTA and GO surfaces through a silanization reaction. This

Table 2 Grapheme based nanocomposites for the removal of toxic metal ions

S. No	Adsorbent	Adsorbate	Adsorption capacity (mg/g)
1	GO/Fe_3O_4	Cu(II)	18.3
2	GO/silica/Fe_3O_4	Pb and Cd (II)	333.3 and 166.7
3	GO/Fe_3O_4/sulfanilic acid	Cd(II)	55.4
4	GO–Mn Fe_3O_4	Pb(II)	673
5	rGo/CO Fe_3O_4	Pb(II)	299.4
6	GO/Mn–doped Fe(III)oxide	Cd and Cu(II)	87.2 and 129.7
7	GO/Fe_3O_4/sulfanilic acid	Cu(II)	50.7 and 56.8
8	GO/$FesO_4$	Cu, Pb and Cd	23.1, 38.5 and 4.4
9	Graphene Oxide (GO)	Pb(II)	35.6
10	TiO_2/GO	Pb(II)	65.6
11	GO–EDTA	Pb(II)	479
12	Graphene nanosheets (GNs)	Pb(II)	22.4
13	GNs–500	Pb(II)	35.2
14	SiO_2–GNs	Pb(II)	113.6
15	MnOi/GNs	Hg(II)	10.8

Reprinted with permission from Santhosh et al. (2016). Copyright 2016 Elsevier

EDTA-GO complex was found to be a perfect adsorbent to remove Pb(II) (Madadrang et al. 2012). Deng et al. (2007) prepared a functionalized graphene and used it as an adsorbent for Pb (II) removal. The adsorption behavior of graphene-based material with different metals is given in Table 2 (Santhosh et al. 2016).

4.3 *Removal of Toxic Metal Ions from Water by Carbon Nanotubes*

The era of Carbon nanotubes (CNTs) has started with the first report of Injima in 1991 (Sadegh et al. 2017). Depending on the number of layers, CNTs are classified into two types: single-walled carbon nanotubes (SWCNTs) and multi-walled carbon nanotubes (MWCNTs) (Sadegh et al. 2017; Iijima 1991, 2002; Arnold et al. 2006). CNTs possess unique chemical, physical and electronic properties which make it a potential candidate to remove various water pollutants (Iijima 2002). Figure 4 highlights some properties of CNTs. CNTs are extensively use to remove inorganic, organic pollutants, toxic heavy metals and industrial effluents (Savage and Diallo 2005; Yang and Xing 2010). CNTs are also very effective in removing microorganisms such as *E. coli*, Salmonella, and viruses (Mostafavi et al. 2009; Akasaka 2009). The theoretical surface area of SWCNTs is very high (3000 m^2/g) which make them superior adsorbents (Yang and Xing 2010). Therefore, CNTs are also called as new

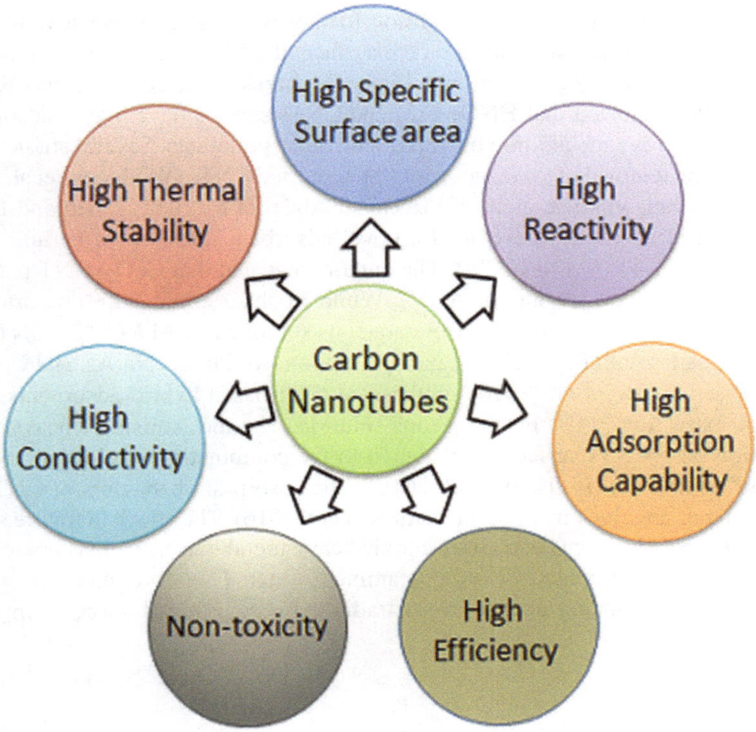

Fig. 4 An overview of properties of CNTs

generation nano-adsorbents. These are capable to remove new emerging pollutants and toxic metal ions even at their lower concentration (μg/L) under varied conditions of pH and temperature (Yang and Xing 2010; Nangmenyi et al. 2009). CNTs have led to new avenues for the removal of heavy metals ions such as As, Cd, Pb, Ni, Cr, Hg, Cu and Zn (Rao et al. 2006, 2007; El-Sheikh 2008; Issa and Meunier 2007; Li et al. 2003). Li et al. studied the competitive adsorption capacities of Cu^{2+}, Pb^{2+}, Cd^{2+} on the surface of MWCNTs and they found the affinity in the following order $Pb^{2+} > Cu^{2+} > Cd^{2+}$ (Li et al. 2003). The absorption capacity of CNTs is significant over a broad pH range. Stafiej et al. (2007) have reported adsorption of Pb(II), Cu(II), Mn(II), Co(II), and Zn(II) on CNTs surface at pH 9.

Although CNTs have great adsorption potential for toxic metal ions, however, the selectivity and sensitivity are the main limitations. The selectivity of CNTs can be improved by its physical and chemical modification. This can be easily achieved by attachment of functional group or suitable chelating agent on its surface. The recent studies have shown enhanced improvement in selectivity and adsorption capacity for toxic metal ions on the modified CNTs surface as compared to the unmodified surface. Oxygen-containing functional groups such as –CO, –COOH, and –OH, can be easily introduced to the CNTs surface by various processes (Yang and Xing 2010; Ihsanullah et al. 2016; Mubarak et al. 2014). In addition, to improve the selectivity and sensitivity, oxidation induces negative charges on the surface of CNTs; consequently, increases the adsorption capacity of heavy metal ions (Mubarak et al. 2014). In this regard, Li et al. (2003) showed that HNO_3 oxidized CNTs surface have higher adsorption capacities for heavy metal ions compared with other pollutants. Several studies have focused on the removal of toxic metal ions by modified CNTs (Wildgoose et al. 2006; Chen et al. 2009; Vukovic et al. 2011; Salam Abdel et al. 2011; Hsieh and Horng 2007). Peng et al. (2005) developed a novel adsorbent made up of cerium oxide supported on CNTs (CeO_2–CNTs). The authors revealed that CeO_2–CNT particles were more effective sorbents for As(V). While Kochkar et al. (2009) reported the removal of Cu(II) and Pb(II) ions from aqueous solutions by MWCNTs. CNTs can form composites with Fe_2O_3, ZrO_2 and TiO_2 to remove Pb, Cr, Ni, As, and Cu ions. In this regard, Zhao et al. (2010) synthesized CNT-based hybrid adsorbents using TiO_2 for extracting Pb(II) from aqueous solution. The mechanism of heavy metal adsorption onto CNTs surface is attributed to the combination of electrostatic and chemical interaction (Bolisetty et al. 2019). The adsorption behaviors of CNTs for various metals are given in Table 3 (Santhosh et al. 2016). The above literature shows that these nano adsorbents can remove toxic heavy metal ions as well as other water pollutant with high efficiency from contaminant water. Therefore, nano adsorbents can become a promising alternative to traditional heavy metal sorbents in water treatment.

5 Nanomaterials and Water Purification: Challenges

The novel and unique properties of nanomaterials make them suitable for many applications including water treatment; therefore, the production of new and novel nanomaterials is also gradually improving day by day (Santhosh et al. 2016; Iijima 1991). Enhanced use of nanomaterials has aroused global concern regarding their unknown toxicity, health hazards, and environmental impact (Colvin 2003; De Volder et al. 2013). Nanoparticles have a greater surface area so their reactivity is also more as compared to the micro-scale particles (Magrez et al. 2006; Lam et al. 2004). Moreover, the size of nanomaterials is smaller than common irritants, so they can easily enter into the cells and can create serious health risks (Mauter and Elimelech 2008; Magrez et al. 2006). Some carbon and metal-containing nanoparticles such as TiO_2 are not safe for skin (Lam et al. 2004; Ray et al. 2009). Many nanomaterials are found to be toxic for cells as well as for the environment (Ray et al. 2009). Another main problem for its application is its high cost of production; therefore, the use of modern nanomaterials in the commercial sector is still marginal. Nanotechnologies for water treatment processes claimed to have an efficiency of more 99%, but these results are only reproducible under ideal conditions of temperature, pressure, pH, pollutant concentration and other experimental parameters. For real field applications, their efficiency can decrease substantially (Bolisetty et al. 2019). It is also important to note that at high concentration nanomaterials may coagulate in water

Table 3 Toxic metal ions adsorption onto carbon nanotubes

S. No	Adsorbents	Metal ions	Adsorption capacity
1	CNTs	Pb^{2+}	17.44 mg/g at pH-7.0
2	CNTs (HNO_3)	Pb^{2+}	49.95 mg/g at pH-7.0
3	MWCNTs	Ni^{2+}	7.53 mg/g at pH-7.0
4	SWCNTs	Ni^{2+}	9.22 mg/g at pH-7.0
5	MWCNTs (HNO_3)	Pb^{2+}	97.08 mg/g at pH-5.0
6	CNTs	Pb^{2+}	1.406 mmol/g
7	CNT-OH	Pb^{2+}	2.07 mmol/g
8	CNT-$CONH_2$	Pb^{2+}	1.907 mmol/g
9	CNT–COO–	Pb^{2+}	4.672 mmol/g
10	CNTs	Cu^{2+}	1.219 mmol/g
11	CNT–OH	Cu^{2+}	1.342 mmol/g
12	CNT–CONH2	Cu^{2+}	1.755 mmol/g
13	CNT–COO^-	Cu^{2+}	3.565 mmol/g
14	CNTs	Cd^{2+}	1.291 mmol/g
15	CNT–OH	Cd^{2+}	1.513 mmol/g
16	CNT–$CONH_2$	Cd^{2+}	1.563 mmol/g
17	CNT–COO^-	Cd^{2+}	3.325 mmol/g
18	CNTs	Hg^{2+}	1.068 mmol/g
19	CNT–OH	Hg^{2+}	1.284 mmol/g
20	CNT–$CONH_2$	Hg^{2+}	1.658 mmol/g
21	CNT–COO^-	Hg^{2+}	3.300 mmol/g

and can lose their nano identity. Furthermore, it was also found that nanomaterials can also aggregate in hard freshwater and seawater. Therefore, the performance of various nanomaterials in more realistic samples needs to be tested.

6 Conclusion

Removal of toxic metals from water sources is the biggest challenge. Conventional water treatment methods are not effective for removing these pollutants at their low concentrations. Nanotechnology is one of the effective and emerging technique for water purification. Nano adsorbents are simple, effective, inexpensive and eco-friendly materials to remove toxic metal ions from water. Therefore, in this book chapter, we have summarized the recent advances and challenges of nanomaterials for the removal of highly toxic metal ions such as As(V), Pb(II), Cd(II), and Hg(II) from contaminated water. Among all, carbon-based nanomaterials are the most popular and promising materials for adsorption of toxic metals ions. However, the extensive use of nanomaterials may have harmful effects on human health and environment especially if they enter into the food chain, therefore, it is important to check their toxicity and environmental risks. Overall, nanomaterials can offer unprecedented opportunities to develop more sensitive, efficient and eco-friendly water-purification systems as compared to the traditional materials but extensive research is required for their commercial applications.

References

Akasaka T, Watari F (2009) Capture of bacteria by flexible carbon nanotubes. Acta Biomater 5:607–612

Alekseeva OV, Bagrovskaya NA, Noskov AV (2016) Sorption of heavy metal ions by fullerene and polystyrene/fullerene film composite. Prot Met Phys Chem Surf 52(3):443–447

Ali I (2012) New generation adsorbents for water treatment. Chem Rev 112:5073–5091

Ali I, Khan TA, Asim M (2011) Removal of arsenic from water by electrocoagulation and electrodialysis techniques. Sep Purif Rev 40:25–42

Ali I, Alothman ZA, Alwarthan A (2017) Supra molecular mechanism of the removal of 17-β-estradiol endocrine disturbing pollutant from water on functionalized iron nano particles. J Mol Liq 441:123–129

Ali I, Alharbi O, Alothman ML, Badjah ZA, Alwarthan AY, Basheer AAA (2018) Artificial neural network modelling of amido black dye sorption on iron composite nano material. Kinetics and thermodynamics studies. J Mol Liq 250:1–8

Amin MT, Alazba AA, Manzoor U (2014) A review of removal of pollutants from water/wastewater using different types of nanomaterials. Adv Mater Sci Eng. https://doi.org/10.1155/2014/825910

Anitha K, Namsani S, Singh JK (2015) Removal of heavy metal ions using a functionalized single-walled carbon nanotube: a molecular dynamics study. J Phys Chem A 119:8349–8358

Arias M, Barral MT, Mejuto JC (2002) Enhancement of coper and cadmium adsorption on kaolin by the presence of humic acids. Chemosphere 48:1081–1088

Arnold MS, Green AA, Hulvat JF, Stupp SI, Hersam MC (2006) Sorting carbon nanotubes by electronic structure using density differentiation. Nat Nanotechnol 1:60–65

Barakat MA (2011) New trends in removing heavy metals from industrial wastewater. Arab J Chem 4:361–377

Binns C (2010) Introduction to nanoscience and nanotechnology. Wiley, New York

Bolisetty S, Mohammad P, Raffaele M (2019) Sustainable technologies for water purification from heavy metals: review and analysis. Chem Soc Rev 48:463–487

Bottero JY, Rose J, Wiesner MR (2006) Nanotechnologies: tools for sustainability in a new wave of water treatment processes. Integr Environ Assess Manag 2:391–395

Cao AM, Monnell JD, Matranga C, Wu JM, Cao LL, Gao D (2007) Hierarchical nanostructured copper oxide and its application in arsenic removal. J Phys Chem C 111:18624–18628

Carpenter AW, de Lannoy CF, Wiesner MR (2015) Cellulose nanomaterials in water treatment technologies. Environ Sci Technol 49:5277–5287

Chen C, Hu J, Shao D, Li J, Wang X (2009) Adsorption behavior of multi-wall carbon nanotube/iron oxide magnetic composites for Ni(II) and Sr(II). J Hazard Mater 164:923–928

Colvin VL (2003) The potential environmental impact of engineered nanomaterials. Nat Biotechnol 21:1160

Coston JA, Fuller CC, Davis JA (1995) Pb^{2+} and Zn^{2+} adsorption by a natural aluminum- and iron-bearing surface coating on an aquifer sand. Geochim Cosmochim Acta 59:3535–3547

Cui L, Wang Y, Gao L, Hu L, Yan L, Wei Q, Du B (2015) EDTA functionalized magnetic graphene oxide for removal of Pb (II), Hg (II) and Cu (II) in water treatment: Adsorption mechanism and separation property. Chem Eng J 281:1–10

Damia B (ed) (2005) Emerging organic pollutants in waste waters and sludge. Springer, New York

De Volder MFL, Tawfick SH, Baughman RH, Hart AJ (2013) Carbon nanotubes: present and future commercial applications. Science 339:535–539

Deng LP, Su YY, Su H, Wang XT, Zhu XB (2007) Sorption and desorption of lead (II) from wastewater by green algae *Cladophora fascicularis*. J Hazard Mater 143:220–225

Diniz CV, Doyle FM, Ciminelli VST (2002) Effect of pH on the adsorption of selected heavy metal ions from concentrated chloride solutions by the chelating resin dowex M-4195. Sep Sci Technol 37:3169–3185

Duffus JH (2002) Heavy metals-a meaningless term? Pure Appl Chem 74:793–807

El-Sheikh AH (2008) Effect of oxidation of activated carbon on its enrichment efficiency of metal ions: comparison with oxidized and non-oxidized multi-walled carbon nanotubes. Talanta 75:127–134

Fahrner W (2005) Nanotechnology and nanoelectronics materials, devices, measurement techniques. Springer, New York

Fawell J, Nieuwenhuijsen MJ (2003) Contaminants in drinking water. Br Med Bull 68:199–208

Forgacs E, Cserháti T, Oros G (2004) Removal of synthetic dyes from wastewaters: a review. Environ Int 30:953–971

Gautam P, Madathil D, Brijesh Nair AN (2013) Nanotechnology in waste water treatment: a review. Int J ChemTech Res 5:2303–2308

Gehrke I, Geiser A, Somborn-Schulz A (2015) Innovations in nanotechnology for water treatment. Nanotechnol Sci Appl 8:1–17

Gollavelli G, Chang CC, Ling YC (2013) Facile synthesis of smart magnetic graphene for safe drinking water: heavy metal removal and disinfection control. ACS Sustain Chem Eng 1:462–472

Gupta K, Ghosh UC (2009) Arsenic removal using hydrous nanostructure Iron(III)-Titanium(IV) binary mixed oxide from aqueous solution. J Hazard Mater 161:884–892

Gupta VK, Nayak A (2012) Cadmium removal and recovery from aqueous solutions by novel adsorbents prepared from orange peel and Fe_2O_3 nanoparticles. Chem Eng J 180:81–90

Gupta VK, Saleh TA (2013) Sorption of pollutants by porous carbon, carbon nanotubes and fullerene—an overview. Environ Sci Pollut Res 20(5):2828–2843

Helmer HR (1997) Water pollution control—a guide to the use of water quality management principles. E & FN Spon, London

Hsieh S, Horng J (2007) Adsorption behavior of heavy metal ions by carbon nanotubes grown on microsized Al_2O_3 particles. J Univ Sci Technol Beijing, Miner, Metall, Mater 14:77–78

Huang ZH, Zheng X, Lv W, Wang M, Yang QH, Kang F (2011) Adsorption of lead (II) ions from aqueous solution on low-temperature exfoliated graphene nanosheets. Langmuir 27:7558–7562

Ihsanullah Abbas A, Al-Amer AM, Laoui T, AlMarri MJ, Nasser MS, Khraisheh M, Atieh MA (2016) Heavy metal removal from aqueous solution by advanced carbon nanotubes: critical review of adsorption applications. Sep Purif Technol 157:141–161

Iijima S (1991) Helical microtubules of graphitic carbon. Nature 354:56

Iijima S (2002) Carbon nanotubes: past, present, and future. Physica B Condens. Matter 323:1–5

Iijima S, Yudasaka M, Yamada R, Bandow S, Suenaga K, Kokai F, Takahashi K (1999) Nano-aggregates of single-walled graphitic carbon nano-horns. Chem Phys Lett 309:165

Issa M, Meunier J (2007) Removal of nickel ions from water by multi-walled carbon nanotubes. J Hazard Mater 46:283–288

John DZ (1990) Handbook of drinking water quality: standards and controls. Van Nostrand Reinhold, New York

Kalhapure RS, Sonawane SJ, Sikwal DR, Jadhav M, Rambharose S, Mocktar C, Govender T (2015) Solid lipid nanoparticles of clotrimazole silver complex: an efficient nanoantibacterial against Staphylococcus aureus and MRSA. Colloids Surf B Biointerfaces 36:651–658

Kochkar H, Turki A, Bergaoui L, Berhault G, Ghorbel AJ (2009) Study of Pd(II) adsorption over titanate nanotubes of different diameters. Colloid Interface Sci 331:27–31

Kroto HW, Heath JR, O'Brien SC, Curl RF, Smalley RE (1985) C60: buckminsterfullerene. Nature 318:162–163

Lam CW, James JT, McCluskey R, Hunter RL (2004) Pulmonary toxicity of single-wall carbon nanotubes in mice 7 and 90 days after intratracheal instillation. Toxicol Sci 77:126–134

Lee YC, Yang JW (2012) Self-assembled flower-like TiO_2 on exfoliated graphite oxide for heavy metal removal. J Ind Eng Chem 18:1178–1185

Leonard P, Hearty S, Brennan J (2003) Advances in biosensors for detection of pathogens in food and water. Enzyme Microbial Technol 32:3–13

Li YH, Wang S, Wei J, Zhang X, Xu C, Luan Z, Wu D, Wei B (2002) Lead adsorption on carbon nanotubes. Chem Phys Lett 357:263–266

Li Y, Ding J, Luan Z, Di Z, Zhu Y, Xu C, Wu D, Wei B (2003) Competitive adsorption of Pb^{2+}, Cu^{2+} and Cd^{2+} ions from aqueous solutions by multiwalled carbon nanotubes. Carbon 41(14):2787–2792

Li J, Zhang S, Chen C, Zhao G, Yang X, Li J, Wang X (2012) Removal of Cu(II) and fulvic acid by graphene oxide nanosheets decorated with Fe_3O_4 nanoparticles. ACS Appl Mater Interfaces 4:4991–5000

Li M, Lv Z, Zheng J, Hu J, Jiang C, Ueda M, Zhang X, Wang L (2017) Positively charged nanofiltration membrane with dendritic surface for toxic element removal. ACS Sustain Chem Eng 5:784–792

Liu Z, Yu R, Dong Y, Li W, Zhou W (2016) Preparation of α-Fe_2O_3 hollow spheres, nanotubes. RSC Adv 6:82854–82861

Madadrang CJ, Kim HY, Gao G, Wang N, Zhu J, Feng H, Gorring M, Kasner ML, Hou S (2012) Adsorption behavior of EDTA-graphene oxide for Pb (II) removal. ACS Appl Mater Interfaces 4:1186–1193

Magrez A, Kasas S, Salicio V, Pasquier N, Seo JW, Celio M, Catsicas S, Schwaller B, Forró L (2006) Cellular toxicity of carbon-based nanomaterials. Nano Lett 6:1121–1125

Maleki Dizaj S, Mennati A, Jafari S, Khezri K, Adibkia K (2015) Antimicrobial activity of carbon-based nanoparticles. Adv Pharm Bull 5:19–23

Mauter MS, Elimelech M (2008) Environmental applications of carbon-based nanomaterials. Environ Sci Technol 42:5843–5859

Mishra AK (2014) Application of nanotechnology in water research. Scrivener publishing. ISBN: 978-1-118-49630-51-552

Monser L, Adhoum N (2002) Modified activated carbon for the removal of copper, zinc, chromium and cyanide from wastewater. Sep Purif Technol 26:137–146

Mostafavi ST, Mehrnia MR, Rashidi AM (2009) Preparation of nanofilter from carbon nanotubes for application in virus removal from water. Desalination 238:271–280

Mubarak NM, Sahu JN, Abdullah EC, Jayakumar NS (2014) Removal of heavy metals from wastewater using carbon nanotubes. Sep Purif Rev 43:311–338

Mueller NC, Nowack B (2010) Nanoparticles for remediation: solving big problems with little particles. Elements 6:395–400

Nangmenyi G, Xao W, Mehrabi S, Mintz E, Economy J (2009) Bactericidal activity of Ag nanoparticle-impregnated fibreglass for water disinfection. J Water Health 7:657

Nemerow N, Dasgupta A (1991) Industrial and hazardous waste treatment. Van Nostrand Reinhold, New York

Novoselov KS, Geim AK, Morozov SV et al (2004) Electric field in atomically thin carbon films. Science 306:666–669

Obare SO, Meyer GJ (2004) Nanostructured materials for environmental remediation of organic contaminants in water. J Environ Sci Health—Part A 39:2549–2582

Ouyang K, Dai K, Walker SL, Huang Q, Yin X, Cai P (2016) Efficient photocatalytic disinfection of Escherichia coli O157:H7 using C70-TiO_2 hybrid under visible light irradiation. Sci Rep 6:25702

Palkar VR (1999) Sol-gel derived nanostructured γ-alumina porous spheres as an adsorbent in liquid chromatography. Nanostruct Mater 11:369–374

Peng X, Luan Z, Ding J, Di Z, Li Y, Tian B (2005) Ceria nanoparticles supported nanotubes for the removal of arsenate from water. Mater Lett 59:399–403

Peydayesh M, Bagheri M, Mohammadi T, Bakhtiari O (2017) Fabrication optimization of polyethersulfone (PES)/polyvinylpyrrolidone (PVP) nanofiltration membranes using Box-Behnken response surface method. RSC Adv 7:24995–25008

Pradeep T (2009) Noble metal nanoparticles for water purification: a critical review. Thin Solid Films 517:6441–6478

Qu X, Alvarez PJJ, Li Q (2013) Applications of nanotechnology in water and wastewater treatment. Water Res 47:3931–3946

Rao MM, Ramesh A, Rao GPC, Seshaiah K (2006) Removal of copper and cadmium from the aqueous solutions by activated carbon derived from ceiba pentandra hulls. J Hazard Mater B 129:123–129

Rao GP, Lu C, Su F (2007) Sorption of divalent metal ions from aqueous solution by carbon nanotubes: a review. Sep Purif Technol 34:224–231

Ravindranath R, Roy P, Periasamy AP, Chen PW, Liang CT, Chang HT (2017) Fe_2O_3/Al_2O_3 microboxes for efficient removal of heavy metal ions. New J Chem 41:7751–7757

Ray PC, Yu H, Fu PP (2009) Toxicity and environmental risks of nanomaterials: challenges and future needs. J Environ Sci Health C Environ Carcinog Ecotoxicol Rev 27:1–35

Ren X, Chen C, Nagatsu M, Wang X (2011) Carbon nanotubes as adsorbents in environmental pollution management: a review. Chem Eng J 170:395–410

Rickerby DG, Morrison M (2007) Nanotechnology and the environment a European perspective. Sci Technol Adv Mater 8:19–24

Ritter L, Solomon K, Sibley et al (2002) Sources, pathways, and relative risks of contaminants in surface water and groundwater: a perspective prepared for the walkerton inquiry. J Toxicol Environ Health—Part A 65:1–142

Rockstrom J (2003) Water for food and nature in drought-prone tropics. vapour shif in rain-fed agriculture. Philos Trans R Soc B: Biol Sci 358:1997–2009

Sadegh H, Ali GAM, Gupta VK, Makhlouf ASH, Shahryari-ghoshekandi R, Nadagouda MN, Sillanpää M, Megiel E (2017) The role of nanomaterials as effective adsorbents and their applications in wastewater treatment. J Nanostructure Chem 7:1–14

Salam Abdel M, Makki MS, Abdelaal M (2011) Preparation and characterization of multi-walled carbon nanotubes/chitosan nanocomposite and its application for the removal of heavy metals from aqueous solution. J Alloys Compd 509:2582–2587

Santhosh C, Velmurugan V, Jacob G, Jeong SK, Grace AN, Bhatnagar A (2016) Role of nanomaterials in water treatment applications: a review. Chem Eng J 306:1116–1137

Santhosh C, Malathi A, Dhaneshvar E, Bhatnagar A, Grace AN, Madhavan J (2019) Nanoscale materials in water purification, micro and nano technologies. https://doi.org/10.1016/b978-0-12-813926-4.00022-7

Savage N, Diallo MS (2005) Nanomaterials and water purification: opportunities and challenges. J Nanopart Res 7:331

Scida K, Stege PW, Haby G, Messina GA, Garcia CD (2011) Recent applications of carbonbased nanomaterials in analytical chemistry. Anal Chim Acta 691:6–17

Sharma Y, Srivastava V, Singh V, Kaul S, Weng C (2009) Nano-adsorbents for the removal of metallic pollutants from water and wastewater. Environ Technol 30:583–609

Sitko R, Zawisza B, Malicka E (2013) Graphene as a new sorbent in analytical chemistry. TrAC. Trends Anal Chem 51:33–43

Stafiej A, Pyrzynska K (2007) Adsorption of heavy metal ions with carbon nanotubes. Sep Purif Technol 58:49–52

Stefaniuk M, Oleszczuk P, Ok PS (2016) Review on nano zerovalent iron (nZVI). from synthesis to environmental applications. Chem Eng J 287:618–632

Tchounwou PB, Yedjou CG, Patlolla AK, Sutton DJ (2012) Heavy metal toxicity and the environment. Exp Suppl 101:133–164

Upadhyayula VKK, Deng S, Mitchell MC, Smith GB (2009) Application of carbon nanotube technology for removal of contaminants in drinking water: a review. Sci Total Environ 408:1–13

Van Benschoten JE, Reed BE, Matsumoto MR, McGarvey PJ (1994) Metal removal by soil washing for an iron oxide coated sandy soil. Water Environ Res 66:168–174

Vörösmarty CJ, Green P, Salisbury J, Lammers RB (2000) Global water resources. Vulnerability from climate change and population growth. Science 289:284–288

Vukovic G, Marinkovic A, Skapin S, Ristic M, Aleksic R, Peric-Gruji A, Uskokovic P (2011) Removal of lead from water by amino modified multi-walled carbon nanotubes. Chem Eng J 173:855–865

Wan X, Huang Y, Chen Y (2012) Focusing on energy and optoelectronic applications: a journey for graphene and graphene oxide at large scale. Acc Chem Res 45:598–607

WHO (2017) Progress on drinking water. Sanitation and hygiene. Available from: https://washdata.org/sites/default/files/documents/reports/2018-01/JMP-2017-report-final.pdf

Wildgoose G, Banks C, Leventis H, Compton R (2006) Chemically modified carbon nanotubes for use in electroanalysis. Microchim Acta 152:187–214

Yang K, Xing B (2010) Adsorption of organic compounds by carbon nanomaterials in aqueous phase: Polanyi theory and its application. Chem Rev https://doi.org/10.1021/cr100059s

Zhao X, Jia Q, Song N, Zhou W, Li Y (2010) Adsorption of Pb(II) from an aqueous solution by titanium dioxide/carbon nanotube nanocomposites: kinetics, thermodynamics, and isotherms. J Chem Eng Data 55(10):4428–4433

Zhao G, Ren X, Gao X, Tan X, Li J, Chen C, Huang Y, Wang X (2011) Removal of Pb (II) ions from aqueous solutions on few-layered graphene oxide nanosheets. Dalton Trans 40:10945–10952

Zhao J, Wang Z, White JC, Xing B (2014) Graphene in the aquatic environment: adsorption, dispersion, toxicity and transformation. Environ Sci Technol 48:9995–10009

Nanoparticles for Drug Delivery

Abu Tariq, Showkat Ahmad Bhawani and Abdul Moheman

Abstract Since ages human kind is using natural and synthetic compounds for the cure of diseases. By large synthetic compounds have edged the natural compounds. Discovery of nanomaterials paved way for smart treatment of diseases which were considered incurable. Nanomedicine and nano drug delivery systems are developing at a very fast pace offering multiple benefits in the treatment of chronic human ailments such as cancer, HIV and many other diseases by target-oriented and site-specific delivery of medicines. A detailed importance of nanomaterials in drug delivery systems is given in this chapter. This chapter also presented a comprehensive scrutiny of the nanomaterials that are handy in targeted and site specific delivery of drugs, their synthesis and applications in the field of drug delivery.

Keywords Nanomaterials · Nanomedicine · Cancer · Drug delivery

1 Introduction

Nanomaterials are organic or inorganic structures with size ranged 1–100 nm, which is comparable to the size of antibodies and DNA plasmids (Whitesides 2003; Lowe 2000; Wang et al. 2008). These materials are responsible for ground breaking change in the field of nanomedicine which includes drug delivery, biosensors and microfluidics (Arayne et al. 2007; Patra and Baek 2014; Joseph and Venkatraman 2017). Mundane and extensive research has been witnessed in past decades in the field of

A. Tariq
School of Distance Education, Universiti Sains Malaysia,
11800 USM Penang, Malaysia
e-mail: tariqabu@gmail.com

S. A. Bhawani (✉)
Department of Chemistry, Faculty of Resource Science and Technology, UNIMAS,
94300 Kota Samarahan, Sarawak, Malaysia
e-mail: sabhawani@gmail.com

A. Moheman
Department of Chemistry, Gandhi Faiz-e-Aam College (Affiliated to M. J. P. Rohilkhand University), Shahjahanpur 242001, India

A. H. Bhat et al. (eds.), *Nanomaterials for Healthcare, Energy and Environment*,
Advanced Structured Materials 118, https://doi.org/10.1007/978-981-13-9833-9_9

nanotechnology leading to fabrication, characterization and modification of functional properties of nanomaterials for medical and biomedical applications (John et al. 2002; Castelvetro and De Vita 2004; Jang and Shea 2003; Shi et al. 2003; Mu and Feng 2001; Jones et al. 2001; Gupta and Gupta 2005). Nanomedicines are developed by using metallic, organic, inorganic and polymeric nanostructures (materials at the atomic or molecular level) including dendrimers, micelles and liposomes at the nanoscale level paving way for the designing of target-specific delivery of drugs (Bala et al. 2004; Rudramurthy et al. 2016). Polymeric nanostructures could be a nanosphere or nanocapsule, both acts as an excellent drug carrier. Liposome, a colloidal nanoparticle similar to red blood cells are efficient, clinically proven nanosystems for targeted drug delivery. These nanospheres are designed in a way that they can move freely in human body exhibiting unique functional properties such as chemical, structural, magnetic, electrical and biological to name a few.

Nanomaterials are considerably used in many applications and the most promising is the potential of targeted, site-specific drug delivery. Bio-nanomaterials are obtained from nanoparticles designed through modification and functionalization which lead to brilliant technological outcome (Wang et al. 2008; Yokoyama et al. 1990, 1991). The primary goal for nanotechnologist in drug delivery include:

- specificity of drug in delivery and targeting,
- enhancing the retentive efficacy and lowering of toxicity,
- biocompatibility and safety, and
- development of new and safe medicines at a faster rate.

To design and develop new efficient nanomaterials one must possess in depth knowledge on the (i) incorporation of drug and its delivery to the target, (ii) stability of formulation, (iii) shelf life of the resultant product, (iv) biocompatibility and biodistribution, and (v) functionality.

2 Types of Nanoparticles for Drug Delivery

There are a large number of drug delivery systems produced, reproduced and successfully employed in the treatment of various chronic ailments worldwide specially tumors, HIV, hypertension, asthma and diabetes. The nanomaterials in quest includes but are not limited to Liposomes (Allen and Cullis 2013), Nanoparticles (NPs) (Wohlfart et al. 2012) which includes solid-lipid nanoparticles, polymeric nanoparticles [nanocapsules, nanospheres and dendrimers (Kesharwani et al. 2014)], metal nanoparticles, quantum dots, nanotubes (CNTs), nanocrystals, nanowires and nanobots (Fig. 1). A detailed insight is drawn during the length of this chapter shedding light on every aspect of the nanomaterials in pursuit.

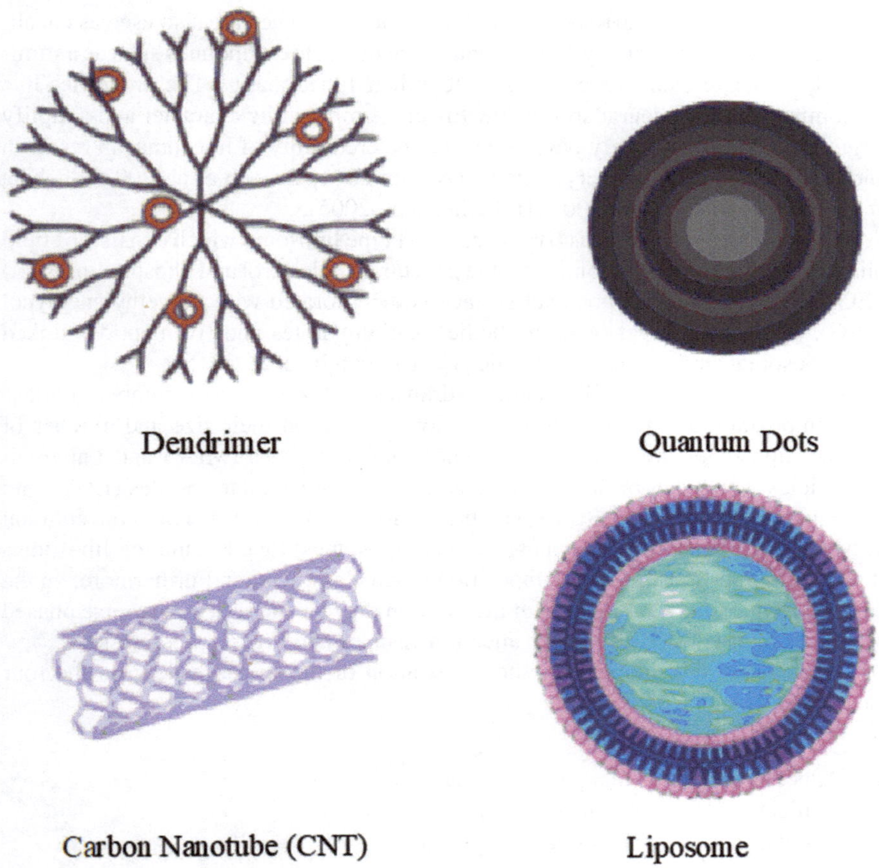

Fig. 1 Nanomaterials; dendrimer, quantum dots, carbon nanotubes and liposomes

2.1 *Liposomes*

Liposomes were discovered in 1960 by Alec Bangham. Liposomes are small concentric bilayered artificial vesicles of spherical shape that can be obtained from cholesterol and natural or synthetic nontoxic phospholipids. The vesicles have aqueous core surrounded by hydrophobic bilayer composed of safe phospholids. The particle size of spherical vesicle ranges from 15 nm to several μm. Liposomes with biocompatibility and amphiphilic character (structure consists of hydrophobic and hydrophilic character) provides a promising systems for delivery of drugs in chronic ailments. The properties of liposomes are size, lipid composition, surface charge and the method of preparation dependent. Furthermore, the composition of number of bilayers and choice of materials determines the rigid or fluidic character of liposomes. Liposomes are extensively used in cosmetic, pharmaceutical, food and farming industries as a carrier for numerous molecules. They are also used as encapsulated liposomal delivery systems to entrap the unstable compounds such as antimicrobial, bioactive elements etc. and protect their functionality. The properties like biocompatibility, biodegradability, low toxicity, amphiphilic character and simplify target-specific drug delivery have enhanced the credibility of liposomes in research and commercial drug delivery systems (Atrooz 2011; Benech et al. 2002; Shehata et al. 2008; Johnston et al. 2007; Hofheinz et al. 2005).

Liposomes are classified as (i) conventional type liposome which consists of lipid bilayer which can make cationic, anionic or neutral cholesterol and phospholipids, (ii) PEGylated type, where liposomal surface is incorporated with polyethyleneglycol (PEG), (iii) ligand-target type; antibodies, carbohydrates and (iv) peptides linked with liposomal surface and theranostic type (Sercombe et al. 2015).

Another classification of liposomes is drawn based on their size, number of bilayers, composition and method of preparations. Based on their size and number of bilayers liposomes are classified as multilamellar vesicles (MLV) and unilamellar vesicles which can be sub categorized as large unilamellar vesicles (LUV) and small unilamellar vesicles (SUV) (Vemuri and Rhodes 1995). Whereas, conventional liposomes, pH sensitive liposomes, cationic liposomes, long circulating liposomes (LCL), immuno-liposome are composition based classification. Furthermore, on the basis of composition and method of preparation they are classified as reverse phased evaporation vesicles (REV), ether injection vesicles and French press vesicles.

General methods adopted for the preparation of liposomes undergoes the four basic steps given below:

1. Drying of lipids from organic solvents
2. Dispersion of lipids in aqueous media
3. Purification of the obtained liposome
4. Characterization of the obtained liposome

The different methods employed for the preparation of the liposomes are;

Mechanical Dispersion Method: Mechanical distribution technique can further be categorized into sonication, membrane extrusion and micro-emulsification to name a few. Following these methods MLV, SUV or LUVs are synthesized (Riaz 1996; Himanshu et al. 2011; Kataria et al. 2011; Mayer et al. 1986; Song et al. 2011; Zhang 2011; Anne and Thomas 2006; Hamilton and Guo 1984; Pick 1981).

Solvent Dispersion Method: Includes method like solvent vaporization (Ether injection) technique, Ethanol injection and reverse phase evaporation method (Deamer and Bangham 1976; Schieren et al. 1978; Batzri and Korn 1973; Szoka and Papahadjopoulos 1978; Handa et al. 2006).

Detergent Removal Method (Riaz 1996; Himanshu et al. 2011): Dialysis, detergent removal of mixed micelles and Gel-permeation chromatography are several methods applied to obtain liposomes.

One disadvantage of liposomes is that the encapsulated drugs are not bioavailable until they are released after enzymatic digestion of walls of liposome vesicle. Drug loading on the resultant liposomes are done via two different pathways, passive where the drug is encapsulated during the preparation of liposome or active where the drug is loaded after the preparation of the liposome (Akbarzadeh et al. 2013).

Liposomes administered either in vivo or in vitro to the target body are different in physicochemical characteristics, thus require a rigorous characterization of the same. This is also due to the different methods followed for the preparation of the liposomes. Characterization after formulation and loading of drug on liposome is a must which could be obtained following below mentioned parameters.

Size and size distribution, percent drug encapsulation, surface charge, vesicle shape and lamellarity and phospholipids identification and assay in liposomal formulations.

2.1.1 Applications of Liposomes

Medicinal and pharmacological applications of liposomes containing various drugs or markers can be categorized as diagnostic and therapeutic. They can be used as a treatment tool, research model, and reagents in the studies of cell interactions, process of recognition and mode of action of certain substances; (Kumari et al. 2012). Liposomal drug delivery systems are considered effective in passive or physiological targeting of tumor tissues providing steady formulation and better pharmacokinetics. It is evident from the results of pre-clinical or clinical research that drugs such as antitumor encapsulated in liposomes showed improved therapeutic activities, retentive enhanced efficacies and reduced toxicities as compared to the free drug. The enhanced pharmacokinetics improves drugs bioavailability to specific target in the infected body. Liposomes are considered as one option out of many for delivery of drug at the target site in the treatment of cancer. Numerous liposomal formulations loaded with anticancer drug had shown low toxic effect as compared to the free drug (Wang et al. 2018; Choi et al. 2012; Gombotz and Wee 1998; Lee and Mooney 2012).

2.2 Polymeric Nanoparticles

Polymeric nanoparticles (PNPs), a submicronic colloidal carriers are more promising and efficient as drug carrier when compared to the liposomes (Akamatsu et al. 2000). The polymeric nanoparticles are resultant product of biodegradable and biocompatible materials and various polymers (natural or synthetic). There are variety of natural polymers such as albumin, gealtin, alginate, collagen and chitosan are used whereas, polylactide-polyglycolide copolymers, polyacrylates and polycaprolactones are several synthetic polymers which are used in synthesis of nanoparticles (Moghimi et al. 2001; Gelperina et al. 2005). The polymeric nanoparticles have attractive physicochemical properties like size, surface potential and amphiphilic character which make them potential drug carriers for bioactive ingredients such as antitumor agents, vaccines etc. Polymers plays a significant role as a host material in metal (Akamatsu et al. 2000; Zeng et al. 2002; Hussain et al. 2003), and semiconductors (Kumar et al. 2001; Sajinovic et al. 2000; Kumar et al. 2000; Yu et al. 2001) nanoparticles. These nanoparticles are designed and synthesized as drug carrier with an aim of delivering the active molecules to the specific target (Yang et al. 2000). The loading and encapsulation of drug material is achieved during the polymerization process. PNPs are generally classified as vesicular systems (nanocapsules) and matrix systems (nanospheres), dendrimers are a unique class of biopolymers.

Nanocapsules: These are the polymeric systems in which drug is encapsulated in a cavity of polymeric membrane.

Nanospheres are spheric vascular matrix systems in which the drug is uniformly distributed throughout the matrix.

Dendrimers are a unique class of polymers, are highly bifurcated branched macromolecules, monodispersed, globular and have well defined 3-D structure. The shape and surface functionalization is well controlled which give rise to properties such as size mono disparity and stability that make these structures attractive and excellent drug carrier systems (Kesharwani et al. 2015; Zhu and Shi 2013; Madaan et al. 2014). Dendrimers can be synthesized from monomers via step growth polymerization, using either of the two approaches, convergent, formation starts from the outside of the dendrimer or divergent, where formation starts inside out or from core extending outwards (Cheng et al. 2008). Drug loading process in dendrimer is achieved via simple encapsulation, or complexation (electrostatic interaction and covalent conjugation) (Tripathy and Das 2013). Dendrimers deliver drugs to specific sites via in vivo degradation of covalent bonds between drug and dendrimers in the presence of certain enzymes or favourable conditions or by release of drug due to physical changes such as change in pH, temperatures etc. (Tripathy and Das 2013). Based on the functionalization moieties dendrimers can be classified as: polyamidoamine (PAMAM), PPI, liquid crystalline, core-shell, chiral, peptide, glycodendrimers and PAMAMOS. The toxicity of dendrimers arises due to the presence of cationic moiety amine, which limits its clinical applications, hence are usually modified to reduce it. Dendrimers are thus applied as transdermal, oral, ocular, pulmonary and also as targeted drug delivery systems (Kesharwani et al. 2014).

There are several other biopolymers such as chitosan, alginate, cellulose and xanthan gum which are efficient drug delivery agents and can be used in many therapeutic applications. Only a few are discussed here;

Chitosan is biopolymer obtained from partial N-deacetylation of chitin which is biodegradable, biocompatible and nontoxic in nature. These unique characteristics make chitosan an eligible drug carrier. Synthesis of chitosan nanoparticles can be achieved through several processes such as ionic gelation, nanoprecipitation, desolvation, emulsion cross linking etc. These chitosan nanoparticles are used as continued drug released systems for buccal, intestinal, nasal, eye and pulmonary epithelia (Portero et al. 2002; Artursson et al. 1994; Fernández-Urrusuno et al. 1999; De Campos et al. 2001; Al-Qadi et al. 2012). These nanoparticles also possess mucoadhesive and antibacterial properties making them suitable for administration of ocular drugs.

Alginate is a natural anionic polysaccharide obtained from marine brown algae. It is a low toxic, biocompatible, biodegradable and mucoadhesive polymer (Draget and Tylor 2011; Downs et al. 1992). Alginate nanoparticles are synthesized by ionic gelation technique using divalent cations such as calcium. These nanoparticles are used in nasal and oral administration due to mucoadhesive properties (Borges et al. 2008; Qurrat ul et al. 2003), also they are used as hydrophilic carriers. As compared to the traditional vaccines these hydrophilic carriers show enhanced immunogenicity due to prolonged antigen release.

Xanthan gum is a naturally occurring high molecular weight polyionic heteropolysaccharide obtained from Xanthomans campestris. It is vastly used in pharmaceutical industries as excipient largely due properties such as non-toxic, non-irritating and bioadhesive (Laffleur and Michalek 2017).

Cellulose is a natural polymer used as carrier in biomedicine. Cellulose possess some unique properties such as high hydrophilicity, excellent biodegradability and biocompatibility, good thermal and chemical stability making it excellent carrier of required drug material. Nano-structured cellulose or nanocellulose are consisted of cellulose nanocrystals (CNC), cellulose nanofibrils (CNF) and bacterial cellulose (BC). Owing to excellent above mentioned properties they are used in tissue repair, regeneration and healing, sensors, smart membrane, antimicrobial nanomaterials etc. (Qiu and Hu 2013; Lin and Dufresne 2014; Jorfi and Foster 2015; Shanmuganathan et al. 2010).

Polymeric micelle is another class of polymeric nanoparticles used in targeted delivery of drugs. These nanostructures are block copolymers under 100 nm size with amphiphilic properties, in aqueous solution it forms a self-assembled core shell structure. The hydrophobic part of the core contains loaded drug hydrophobic in nature such as paclitaxel, docetaxel and camptothecin, whereas hydrophilic part of shell makes whole system soluble stabilizing the core. The polymeric shell restrain itself from nonspecific interaction with the surrounding biological components. Synthesis of polymeric micelles can be achieved via convenient solvent based direct dissolution of polymer followed by dialysis or by precipitation of one block by adding a solvent (Xu et al. 2013; Kulthe et al. 2012). The assembly creation starts when the amphiphilic molecules reaches the Critical Micellar Concentration or CMC (Kulthe et al. 2012). Solvent system, hydrophobic chain size of amphiphilic molecules, amphiphilic concentration and temperature are certain factors which govern the formation of micelle (Devarajan and Jain 2016).

2.2.1 Applications of Polymeric Nanoparticles

In comparison to liposomes, polymeric nanoparticles are far more impressive and efficient as drug carrier. This all is due to the strong characteristic properties such as low toxicity, biodegradability, biocompatibility, amphiphilic character, mucoadhesiveness and monodisparity. The other aspect of its extensive use is its stability, specificity and cheap fabrication by easy to go synthetic approaches. Polyactides and poly(DL-lactide-co-glycolide) polymers are commonly investigated for their use in drug delivery, upon implantation in the body, it undergoes hydrolysis resulting in the formation of biocompatible fragments (Jain 2000). MCF-7, breast cancer cells easily absorb Doxorubicin encapsulated in these nanoparticles and release it into the cytoplasm (Jie et al. 2013). In another study, folic acid conjugated chitosan-polylactide copolymers were synthesized and used as active targeting drug carrier for paclitaxel, furthermore, it was confirmed with folic acid receptor-expressed MCF-7 breast cancer cells (Shengtang et al. 2013). Dendrimers on the other hand, are extensively studied for both drug delivery and gene delivery, as penicillin carrier and in anticancer treatments (Venkatesan et al. 2009, 2011; Sundar et al. 2010; Hussain et al. 1997; Arunkumar 2016a, b). Jain et al. (2014) studied doxorubicin-folate conjugated poly-1-lysine dendrimers and described the formulation as very much capable of preventing cancer when compared with the free doxorubicin. The release of drug in dendrimer formulation here in study was pH dependent. Another pH sensitive drug released system folate-conjugated polypropylene imine dendrimers (FA-PPI) was developed by Kaur et al. (2017). FA-PPI was used as nanocarrier of methotrexate (MTX) for specific targeting of cancer cells and its treatment. Further in vitro studies on MCF-7 cell lines show sustained release, increase cell uptake and low cytotoxicity (Kaur et al. 2017). Furthermore, the new formulations, methotrexate (MTX)-loaded and folic acid-conjugated 5.0G PPI (MTX-FAPPI) was compared with free drug methotrexate (MTX) and found that the formulations were selectively taken up by the tumor cells. Silva with his co-workers synthesized and studied the efficacy of a 0.75% w/w isotonic solution of hydroxylpropyl methylcellulose (HPMC) containing chitosan/sodium tripolyphosphate/hyaluronic acid nanoparticles. This nanoparticle formulation was used to deliver ceftazidime antibiotic to the eye (Silva et al. 2017). Furthermore, the cytotoxicity of nanoparticles was studied for two cell lines ARPE-19 and HEK239T and the end result was negative in both cases. The nanoparticles also showed positive response in preserving the antibacterial activity, thus making the formulation formidable with improved mucoadhesive properties for administration of ocular drugs. Carboxymethyl chitosan nanoparticles were prepared for the release of intra-nasal carbamazepine (CBZ) to bypass the blood-brain barrier (BBB) membrane. The treatment efficacy was refined, and increase in medication in brain was obtained thereby reducing the systemic drug exposure (Liu et al. 2018). An insulin containing alginate nanoparticles with nicotinamide was developed by Patil and Devarajan (2016). The obtained formulation was applied on diabetic rats as a permeation agent in order to lower the levels of glucose serum and raise insulin levels. Also, Haque et al. (2014) prepared venlafaxine (VLF) encapsulated alginate nanoparticles. The drug VLF was released via intranasal for treatment of depression, thus proving the vitality of the nanoparticles for the treatment of depression. Another example is of alginate microcapsules prepared by Roman and co-worker containing epidermal growth factor bound on its outer region to target the non-small cell lung cancer cells (Román et al. 2016). Cisplatin is another carcinogenic drug which is loaded in the nanoparticles for the treatment of cancer. Elseoud et al. (2018) studied the utilization of cellulose nanocrystals and chitosan nanoparticles for the oral releasing of replinide. For the treatment of sialorrhea, Laffleur and Michalek (2017) synthesized a xanthan gum based carrier with l-cysteine to release tannin in buccal mucosa. Increased adhesion on the buccal mucosa can be achieved by the thiolation of xanthan gum. Polymeric micelles are considered useful for both against cancer and ocular drug delivery (Kulthe et al. 2012; Mandal et al. 2017). The polymeric micelle is used for reaching the posterior ocular tissues. Drug dasatinib encapsulated nanoparticle obtained from micellation of PEG-b-PC was reported for the treatment of proliferative vitreoretinopathy (PVR) (Li et al. 2016). The size of resultant nanoparticles was found to be 55 nm with a narrow distribution they turned out to be noncytotoxic to ARPE-19 cells. In comparison to the free drug this micellar formulation balefully oppressed the attachment, cell proliferation and relocation (Li et al. 2016).

2.3 *Solid Lipid Nanoparticles*

In 1990, an era of colloidal drug carrier systems such as emulsions, liposomes and polymeric nanoparticles, for controlled drug delivery emerged Solid lipid nanoparticles (SLN). They were developed due to their unique properties such as small size, large surface area, high drug loading and the interaction of phases at the interface which are considered as an advantage over the properties of liposomes and polymeric nanoparticles. They are considered as nontoxic, biodegradable, biocompatible, stable against coalescence, physically stable and excellent carrier of lipophilic drugs (Cavalli et al. 2002). SLN are colloidal nanoparticle carriers in the range of 50–1000 nm, mainly composed of solid lipid matrix dispersed in water or aqueous surfactant solution (Mozafari 2005; Houli et al. 2009; Melike and Gulgun 2007). To avoid aggregation and achieve stabilized dispersion, different kind of surfactants were applied under GRAS (Generally Recognized as Safe) status. They are composed of solid hydrophobic core with dissolved or dispersed drug generally coated with mono layered phospholipids. The solid fat matrix of core contains embedded hydrophobic chains of phospholipids with potential to carry both lipophilic and hydrophilic drugs (Shah et al. 2011). The developed SLNs were studied in the area of parental, pulmonal and dermal application routes. For gene transfer, new formulations were derived using same cationic lipids as for liposomal transfection.

Several methods are applied to prepare SLNs from lipid, emulsifier and water/solvent. Following are the methods;

High pressure homogenization (Hot homogenization and Cold homogenization)
Ultrasonic/high speed homogenization
Solvent evaporation method
Solvent emulsification-diffusion method
Super critical fluid method
Microemulsion based method
Spray drying method
Double emulsion method
Precipitation technique
Film-ultrasound dispersion

Drug incorporation was achieved through the different model as mentioned here in; Homogenious matrix model (Drug molecularly dispersed in the lipid matrix), Drug enrich core—core shell model and Drug enrich shell—core shell model (Rabinarayan and Padilama 2010). In homogeneous matrix model, also termed as solid solution model molecularly dispersed drug or amorphous clusters is mainly obtained when highly lipophilic drugs are incorporated into the SLN. To obtain the resultant SLN, hot homogenization or cold homogenization technique is applied, could be achieved via avoiding potentially drug solubilizing surfactants. During the production of SLN, the drug enriched shell with core shell model is obtained, where the drug partitioned to water phase. Precipitation of lipid is observed upon cooling giving rise to drug free lipid core due to phase separation. At the same point of time, the repartitioning of the drug into the lipid-lipid phase in the outer shell is observed with gradual increase in the formation of the same giving rise to drug enriched crystalized shell. Tetracaine SLN is a good example of hot homogenization model with drug enriched shell. To obtain a drug enriched core, drug for example prednisolone is dissolved in melted lipid. The nanoemulsion obtained left for cooling lead to supersaturation of drug in the melted lipid and precipitation of the drug afterwards. Further cooling of the microemulsion will lead to precipitation of lipid forming a lipid membrane shell around the drug enriched core.

2.3.1 Applications of SLNs

A large number of applications is widely known for the SLNs to establish it as a potential carrier of drug in the field of pharmaceuticals and cosmoceuticals. SLNs are effective drug and gene carrier for treatment of various ailments such as cancer, vaccination, parasitic diseases etc. Tamoxifen-SLN is a formulation used in breast cancer treatment, the resultant formulation prolongs the release of the drug Tamoxifin after IV administration. Similarly, methotrexate and camptothecin loaded SLNs are used in different tumor targets (Müller et al. 2000). Another class of drug Mitoxantrone is encapsulated in SLN for the formulation of injections with reduced toxicity and improved safety and bioavailability of drug (Bin et al. 2006). The penetration of drug into brain is one of the most difficult task in treatment of diseases. To improve the drug penetrations through blood-brain barriers (BBB), SLNs are promising contenders, they improve drug targeting system for treatment of central nervous system disorders. In case of limited access of the drug 5-fluoro-2′-deoxyuridine (FUdR) into the brain, 3′,5′-dioctanoyl-5-fluoro-2′-dexoyuridine (DOFUdR) incorporated SLN (DOFUdR-SLN) was synthesized to overcome the BBB (Antonio and Eliana 2007). SLN systems with encapsulated drugs rifampicin, isonizide, pyrazinamide was used for the treatment of tuberculosis and improves the patient compliance by decreasing dosing frequency (Pandey et al. 2005).

2.4 Inorganic Nanoparticles

Various class of inorganic nanoparticles offer important multifunctional platforms for biomedical applications. These includes silica nanoparticles (Tan et al. 2004), metal nanoparticles (Daniel and Astruc 2004), quantum dots (Stroh et al. 2005; Michalet et al. 2005) and lanthanide nanoparticles (Nichkova et al. 2005; Chen et al. 2007a, b). These nanoparticles showed several advantages when it comes to surface functionalization, good biocompatibility and versatility are two such example advantages over liposomes etc. Silver and gold nanoaparticles show surface plasmon resonance (SPR) which is significantly absent in liposomes, dendrimers and micelles. When it comes to antimicrobial activity, silver nanoparticles exhibit great tendency, however very few research had been conducted for drug delivery and release.

2.4.1 Gold Nanoparticles

Gold on other hand is a versatile element which is used in medicine for centuries covering both antimicrobial and anticancer treatment due to the presence of unique properties. Gold nanoparticles (Au-NPs) and nanorods are excellent carrier of drugs in various treatments of cancer and are considered for different potential bio-molecular potential applications (Kim et al. 2006). Gold nanoparticles edged over other metallic nanoparticles in term of biocompatibility and non-cytotoxicity, approved by FDA, thus considered as favourable for drug delivery (Dhar et al. 2008). The preparation of Au-NPs is easy and the size of the NPs can be easily controlled. The Au-NPs are selective and specific in recognizing the tumor cells. To enhance its properties, Au-NPs can be conjugated with amino acids and proteins (Selvakannan et al. 2004; Niemeyer 2003). The functionalization of Au-NPs can be easily achieved which make Au-NPs a promising carrier for drug delivery as biomarkers, specifically for drug resistant cancer cell (Hwu et al. 2009). There are several reported methods for the preparation of Au-Nps, one such is reported by Dhar et al. (2008). Au-Nps are prepared by using natural gellan gum and delivery of doxorubicin hydrochloride was successfully achieved, they have also suggested a successful loading method of doxorubicin onto Au-NPs. A well-defined method of preparation of Drug-Au-NPs system by Gibson et al. (2007) was reported for accurate measurement of biological activities. Au-NPs can also be synthesized by reduction of $AuCl_4$ salts by $NaBH_4$ in the presence of thiols which subsequently forms monolayer around the core gold atom (Huang et al. 2007). Au-NPs have low toxicity, and have tendency to protect drugs from potential transformation in tissues and organs which allows Au-NPs to be actively used in various treatments. Applications of Au-NPs includes delivery of insulin via nasal route (Joshi et al. 2006), antimicrobial activity against E. coli strains (Gu et al. 2003), and Au-NP encapsulated by ciprofloxacin for improved drug delivery (Tom et al. 2004). Three paclitaxel-conjugated NPs was synthesised using Fe_3O_4 and gold as core, making a new class of drugs for treatment of cancer (Hwu et al. 2009). Gold nanorods are another carrier of drugs whose cytotoxicity could be attributed to the presence of the stabilizer CTAB. Similarly, gold nanoshells are prepared from gols solutions consisting gold and copper or gold and silver to function as contrast agents in Magnetic Resonance Imaging (MRI) (Su et al. 2006). Furthermore, gold-silica nanoshell is used for photothermal ablation of tumor cells (Bernardi et al. 2008). Au-NPs are also used as nanobioconjugates, for example, it can be formulated with polyethylene glycol (PEG) or antibodies to form nanobioconjugates. These bionanoconjugates penetrates the bloodstream and remain there to get penetrated in tumor cells (Verma et al. 2008). Literature also suggest that the multifunctional Au-NPs are potentially used for both imaging and treatment of tumor (Llevot and Astruc 2012). Synthesis of nanocages of gold was achieved by Xie et al. (2007) and showed its potential for the treatment of tumor by targeted inhibition of tumor cells.

2.4.2 Silica Nanoparticles

Another inorganic material which is commonly investigated for drug delivery, obtained from silicon-based materials are porous silicon and silica or silicon dioxide. Several architectures are obtained which include calcifies nanopores, platinum containing nanopores, porous nanoparticles and nanoneedles (Prinz et al. 2003; Desai 1996). The nanopores are designed in a way to achieve constant drug deliver rate through the pores via controlled density and diameter. Porous hollow silica nanoparticles (PHSNP) are another class of silica based NPs which are fabricated in a suspension containing calcium carbonate sacrificial nanoscale templates (Chen et al. 2004). Furthermore, sodium silicate is added to the suspension as precursors, the resultant suspension is dried and calcinated to obtain a core of the template material coated with porous silica shell. On dissolving the material into etch bath porous silica shell is obtained. These synthesised PHSNPs are mixed with drug molecules subsequently dried to obtain silica nanoparticles encapsulated with drug molecules (Chen et al. 2004). There are several other examples of silica NPs being tested for drug delivery. Li et al. (2000) reported the use of silicon-based delivery system which includes platinum embedded porous silicon for treatment of tumor (Li et al. 2004). Calcified porous silicon is considered as an artificial growth factor agent (Weis et al. 2002), silicon nanopores as antibody carrier (Tao and Desai 2003) and porous silica nanoparticles loaded with antibiotics (Chen et al. 2004), enzymes (Jain et al. 1998) and DNA (Sameti et al. 2003).

2.4.3 Quantum Dots

Quantum dots (QDs) are another class of drug carrier nanomaterials derived from semiconductors. The semiconductor nanocrystals and coreshell nanocrystals contains interface between different semiconductors. The QD particle diameter ranges between 2 and 10 nm, which increases to 5–20 nm after the drug encapsulation process. QD nanocrystlas have unique and fascinating optical properties, such as photoluminescence and absorbance, which are size dependent (Prasad 2004). These properties make QDs as an indispensable tool in biomedical field, specifically for multiplexed, quantitative and long term fluorescence imaging and detection (Michalet et al. 2005). The QDs falls in the emission range of near-infrared region (<650 nm), unlike conventional organic dyes. This emission range is considered as very desirable characteristic in the field of biomedical imaging, due to the low absorption by the tissues and light scattering (Prasad 2004; Volkov 2015). QDs are also considered a promising agents in targeted drug delivery, sensors and bioimaging, also as contrast agents for in vivo imaging (Volkov 2015; Shi et al. 2015; Han et al. 2015; So et al. 2006). A multifunctional biocompatible graphene oxide QD covered with luminescent magnetic nanoplatform was synthesized by Shi and co-worker for diagnostic of glypican-3-expressing Hep G2, a specific liver cancer tumor cells (Shi et al. 2015). In sustained and/or controlled release of therapeutic molecules, QDs improve its efficacy via external stimulation by light, heat, radio frequency or magnetic fields for controlled release (Xu et al. 2016; Huang et al. 2015). A theranostic system was developed by Olerile et al. (2017) based on co-loaded of QDs and anticancer drug in nano-structured lipid carrier. A pH responsive QD based on ZnO QD decorated with PEG and hyaluronic acid was synthesized by Cai et al. (2016) for targeting specific cells with HA-receptor CD44. This QD nanocarrier was combined with doxorubicin (DOX) and studied for its sustained release efficiency. The author reported enhanced antitumor activity for this combination of DOX and ZnO QDs. Polymer and co-polymer materials are used to conjugate with QDs to obtain bioconjugated QDs of variable sizes. Some contains an amphiphilic triblock copolymer layer for in vivo protection and multiple PED molecules. The resultant materials are highly stable with improved biocompatibility and circulation and able to produce bright signals. Viral diagnosis is another classic application of QDs. Conjugated QD with different biological molecules such as antibodies is another promising combination for detecting and quantifying a wide variety of cancer-associated molecules (Rizvi and Saleh 2018). The toxicity of QDs could be reduced after modifying the surface of the QD with N-acetylcysteine (Choi et al. 2007). Lovric et al. (2005) showed the cytotoxicity of QD by the induction of reactive oxygen which damages the plasma membranes, mitochondria and nucleus.

2.4.4 Nanotubes—CNTs

Carbon have many well-known allotropes, among them is carbon nanotubes (CNTs), a cylindrical nanostructure, also known as buckytubes. The length-to-diameter ratio of the constructed nanotubes is found to be 132,000,000:1, significantly large when compared to other materials (Velavan et al. 2015). The cylindrical carbon tubes consist of several unique properties such as electrical, thermal and stability, making them promising agents in the field of nanotechnology, electronics, optics and other material sciences. CNT belongs to the family of fullerenes which can be obtained by polymerization of carbon atoms under specific conditions (Terrones and Terrones 2003). The length of nanotubes as reported till 2010 is 18 cm and a diameter of few nanometers approximately 1/50,000th width of human hair. The CNTs can be classified as single-walled nanotubes (SWNTs) and multi-walled nanotubes (MWNTs) with 0.7–1.5 nm and 2–50 nm diameter, respectively. The nanotubes are composed entirely of sp^2 c–c bonds and orbital hybridization, which provide nanotubes with unique strength. Moreover, these nanotubes are held together by Van der Waals forces aligning them to form a "rope". These tubes have potential to act as biopersistent fibres. CNTs are promising agents when comes to theranostics, they exhibit properties which is highly recommended in the same. It is due to the fact that NIR absorption possessed by CNTs is effectively used in both high-resolution imaging and photothermal therapy (Lee et al. 2012). Direct therapeutic effect of CNTs based on strong optical absorbance, NIR absorption effect was reported, herein the CNTs induces the destruction of tumor tissues by PTT (Heath and Davis 2008). The biodegradability of CNT is a matter of study, which could be achieved by application of appropriate chemical modification (Bianco et al. 2011). Large surface area of CNTs allows the functionalization of outer surface for specific cancer receptors and contrast agents as well (Dinesh et al. 2016). CNTs in addition to high surface area possess a very hydrophobic hollow interior which allow loading of water soluble drugs or other bioactive molecules and their safe delivery to cellular environment (Li et al. 2012). In some cases, these tubes acts as "cell membrane needles" (Fabbro et al. 2012; Al-Jamal et al. 2012). The unique properties of CNTs offer valuable support in the treatment of various diseases as smart nanomedicines, however, their application in the field have to go miles (Kostarelos et al. 2009). The phagocytosis of dendritic cells can be achieved with CNTs at the tumor cites (Shen et al. 2006). An extensive research had been made to show the credibility of CNTs in the treatment of the malignant tumors. Also the CNTs can be considered an excellent class of tumor PTT system and therefore are highly promising for clinical trials.

2.4.5 Fullerenes

Fullerene, the third carbon allotrope, like graphite is a molecule composed of entirely carbon, which is stacked grapheme sheets of linked hexagonal rings, additionally comprises of pentagonal or sometimes heptagonal tings to give promising porous molecules (Holister et al. 2003). Fullerenes are present in the form of hollow sphere, ellipsoid, or tube. The spherical form of fullerenes is called buckballs whereas the cylindrical ones are known as carbon nanotubes or buckytubes. Buckyball clusters are comprised of less than 300 carbon atoms and are commonly termed as endohedral fullerenes which includes famous buckminsterfullerene, C60. Fullerenes are being extensively studied as potential and promising agent for treatment of several ailments. It has been explored as antimicrobial agents owing to its potency for induction of reactive oxygen species after photoexcitation (Yamakoshi et al. 2003). Fullerenes and their functionalized derivatives plays versatile role in the field of biomedical applications such as HIV, liver toxicity, neurological diseases etc.

Unique chemical and physical properties owing to three dimensional structure of fullerenes lead to great number of applications in the field of medical sciences. Also chemical functionalization plays a very important role in the alteration of properties of new compounds resulting in some effective results in treatment. The fullerenes and its derivative compositions could be handy in the treatment of HIV, cancer, imaging and gene therapy. A large number of fullerene derivatives were synthesized and tested for HIV-1 protease showing potent and selective activity (Prato 1999; Bosi et al. 2003; Pantarotto et al. 2004). Conjugated perfluorinated fullerenes was tested for fluorine-19 imaging (Wilson et al. 1999), also other studies were made for molecular imaging, including detection, identification and/or sequencing of biomolecules like nucleic acids or proteins using scanning probe microscopy. Metallofullerenes such as $Gd@C_{82}(OH)_{22}$ was studied by Yin et al (2010), it can circumvent tumor resistance to cisplatin by reacting endocytosis. A large number studies were conducted to establish fullerene and its derivatives as an agent of cancer therapy. For example, pyrrolidinium fullerene derivative induces mitochondrion and caspase dependent apoptosis through generation of ROS in treatment of promyrloleukaemia (HL-60) cells in human beings, a chitosan/fullerene nanogel was reported effective for tumor therapy (Castro et al. 2017). Fullerenes were also studied for nanoparticle based anticoagulants, for the acceleration of blood plasma clot lysis and gene therapy via DNA vectorization (Moussa 2018).

3 Conclusion and Future Perspective

The present chapter discusses the progressive advancement and applications of nanomaterials in the field of nanomedicines, especially in drug delivery. Nanoparticles are considered as one of the best agent for targeted drug delivery. They are easily available via simple route of synthesis, are easily functionalized, and have the unique characteristic of their size, stability, low toxicity and biocompatibility. With the promising applications of nanomaterials in the future, may be used extensively for the diagnosis and treatment of cancer. There is a range of nanomaterials from liposomes, polymer nanomaterials to fullerenes that are discussed for their applicability to diagnose, specific delivery to targets and sense or activate materials in live systems. The enhanced efficacy of the naturally occurring bioactive compounds is due to the advancement in the field of nanotechnology, which is mainly achieved through these nanocarriers formulated with liposomes, gold, silver, solid lipid nanoparticles, micelles or dendrimers. Several easy and hands-on techniques are applied to formulate these nanomaterials. For example, polymeric nanoparticles can be synthesized by using solvent evaporation, emulsion polymerization or surfactant free polymerization. Integration of therapy and diagnosis or theranostic is considered as one of the promising development in the field of nanomedicines by putting cancer as a model disease. It is evident for the ongoing researches and advancement that application of nanotechnology in drug delivery and medicine have created a lot of opportunities for safe and customized treatment of cancer, HIV etc.

However, with the advancement in nanomaterials and its applications in the treatment of many disease, comes with challenges. One of the problem is chronic and acute toxicity of the nanomaterials which cannot be ignored and is of more concern. For example, NPs and CNTs may damage the cardiovascular and respiratory systems (Proffitt 2004); Lam et al. 2004), the CNTs are reported as lethal for the lung tissues (Renwick et al. 2004) and are toxic to immune system in mice. Also the biocompatibility and safety of CNTs is a matter of concern. Due to presence of heavy metals the toxicity of QDs can also not be denied. Therefore, the safety/toxicity issue of the nanomaterials for living systems should intensively have studied in the future. However, so far, a conceptual understanding of nanomaterials and the biological responses is needed for the development and application of safe nanomaterials in drug delivery in the future. Furthermore, a deep understanding between the two domain, drug delivery and particle toxicology is necessary for the development and advancement of new and novel ideas in the field of nanomedicines.

References

Akamatsu K, Takei S, Mizuhata M et al (2000) Preparation and characterization of polymer thin films containing silver and silver sulfide nanoparticles. Thin Solid Films 359:55–60

Akbarzadeh A, Rezaei-Sadabady R, Davaran S et al (2013) Liposome: classification, preparation, and applications. Nanoscale Res Lett 8:102

Al-Jamal KT, Nunes A, Methven L et al (2012) Degree of Chemical Functionalization of Carbon Nanotubes Determines Tissue Distribution and Excretion Profile. Angew Chem Int Ed 51:6389–6393

Allen TM, Culli PR (2013) Liposomal drug delivery systems: from concept to clinical applications. Adv Drug Delivery Rev 65:36–48

Al-Qadi S, Grenha A, Carrión-Recio D et al (2012) Microencapsulated chitosan nanoparticles for pulmonary protein delivery: in vivo evaluation of insulin-loaded formulations. J Control Release 157:383–390

Anne S, Thomas R (2006). In Mozafari MR (eds) Nanocarrier technologies: frontiers of nanotherapy, pp 41–50

Antonio AJ, Eliana S (2007) Solid lipid nanoparticles as a drug delivery system for peptides and proteins. Adv Drug Deliv Rev 59:478–490

Arayne MS, Sultana N, Qureshi F (2007) Nanoparticles in delivery of cardiovascular drugs. Pak J Pharm Sci 20:340–348

Artursson P, Lindmark T, Davis SS et al (1994) Effect of chitosan on the permeability of monolayers of intestinal epithelial cells (Caco-2). Pharm Res 11:1358–1361

Arunkumar A et al (2016a) Formulation, evaluation and optimization of sustained release Bilayer tablets of Niacin and Green Tea extract by employing Box-Behnken design. J Sci Res Phar 5(2):23–28

Arunkumar A et al (2016b) Development and Validation of New Analytical methods for Simultaneous estimation of Epigallocatechin gallate, a component of Green Tea extract and Niacin in a Pharmaceutical dosage form. J Pharm Res 5(2):21–24

Atrooz OM (2011) Effects of alkylresorcinolic lipids obtained from acetonic extract of Jordanian wheat grains on liposome properties. Int J Biol Chem 5(5):314–321

Bala I, Hariharan S, Kumar MN (2004) PLGA nanoparticles in drug delivery: the state of the art. Crit Rev Ther Drug Carrier Syst 21:387–422

Batzri S, Korn ED (1973) Single bilayer liposomes prepared without sonication. Biochim Biophy Acta 298(4):1015–1019

Benech RO, Kheadr EE, Laridi R et al (2002) Inhibition of Listeria innocua in cheddar cheese by addition of nisin Z in liposomes or by in situ production in mixed culture. Appl Environ Microbiol 68:3683–3690

Bernardi RJ, Lowery AR, Thompson PA et al (2008) Immunoshells for targeted photothermal ablation in medulloblastoma and glioma: an in vitro evaluation using human cell lines. J Neurooncol 86(2):165–172

Bianco A, Kostarelos K, Prato M (2011) Making carbon nanotubes biocompatible and biodegradable. Chem Commun 47:10182–10188

Bin L, Su-Bin X, Hong Y et al (2006) Solid lipid nanoparticles of mitoxantrone for local injection against breast cancer and its lymph node metastases. Eur J Pharm Sci 28(1–2):86–95

Borges O, Cordeiro-da-Silva A, Tavares J et al (2008) Immune response by nasal delivery of hepatitis B surface antigen and codelivery of a CpG ODN in alginate coated chitosan nanoparticles. Eur J Pharm Biopharm 69:405–416

Bosi S, Da Ros T, Spalluto G et al (2003) Fullerene derivatives: an attractive tool for biological applications. Eur J Med Chem 38(11):913–923

Cai X, Luo Y, Zhang W et al (2016) pH-Sensitive ZnO quantum dots–doxorubicin nanoparticles for lung cancer targeted drug delivery. ACS Appl Mater Interfaces 8:22442–22450

Castelvetro V, De Vita C (2004) Nanostructured hybrid materials from aqueous polymer dispersions. Adv Colloid Interface Sci 108–109:167–185

Castro E, Garcia AH, Zavala G et al (2017) Fullerenes in biology and medicines. J Mat Chem B 5:6523–6535

Cavalli R, Gasco MR, Chetoni P et al (2002) Solid lipid nanoparticles (SLN) as ocular delivery system for tobramycin. Int J Pharm 238:241–245

Chen JF, Ding HM, Wang JX et al (2004) Preparation and characterization of porous hollow silica nanoparticles for drug delivery application. Biomaterials 25:723–727

Chen JY, Wang DL, Xi JF et al (2007a) Immuno gold nanocages with tailored optical properties for targeted photothermal destruction of cancer cells. Nano Lett 7(5):1318–1322

Chen Y, Chi Y, Wen H et al (2007b) Sensitized luminescent terbium nanoparticles: preparation and time-resolved fluorescence assay for DNA. Anal Chem 79:960–965

Cheng Y, Xu Z, Ma M et al (2008) Dendrimers as drug carriers: applications in different routes of drug administration. J Pharm Sci 97:123–143

Choi AO, Cho SJ, Desbarats J et al (2007) Quantum dot-induced cell death involves Fas upregulation and lipid peroxidation in human neuroblastoma cells. J Nanobiotechnol 12:1

Choi KY, Jeon EJ, Yoon HY et al (2012) Theranostic nanoparticles based on PEGylated hyaluronic acid for the diagnosis, therapy and monitoring of colon cancer. Biomaterials 33:6186–6193

Daniel MC, Astruc D (2004) Gold nanoparticles: assembly, supramolecular chemistry, quantum-size-related properties, and applications toward biology, catalysis, and nanotechnology. Chem Rev 104:293–346

De Campos AM, Sánchez A, Alonso MJ (2001) Chitosan nanoparticles: a new vehicle for the improvement of the delivery of drugs to the ocular surface. Application to cyclosporin A. Int J Pharm 224:159–168

Deamer D, Bangham AD (1976) Large volume liposomes by an ether vaporization method. Biochim Biophys Acta 443(3):629–634

Desai MP (1996) Gastrointestinal uptake of biodegradable microparticles: effect of particle size. Pharm Res 13:1838–1845

Devarajan PV, Jain S (2016) Targeted drug delivery: concepts and design. Springer, Berlin

Dhar S, Reddy EM, Shiras A et al (2008) Natural gum reduced/stabilized gold nanoparticles for drug delivery formulations. Chem Eur J 14:10244–10250

Dinesh B, Bianco A et al (2016) Designing multimodal carbon nanotubes by covalent multi-functionalization. Nanoscale 8(44):18596–18611

Downs EC, Robertson NE, Riss TL et al (1992) Calcium alginate beads as a slow-release system for delivering angiogenic molecules in vivo and in vitro. J Cell Physiol 152:422–429

Draget KI, Tylor C (2011) Chemical, physical and biological properties of alginates and their biomedical implications. Food Hydrocoll 25:251–256
Elseoud WSA, Hassan ML, Sabaa MW et al (2018) Chitosan nanoparticles/cellulose nanocrystals nanocomposites as a carrier system for the controlled release of repaglinide. Int J Biol Macromol 111:604–613
Fabbro C, Ali-Boucetta H, Da Ros T et al (2012) Targeting carbon nanotubes against cancer. Chem Commun 48:3911–3926
Fernández-Urrusuno R, Calvo P, Remuñán-López C et al (1999) Enhancement of nasal absorption of insulin using chitosan nanoparticles. Pharm Res 16:1576–1581
Gelperina S, Kisich K, Iseman MD et al (2005) The potential advantages of nanoparticle drug delivery systems in chemotherapy of tuberculosis. Am J Respir Crit Care Med 172:1487–1490
Gibson JD, Bshnu PK, Eugene RZ (2007) Paclitaxel-functionalized gold nanoparticles. J Am Chem Soc 129:11653–11661
Gombotz WR, Wee S (1998) Protein release from alginate matrices. Adv Drug Deliv Rev 31:267–285
Gu HW, Ho PL, Tong E et al (2003) Presenting vancomycin on nanoparticles to enhance antimicrobial activities. Nano Lett 3:1261–1263
Gupta AK, Gupta M (2005) Synthesis and surface engineering of iron oxide nanoparticles for biomedical applications. Biomaterials 26:3995–4201
Hamilton RL, Guo LSS (1984) Liposomes preparation methods. J Clin Biochem Nut 7:175
Han H-S, Niemeyer E, Huang Y et al (2015) Quantum dot/antibody conjugates for in vivo cytometric imaging in mice. Proc Natl Acad Sci 112:1350–1355
Handa T, Naito S, Hiramatsu M et al (2006) Thermal SiO and $H_{13}CO^+$ line observations of the dense molecular cloud G0.11-0.11 in the Galactic Center Region. Astrophys J 636:261–266
Haque S, Md S, Sahni JK et al (2014) Development and evaluation of brain targeted intranasal alginate nanoparticles for treatment of depression. J Psychiatr Res 48:1–12
Heath JR, Davis ME (2008) Nanotechnology and cancer. Annu Rev Med 59:251–256
Himanshu A, Sitasharan P, Singhai AK (2011) Liposomes as drug carriers. IJPLS 2(7):945–951
Hofheinz RD, Gnad-Vogt SU, Beyer U et al (2005) Liposomal encapsulated anti-cancer drugs. Anticancer Drugs 16:691–707
Holister P, Cristina RV, Fullerenes HT (2003) Nanoparticles, technology white papers nr. Cientifica 3:1–12
Huang X, Jain PK, El-Sayed IH et al (2007) Gold nanoparticles: interesting optical properties and recent applications in cancer diagnostics and therapy. Nanomedicine 2:681–693
Huang CL, Huang CC, Mai FD et al (2015) Application of paramagnetic graphene quantum dots as a platform for simultaneous dual modality bioimaging and tumor-targeted drug delivery. J Mater Chem B 3:651–664
Hussain N, Jani PU, Florence AT (1997) Enhanced oral uptake of tomato lectin-conjugated nanoparticles in the rat. Pharm Res 14:613–618
Hussain I, Brust M, Papworth AJ et al (2003) Preparation of acrylate-stabilized gold and silver hydrosols and gold-polymer composite films. Langmuir 19:4831–4835
Hwu JR, Lin YS, Josephrajan T et al (2009) Targeted paclitaxel by conjugation to iron oxide and gold nanoparticles. J Am Chem Soc 131:66–68
Jain RA (2000) The manufacturing techniques of various drug loaded biodegradable poly(lactide-co-glycolide) (PLGA) devices. Biomaterials 21:2475–2490
Jain TK, Roy I, De TK et al (1998) Nanometer silica particles enacapsulating active compounds: a novel ceramic drug carrier. J Am Chem Soc 120:11092–11095
Jain K, Gupta U, Jain NK (2014) Dendronized nanoconjugates of lysine and folate for treatment of cancer. Eur J Pharm Biopharm 87:500–509
Jang JH, Shea LD (2003) Controllable delivery of non-viral DNA from porous scaffolds. J Controlled Release 86:157–168
Jie Y, Wenfeng L, Chang Y et al (2013) Amphiphilically modified chitosan cationic nanoparticles for drug delivery. J Nanopart Res 15:2123

John VT, Simmons B, McPherson GL et al (2002) Recent developments in materials synthesis in surfactant systems. Curr Opin Colloid Interface Sci 7:288

Johnston MJ, Semple SC, Klimuk SK et al (2007) Characterization of the drug retention and pharmacokinetic properties of liposomal nanoparticles containing dihydrosphingomyelin. Biochim Biophys Acta 1768:1121–1127

Jones CD, Fidalgo MM, Wiesner MR et al (2001) Alumina ultrafiltration membranes derived from carboxylate-alumoxane nanoparticles. J Membr Sci 193:175–184

Jorfi M, Foster EJ (2015) Recent advances in nanocellulose for biomedical applications. J Appl Poly Sci 132:41719

Joseph RR, Venkatraman SS (2017) Drug delivery to the eye: what benefits do nanocarriers offer? Nanomedicine 12:683–702

Joshi HM, Bhumkar DR, Joshi K et al (2006) Gold nanoparticles as carriers for efficient transmucosal insulin delivery. Langmuir 22:300–305

Kataria S, Sandhu P, Bilandi A, Akanksha M et al (2011) Stealth liposomes: a review. IJRAP 2(5):1534–1538

Kaur A, Jain K, Mehra NK (2017) Development and characterization of surface engineered PPI dendrimers for targeted drug delivery. Artif Cells Nanomed Biotechnol 45:414–425

Kesharwani P, Jain K, Jain NK (2014) Dendrimer as nanocarrier for drug delivery. Progr Polym Sci 39:268–307

Kesharwani P, Xie L, Banerjee S et al (2015) Hyaluronic acid-conjugated polyamidoamine dendrimers for targeted delivery of 3, 4-difluorobenzylidene curcumin to CD44 overexpressing pancreatic cancer cells. Colloids Surf B 136:413–423

Kim CK, Kalluru RR, Singh JP et al (2006) Gold-nanoparticle-based miniaturized laser-induced fluorescence probe for specific DNA hybridization detection: studies on size-dependent optical properties. Nanotech 17:3085

Kostarelos K, Bianco A, Prato M (2009) Promises, facts and challenges for carbon nanotubes in imaging and therapeutics. Nat Nanotechnol 4:627–633

Kulthe SS, Choudhari YM, Inamdar NN et al (2012) Polymeric micelles: authoritative aspects for drug delivery. Design Monomers Polym 15:465–521

Kumar RV, Koltypin Y, Cohen YS et al (2000) Preparation of amorphous magnetite nanoparticles embedded in polyvinyl alcohol using ultrasound radiation. J Mater Chem 10:1125–1129

Kumar RV, Elgamiel R, Diamant Y et al (2001) Sonochemical preparation and characterization of nanocrystalline copper oxide embedded in poly(vinyl alcohol) and its effect on crystal growth of copper oxide. Langmuir 17:1406–1410

Kumari A, Kumar V, Yadav S (2012) Nanotechnology: a tool to enhance therapeutic values of natural plant products. Trends Med Res 7:34–42

Laffleur F, Michalek M (2017) Modified xanthan gum for buccal delivery—a promising approach in treating sialorrhea. Int J Biol Macromol 102:1250–1256

Lam CW, James JT, McCluskey R et al (2004) Pulmonary toxicity of single wall carbon nanotubes in mice 7 and 90 days after intratracheal instillatioin. Toxicol Sci 77(1):126–134

Lee KY, Mooney DJ (2012) Alginate: properties and biomedical applications. Prog Polym Sci 37:106–126

Lee DE, Koo H, Sun IC et al (2012) Multifunctional nanoparticles for multimodal imaging and theragnosis. Chem Soc Rev 41:2656–2672

Li X, St John J, Coffer JL et al (2000) Porosified silicon wafer structures impregnated with platinum anti-tumor compounds: fabrication, characterization, and diffusion studies. Biomed Microdevices 2:265–272

Li ZZ, Wen LX, Shao L et al (2004) Fabrication of porous hollow silica nanoparticles and their applications in drug release control. J Controlled Release 98:245–254

Li Houli, Xiaobin Z, Yukun M et al (2009) Enhancement of gastrointestinal absorption of quercetin by solid lipid nanoparticles. J Controlled Release 133:238–244

Li J, Yap SQ, Yoong SL et al (2012) Carbon nanotube bottles for incorporation, release and enhanced cytotoxic effect of cisplatin. Carbon 50:1625–1634

Li Q, Lai KL, Chan PS et al (2016) Micellar delivery of dasatinib for the inhibition of pathologic cellular processes of the retinal pigment epithelium. Colloids Surf B 140:278–286
Lin N, Dufresne A (2014) Nanocellulose in biomedicine: current status and future prospect. Eur Poly J 59:302–325
Liu S, Yang S, Ho PC (2018) Intranasal administration of carbamazepine-loaded carboxymethyl chitosan nanoparticles for drug delivery to the brain. Asian J Pharm Sci 13:72–81
Llevot A, Astruc D (2012) Applications of vectorized gold nanoparticles to the diagnosis and therapy of cancer. Chem Soc Rev 41(1):242–257
Lovric J, Cho SJ, Winnik FM et al (2005) Unmodified cadmium telluride quantum dots induce reactive oxygen species formation leading to multiple organelle damage and cell death. Chem Biol 12:1159–1161
Lowe CR (2000) Nanobiotechnology: the fabrication and applications of chemical and biological nanostructures. Curr Opin Chem Biol 10:428–434
Madaan K, Kumar S, Poonia N et al (2014) Dendrimers in drug delivery and targeting: drug-dendrimer interactions and toxicity issues. J Pharm Bioallied Sci 6:139
Mandal A, Bisht R, Rupenthal ID et al (2017) Polymeric micelles for ocular drug delivery: from structural frameworks to recent preclinical studies. J Controlled Release 248:96–116
Mayer LD, Bally MB, Hope MJ (1986) Techniques for encapsulating bioactive agents into liposomes. Chem Phys Lipids 40:333–345
Melike U, Gulgun Y (2007) Importance of solid lipid nanoparticles (SLN) in various administration routes and future perspectives. Int J Nanomed 2(3):289–300
Michalet X, Pinaud FF, Bentolila LA et al (2005) Quantum dots for live cells, in vivo imaging, and diagnostics. Science 307:538–544
Moghimi SM, Hunter AC, Murray JC (2001) Long-circulating and target-specific nanoparticles: theory to practice. Pharmacol Rev 53:283
Moussa F (2018) [60] Fullerene and derivatives for biomedical applications. In Nanobiomaterials: nanostructured materials for biomedical applications. Woodhead Publishing, Elsevier, pp 113–136
Mozafari MR (2005) Liposomes: an overview of manufacturing techniques. Cell Mol Biol Lett 10(4):711–719
Mu L, Feng SS (2001) Fabrication, characterization and in vitro release of paclitaxel (taxol) loaded poly (lactic-co-glycolic acid) microspheres prepared by spray drying technique with lipid/cholesterol emulsifiers. J Controlled Release 76:239–254
Müller RH, Mäder K, Gohla S (2000) Solid lipid nanoparticles (SLN) for controlled drug delivery—a review of the state of the art. Eur J Pharm Biopharm 50:161–177
Nichkova M, Dosev D, Gee SJ et al (2005) Microarray immunoassay for phenoxybenzoic acid using polymer encapsulated Eu:Gd_2O_3 nanoparticles as fluorescent labels. Anal Chem 77:6864–6873
Niemeyer CM (2003) Functional hybrid devices of proteins and inorganic nanoparticles. Angew Chem 42:5796–5800
Olerile LD, Liu Y, Zhang B et al (2017) Near-infrared mediated quantum dots and paclitaxel co-loaded nanostructured lipid carriers for cancer theranostic. Colloids Surf B 150:121–130
Pandey R, Sharma S, Khuller GK (2005) Oral SLN Based antitubercular chemotherapy. Tuberculosis (Edinb) 85:415–420
Pantarotto D, Tagmatarchis N, Bianco A et al (2004) Synthesis and biological properties of fullerene containing amino acids and peptides. Mini Rev Med Chem 4:805–814
Patil NH, Devarajan PV (2016) Insulin-loaded alginic acid nanoparticles for sublingual delivery. Drug Deliv 23:429–436
Patra JK, Baek K-H (2014) Green nanobiotechnology: factors affecting synthesis and characterization techniques. J Nanomater 2014:219
Pick U (1981) Liposomes with a large trapping capacity prepared by freezing and thawing of sonicated phospholipid mixtures. Arch Biochem Biophys 212:186–194
Portero A, Remunan-Lopez C, Criado M et al (2002) Reacetylated chitosan microspheres for controlled delivery of anti-microbial agents to the gastric mucosa. J Microencapsul 19:797–809

Prasad PN (2004) Nanophotonics. Wiley, New York
Prato M (1999) Fullerenes and related structures. Springer, Berlin, pp 173–187
Prinz AV, Ya Prinz V, Seleznev VA (2003) Semiconductor micro- and nanoneedles for microinjections and ink-jet printing. Microelectron Eng 67–68:782–788
Proffitt F (2004) Yellow light for nanotech. Science 305(5685):762–765
Qiu XY, Hu SW (2013) "Smart" materials based on cellulose: a review of the preparations, properties, and applications. Materials 6:738–781
Qurratul A, Sharma S, Khuller GK et al (2003) Alginate-based oral drug delivery system for tuberculosis: pharmacokinetics and therapeutic effects. J Antimicrob Chemother 51:931–938
Rabinarayan P, Padilama S (2010) Production of solid lipid nanoparticles-drug loading and release mechanism. J Chem Pharm Res 2(1):211–227
Renwick LC, Brown D, Clouter A et al (2004) Increased inflammation and altered macrophage chemotactic responses caused by two ultrafine particles. Occup Environ Med 61(5):442–447
Riaz M (1996) Liposome preparation method. Pak J Pharm Sci 9(1):65–77
Rizvi SAA, Saleh AM (2018) Applications of nanoparticle systems in drug delivery technology. Saudi Pharm J 26:64–70
Román JV, Galán MA, del Valle EMM (2016) Preparation and preliminary evaluation of alginate crosslinked microcapsules as potential drug delivery system (DDS) for human lung cancer therapy. Biomed Phys Eng Express 2:035015
Rudramurthy GR, Swamy MK, Sinniah UR et al (2016) Nanoparticles: alternatives against drug-resistant pathogenic microbes. Molecules 21:836
Sajinovic D, Saponjic ZV, Cvjeticanin N et al (2000) Synthesis and characterization of CdS quantum dots–polystyrene composite. Chem Phys Lett 329:168–172
Sameti M, Bohr G, Ravi Kumar MN et al (2003) Stabilisation by freeze-drying of cationically modified silica nanoparticles for gene delivery. Int J Pharm 266:51–60
Schieren H, Rudolph S, Findelstein M et al (1978) Comparison of large unilamellar vesicles prepared by a petroleum ether vaporization method with multilamellar vesicles: ESR, diffusion and entrapment analyses. Biochim Biophys Acta 542(1):137–153
Selvakannan PR, Mandal S, Phadtare S et al (2004) Water-dispersible tryptophan-protected gold nanoparticles prepared by the spontaneous reduction of aqueous chloroaurate ions by the amino acid. J Colloid Interface Sci 269:97–102
Sercombe L, Veerati T, Moheimani F (2015) Advances and challenges of liposome assisted drug delivery. Front Pharm 6:286
Shah C, Shah V, Upadhyay U (2011) Solid lipid nanoparticles: a review. Curr Pharm Res 1(4):351–368
Shanmuganathan K, Capadona JR, Rowan SJ (2010) Biomimetic mechanically adaptive nanocomposites. Prog Poly Sci 35:212–222
Shehata T, Ogawara K, Higaki K et al (2008) Prolongation of residence time of liposome by surface-modification with mixture of hydrophilic polymers. Int J Pharm 359:272–279
Shen H, Ackerman AL, Cody V et al (2006) Enhanced and prolonged cross-presentation following endosomal escape of exogenous antigens encapsulated in biodegradable nanoparticles. Immunology 117(1):78–88
Shengtang H, Ying W, Zheng W et al (2013) Folate-conjugated chitosan–polylactide nanoparticles for enhanced intracellular uptake of anticancer drug. J Nanopart Res 15:2096
Shi M, Yang YY, Chaw CS, Goh SH et al (2003) Double walled POE/PLGA microspheres: encapsulation of water-soluble and water-insoluble proteins and their release properties. J Controlled Release 89:167–177
Shi Y, Pramanik A, Tchounwou C et al (2015) Multifunctional biocompatible graphene oxide quantum dots decorated magnetic nanoplatform for efficient capture and two-photon imaging of rare tumor cells. ACS Appl Mater Interfaces 7:10935–10943
Silva MM, Calado R, Marto J et al (2017) Nanoparticles as a mucoadhesive drug delivery system for ocular administration. Mar Drugs 15:370

So M-K, Xu C, Loening AM et al (2006) Self-illuminating quantum dot conjugates for in vivo imaging. Nat Biotechnol 24:339

Song H, Geng HQ, Ruan J et al (2011) Development of polysorbate 80/phospholipid mixed micellar formation for docetaxel and assessment of its in vivo distribution in animal models. Nanoscale Res Lett 6:354

Stroh M, Zimmer JP, Duda DG et al (2005) Quantum dots spectrally distinguish multiple species within the tumor milieu in vivo. Nat Med 11:678–682

Su CH, Sheu HS, Lin CY et al (2006) Nanoshell resonance imaging contrast agents. J Am Chem Soc 129:2139–2146

Sundar RS, Murugesan A, Venkatesan P et al (2010) Formulation development and evaluation of carprofen Microspheres. Int J Pharm Tech Res 2(3):1674–1676

Szoka F Jr, Papahadjopoulos D (1978) Procedure for preparation of liposomes with large internal aqueous space and high capture by reverse-phase evaporation. Proc Natl Acad Sci USA 75(9):4194–4198

Tan W, Wang KM, He X et al (2004) Bionanotechnology based on silica nanoparticles. Med Res Rev 24:621–638

Tao SL, Desai TA (2003) Microfabricated drug delivery systems: from particles to pores. Adv Drug Delivery Rev 55:315–328

Terrones M, Terrones H (2003) The carbon nanocosmos: novel materials for the twenty first century. Philos Trans A Math Phys Eng Sci 361(1813):2789–2806

Tom RT, Suryanarayanan V, Reddy PG et al (2004) Ciprofloxacin-protected gold nanoparticles. Langmuir 20:1909–1914

Tripathy S, Das M (2013) Dendrimers and their applications as novel drug delivery carriers. J Appl Pharm Sci 3:142–149

Velavan P, Karuppusamy K, Venkatesan P (2015) Nanoparticles as drug delivery systems. J Pharm Sci Res 7(12):1118–1122

Vemuri S, Rhodes CT (1995) Preparation and characterization of liposomes as therapeutic delivery systems: a review. Pharm Acta Helv 70(2):95–111

Venkatesan P, Manavalan R, Valliappan K (2009) Microencapsulation: a vital technique in novel drug delivery system. J Pharm Sci Res 1(4):26–35

Venkatesan P, Janardhan VS, Manavalan R et al (2011) Preformulation parameters characterization to design, development and formulation of loxoprofen loaded microspheres. Int J Pharm Biomed Res 2(3):107–117

Verma A, Uzun O, Hu Y et al (2008) Surface-structured-regulated cell-membrane penetration by monolayer protected nanoparticles. Nat Mater 7(7):588–595

Volkov Y (2015) Quantum dots in nanomedicine: recent trends, advances and unresolved issues. Biochem Biophys Res Commun 468:419–427

Wang L, Zhao W, Tan W (2008) Bioconjugated silica nanoparticles: development and applications. Nano Res 1:99–115

Wang G, Gao S, Tian R et al (2018) Theranostic hyaluronic acid-iron micellar nanoparticles for magnetic-field-enhanced in vivo cancer chemotherapy. ChemMedChem 13:78–86

Weis RP, Montchamp JL, Coffer JL et al (2002) Calcified nanostructured silicon wafer surfaces for biosensing: effects of surface modification on bioactivity. Dis Markers 18:159–165

Whitesides GM (2003) The 'right' size in nanobiotechnology. Nat Biotechnol 21:1161–1165

Wilson LJ, Cagle DW, Thrash TP et al (1999) Metallofullerene drug design. Coord Chem Rev 190–192:199–207

Wohlfart S, Gelperina S, Kreuter J (2012) Transport of drugs across the blood brain barrier by nanoparticles. J Controlled Release 161:264–273

Xu W, Ling P, Zhang T (2013) Polymeric micelles, a promising drug delivery system to enhance bioavailability of poorly water-soluble drugs. J Drug Deliv 2013:340315

Xu G, Zeng S, Zhang B et al (2016) New generation cadmium-free quantum dots for biophotonics and nanomedicine. Chem Rev 116:12234–12327

Yamakoshi Y, Umezava N, Ryu A et al (2003) Active oxygen species generated from photoexcited fullerene C-60 as potential medicines O^{2-} vs $^1O_2$. J Am Chem Soc 125:12803–12809
Yang SC, Ge HX, Hu Y et al (2000) Doxorubicin-loaded poly(butylcyanoacrylate) nanoparticles produced by emulsifier-free emulsion polymerization. J Appl Polym Sci 78:517–526
Yin JJ et al (2010) Metallofullerene nanoparticles circumvent tumor resistance to cisplatin by reactivating endocytosis. Proc Natl Acad Sci USA 107:7449–7454
Yokoyama M, Miyauchi M, Yamada N et al (1990) Polymer micelles as novel drug carrier: adriamycin-conjugated poly(ethylene glycol)-poly(aspartic acid) block copolymer. J Controlled Release 11:269
Yokoyama M, Okano T, Sakurai Y et al (1991) Toxicity and antitumor activity against solid tumors of micelle-forming polymeric anticancer drug and its extremely long circulation in blood. Cancer Res 51:3229–3236
Yu SH, Yoshimura M, Moreno JMC et al (2001) In situ fabrication and optical properties of a novel polystyrene/semiconductor nanocomposite embedded with CdS nanowires by a soft solution processing route. Langmuir 17:1700
Zeng R, Rong MZ, Zhang MQ et al (2002) Laser ablation of polymer based silver nanocomposites. Appl Surf Sci 187:239–247
Zhang Y (2011) Relations between size and function of substance particles. Nano Biomed Eng 3(1):1–16
Zhu J, Shi X (2013) Dendrimer-based nanodevices for targeted drug delivery applications. J Mater Chem B 1:4199–4211

Overview of Nanofluids to Ionanofluids: Applications and Challenges

Imran Khan, **Aamir Hussain Bhat**, **Dhananjay K. Sharma**, **Mohd Amil Usmani** and **Farah Khan**

Abstract Nanofluids are a new class of fluids engineered by dispersing nanometer-size structures (particles, fibers, tubes, droplets) in base fluids. In the last decade though the researcher has focused on the development in the field of nanofluids science with the emphasis on heat transfer but attention should also be keen on nanofluids preparation since the final properties of nanofluids are dependent on the stability of the dispersion. The current chapter focus on the recent progress on the study of nanofluids, such as the preparation methods, stability of nanofluids, methods to enhance the stability for nanofluids, the stability mechanisms of nanofluids, and presents the broad range of current and future applications in various fields including energy and mechanical and biomedical fields. This chapter also identifies the opportunities for future research. In addition to this, the chapter also focuses on a new class of nanofluids termed as "Ionanofluids". Ionanofluids represent a new and innovative class of heat transfer fluids that encompass multiple disciplines like nanoscience, mechanical, and chemical engineering. This ionanofluids exhibit enhanced thermal properties compared to their base ionic liquids (ILs) which further increase with increasing concentration of carbon nanotubes (CNTs) as well as fluid temperature to some extent. This chapter concludes that ionanofluids (INFs) show great promises to be used as innovative heat transfer fluids and novel media for the exploitation of green energy technology.

I. Khan (✉)
Department of Chemistry, College of Science, Sultan Qaboos University, PO Box 36, PC 123 Muscat, Oman
e-mail: imran.pchr@gmail.com; imrank@squ.edu.om

A. H. Bhat
Department of Applied Sciences, Higher College of Technology, AL-Khuair, Muscat, Oman

D. K. Sharma
Department of Mechanical Engineering, TEMA, University of Aveiro, 3810-193 Aveiro, Portugal

M. A. Usmani
Department of Chemistry, Gandhi Faiz-E-Aam College, P.O.Box 12676, Shahjahanpur, India

F. Khan
Independent Researcher, Muscat, Oman

A. H. Bhat et al. (eds.), *Nanomaterials for Healthcare, Energy and Environment*, Advanced Structured Materials 118, https://doi.org/10.1007/978-981-13-9833-9_10

Keyword Nanofluids · Nanoparticles · Thermal conductivity · Ionanofluids · Ionic liquids

1 Introduction

Nanofluids are a relatively new class of fluids which consist of a base fluid with nano-sized particles (1–100 nm) suspended within them. These particles, generally a metal or metal oxide, increase conduction and convection coefficients, allowing for more heat transfer out of the coolant (Choi 1995). The most common liquids used as base fluid are conventional coolants, such as deionized water, engine oil, acetone, ethylene glycol. Whereas the most common nanoparticle materials used are aluminum (Al), aluminum oxide (Al_2O_3), copper (Cu), copper oxide (CuO), gold (Au), silver (Ag), silica dioxide (SiO_2), titanium dioxide (TiO_2) and carbon nanotubes (CNTs either single-walled, double-walled or multi-walled). Nanofluids are a very important area of emerging technology and are playing an increasingly important role in the continuing advances of nanotechnology and biotechnology worldwide. They have enormously exciting potential applications and may revolutionize the field of heat transfer. With powerful microfluidic technology, scaling up techniques, thermal-wave theory and constructal theory, research and engineering practice in nanofluids is entering a new era. Nanofluids are research challenges of rare potential but daunting difficulty. The potential comes from both scientific and practical opportunities in many fields. The purpose of this chapter is to focus on the recent advances and development of nanofluids showing the specific area of application with the emphasis on preparation and characterization. And finally concluding with the promising nanofluids called 'ionanofluids' as innovative heat transfer fluids.

1.1 From Lab to Industry

Nanofluids being colloidal suspensions of nanosized solid particles in a liquid. In the recent year the number of publication (Yu et al. 2007; Vasu et al. 2009; Bi et al. 2008; Kuo et al. 2003; Murshed et al. 2008a; Natarajan and Sathish 2009) on nanofluids are increasing significantly which clearly specify industrial importance of this fluids. There are many advantage of this nanofluids which are worth to mention (Chol 1995; Saidur et al. 2011) such as high specific surface area, low pumping power required to accomplish the equivalent heat transfer, high stability of the colloidal suspension, reduced particle clogging compared to conventional colloids, and higher level of control of the thermodynamics and transport properties by varying the particle material, concentration, size, and shape. Over years, the experiment conducted have clearly shown that nanofluids tend to have substantially higher thermal conductivity than the base fluids (Keblinski et al. 2005). To explain the thermal enhancement in nanofluids, various reason has been given in the literature but still there are many noticeable contradictions (Keblinski et al. 2008). Application of nanofluids from laboratory to real application in industries is still inaccurate and for which at least four reasons are given in the literature why a definitive theory on nanofluids still does not exist (Das et al. 2007; Wang and Fan 2010):

- The thermal behavior is too different from solid–solid composites or standard solid–liquid suspensions.
- The thermal transport in nanofluids, besides being surprisingly efficient compared to standard solid–liquid suspensions, depends on nontraditional variables, such as particle size, shape, and surface treatment.
- The understanding of the physics behind nanofluids requires a multidisciplinary approach.
- Probably, the most daunting difficulty is related to multiscale issues.

In the year 2010, author Wang and Fan (2010) discussed some key issue regarding nanofluids such as the molecular scale, the microscale, the mesoscale, and the macroscale. The main difficulty is in the methods chosen to correlate and optimize the interplay among these scales. In one of the recent method i.e. one-step chemical solution method (CSM) has the benefit of controlling atoms and molecules through chemical reactions in the liquid phase. Though, the problem of controlling the microscale while operating at the macroscale is impossible. Therefore, by changing the batch-based macroreactors in the CSM by continuous-flow microfluidic microreactors of microchannels, droplets and slugs, a novel microfluidic one-step CSM is proposed for effective synthesis of high-quality nanofluids with controllable microstructures. And in the coming future, the poetical of this synthesis technique against an important target of controlling reactions accurately with a rapid and precise mixing would be of excessive demand. The success of this technology may fill the gap between the lab experiments to real application in industries with the economical investment at a marketable scale.

1.2 Scientific and Engineering Significance

Three possible approaches have been pursued for the study of nanofluids: experimental, empirical, and numerical. In terms of practical application this nanofluids, were also found to be even more promising as their convective heat transfer coefficient and critical heat flux were reported to be substantially higher as compared to those of their base fluids. Though literature shows that number of paper based on nanofluids experiment has been increasing significantly since 1993 but very few works have been published on empirical or numerical studies of nanofluids (Das et al. 2007). The current lack of understanding of the basic mechanism of energy transport at the nanoscale makes the published empirical works on nanofluids extremely case dependent. The main reason behind this that mostly authors are focusing more on the experiment and fitting the data to the model rather than understanding the main mechanism involved. It is also true that some author has been constantly giving

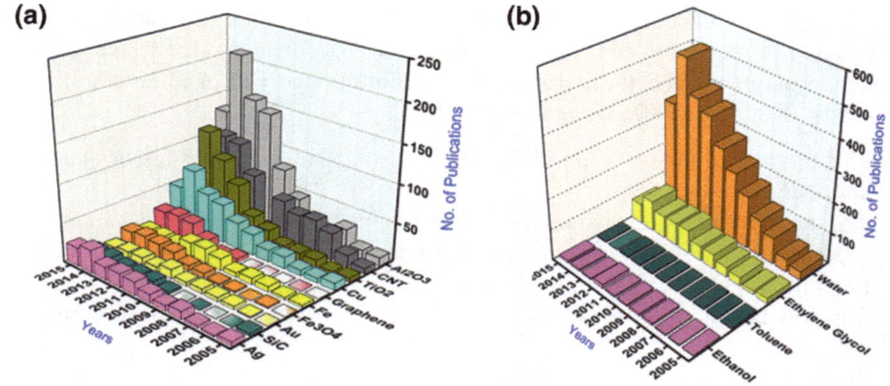

Fig. 1 Growth of publications on nanofluids from year 2005 to 2015 searched by topic "nanofluids" and refined by "nano-particles" in web of science on August 18, 2015, **a** number of publication on nanofluids versus various nanoparticles; **b** number of publication versus various base fluids

the basic concept which simplifies the chemistry and physics behind (Ghosh et al. 2011; Li et al. 2008; Sankar et al. 2008; Sarkar and Selvam 2007), but the more conceptual work were published after a boom came in the computer simulation work. More specifically, two of these works use molecular dynamics (MD) with a simple Lennard-Jones (LJ) potential to study the concept behind this. There are numerous studies which makes impossible to study and understand at the lab level but simulation has made still made the hope alive. Figure 1 shows the rapid growth of number of publications on nanofluids using the keyword "nanofluid" and refining it with key word "nano-particles" in the web of science from the year 2005–2015.

2 Preparation of Nanofluids and Its Limitation

2.1 *Preparation*

There are numerous review papers available in the literature regarding preparation of nanofluids (Murshed and Nieto de Castro 2014; Yu and Xie 2012; Haddad et al. 2014). So in this regards here we described a very short description for the preparation of nanofluids. In order to prepare stable nanofluids by suspending nanoparticles into base fluids, some special requirements are necessary such as even suspension, durability and stable suspension, low agglomeration of particles and no chemical change in fluids. In practical, there are two types of techniques commonly used to synthesize nanofluids and they are the one-step method and the two-step method.

Two-step method is the most widely used method for preparing nanofluids. In this method nanoparticle, nanofibers, nanotubes, or other nanomaterials used are firstly formed as dry powders by using chemical or physical methods, latter this nanosized powder are dispersed into a fluid in the second processing step using intensive magnetic force agitation, ultrasonic agitation, high-shear mixing, homogenizing, and ball milling. The most economical method is the two-step method which produces in large scale. Because of the high surface area and surface activity these nanoparticles have the tendency to aggregate which tend to use surfactant to improve the stability of nanofluids i.e. nanoparticle in the fluid. The important technique to enhance the stability of nanoparticles in fluids is the use of surfactants. However, the functionality of the surfactants under high temperature is also a big concern, especially for high-temperature applications. Due to the difficulty in preparing stable nanofluids by two-step method, several advanced techniques are developed to produce nanofluids, including one-step method. Although the one-step method yields better dispersion and stability of nanofluids, it is less popular in preparing nanofluids mainly due to the complexity in directly synthesizing suspensions of nanoparticles. Two-step method is the most widely used method for preparing nanofluids. The most common and popular method i.e. two step method of nanofluids preparation process is depicted in Fig. 2.

These methods can change the surface properties of the suspended particles and can be used to suppress the formation of particle clusters in order to obtain stable suspensions. The use of these techniques depends on the required application of the nanofluid. A summary of preparation process of nanofluids used by various researchers is summarized in Table 1.

The one-step process consists of simultaneously making and dispersing the particles in the fluid. In this method, the processes of drying, storage, transportation, and dispersion of nanoparticles are avoided, so the agglomeration of nanoparticles is minimized, and the stability of fluids is increased (Li et al. 2009). The one-step processes can prepare uniformly dispersed nanoparticles, and the particles can be

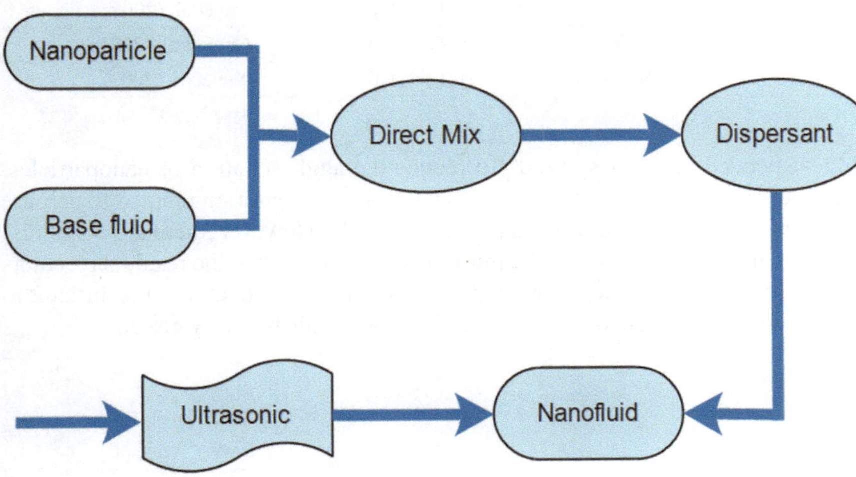

Fig. 2 Two-step preparation process of nanofluid

Table 1 Preparation methods of different nanofluids

Nanofluids system		Methods used	References
Nanofluids	Nano-particles		
Water	Ag	Two step	Godson et al. (2010)
	Al_2O_3	Two step	Eastman et al. (1996)
	Al_2O_3	Two step	Xie et al. (2002a)
	Cu	Two step	Liu et al. (2006a)
	Cu	Two step	Xuan and Li (2000)
	Cu	Two step	Xuan and Li (2000)
	CuO	Two step	Zhang et al. (2006)
	Fe_3O_4	Single step	Zhu et al. (2006)
	FC-72	Two step	Yang and Han (2006)
	MWCNT	Two step	Hong et al. (2005)
	NCTs	Two step	Xie et al. (2003)
	SiC	Two step	Huaqing et al. (2001)
	TiO_2	Two step	Murshed et al. (2005)
	TiO_2	Two step	Murshed et al. (2005)
Ethylene Glycol	Al_2O_3	Single step	Xie et al. (2002)
	Cu	Single step	Eastman et al. (2001)
	Fe	Single step	Hong et al. (2005)
	NCTs	Two step	Xie et al. (2003)
Toluene	Ag	Two step	Patel et al. (2003)
	Au	Two step	Patel et al. (2003)
Ethanol	Au	Two step	Putnam et al. (2006)
Engine oil	NCTs	Two step	Liu et al. (2005)
Poly oil	NCTs	Two step	Choi et al. (2001a)
Decene	NCTs	Two step	Xie et al. (2003)

stably suspended in the base fluid. To reduce the agglomeration of nanoparticles, Eastman et al. developed a one-step physical vapor condensation method to prepare Cu/ethylene glycol nanofluids (Eastman et al. 2001). However, there are some disadvantages in one-step method. The most important one is that the residual reactants are left in the nanofluids due to incomplete reaction or stabilization. It is difficult to elucidate the nanoparticle effect without eliminating this impurity effect.

2.2 *Drawbacks*

The use of nanofluids seems attractive in a broad range of applications as reported in the previous section. But the development in the area of nanofluid application is hindered by many factors in which long term stability of nanofluid in suspension is major reason. Keeping this in mind, this paper focuses various challenges that should be resolved in the near future. The following are the most pressing issues.

2.2.1 Stability of Suspension

Long term physical and chemical stability of nanofluids is an important practical issue because of aggregation of nanoparticles due to very strong van der walls interactions so the suspension is not homogeneous. Physical or chemical methods have been applied to get stable nanofluids such as (i) an addition of surfactant; (ii) surface modification of the suspended particles; (iii) applying strong force on the clusters of the suspended particles. In the year 1997, Lee and Choi (1997) found that Al_2O_3 nanofluids kept after 30 days exhibit some settlement compared to fresh nanofluids. Particles settling must be examined carefully since it may lead to clogging of coolant passages.

2.2.2 Increased Pressure Drop and Pumping Power

Pressure drop development and required pumping power during the flow of coolant determines the efficiency of nanofluid application. It is known that higher density and viscosity leads to higher pressure drop and pumping power. There are many studies showing significant increase of nanofluids pressure drop compared to base fluid. One of the experimental study by Choi et al. (2009) calculated 40% increase of pumping power compared to water for a given flow rate.

2.2.3 Lower Specific Heat

An ideal heat transfer fluid should possess higher value of specific heat so the fluid can exchange more heat. Previous studies show that nanofluids exhibit lower specific heat than base fluid, which results the limiting use of nanofluid application.

2.2.4 Cost Efficiency of Nanofluids

As discussed in previous section nanofluids are prepared by either one step or two step methods. Both methods require advanced and sophisticated equipment's. This leads to higher production cost of nanofluids resulting the higher cost of nanofluids which is a major drawback of nanofluid applications.

2.2.5 Future Research

For the enhancement of nanofluids, further researches are required. An important focus for future research should be determining the key energy transport mechanism in nanofluids. Mostly heat transfer depends on thermal conductivity of nanofluids. The thermal conductivity of nanofluids can be a function of parameters such as particle shapes, particle agglomeration etc. therefore future research should be focused on outcome of the main parameters affecting the thermal conductivity of nanofluids. Theoretical predictions should be evaluated in terms of agreement with experiments regarding concentration, particle size and temperature dependence. In the currently scenario, the available nanoparticles are limited and their specifications cannot be determined accurately. The most puzzling point here is to obtain the desirable nanoparticle product. The development of the nanoparticle production technique will be very helpful for the nanofluids research. We have to face public concern about their safety both in production and in use. Nanofluids engineers would be prudent to pursue green designs by choosing nontoxic or biodegradable nanoparticles. It can be concluded that low cost, high volume production of stable green nanofluids is one of the most challenging directions for future applied research.

3 Characteristics of Nanofluids

This section exemplifies the most common nanofluid preparation methods by providing information about the last stages of the fluid creation. Note that the "Quality" of a nanofluid represents the extent of achievability of the desired properties of the mixture. The desired properties of nanofluids can be stated as,

- To achieve the even, durable and stable suspension of the solid nanoparticles in the host fluid (Base fluid).
- Low or no formation of agglomerates.
- No chemical change of the base fluid (i.e. the solid particles must not chemically react with the host fluid).

Nanofluids follow either single or multi-step creation methods. The single-step creation approach refers to a direct evaporation method (Vacuum Evaporation onto a Running Oil Substrate-VEROS). This method attains the best quality nanofluids; however, there are substantial limitations on the flexibility to create customized nanoparticle volumetric concentrations and base fluid type samples. The multi-step method provides more flexibility, but, can lead to low quality of the attained mixture. It can be created either by diluting a very dense solution of the required nanofluids with the matching base fluid or by mixing directly the nanoparticles of choice with the desired base fluid. This step provides more flexibility than the single-step (one step) method as the nanoparticles volumetric concentration can be made to order; however, the quality of the resulting nanofluids is lower than the single-step method. The second approach of the multi-step method is the most widely used amongst researchers, since it provides maximum flexibility to control the volumetric concentration of the nanoparticles, along with the base fluid type to be customized given the nanoparticle material, shape and size. On the other hand, this procedure delivered the lowest quality of nanofluids in comparison to all the other methods (Das et al. 2007).

4 Calculation and Overview of Thermal Conductivity of Nanofluids

The suspended particles (nanoparticles) remarkably enhances the thermal conductivities of the nanofluids which is the main reasons to exhibit superior heat transfer as compared to conventional heat transfer fluids. Furthermore the thermal conductivities are strongly dependent on volume fraction dimensions and properties of nanoparticles. The theoretical approach for the measurement of thermal conductivities are very difficult so experimental approach named hot-wire apparatus was introduced by Wakeham et al. which was later supported by Masuda et al. in the year 1993 (used to measure the thermal conductivities of liquid containing ultra-fine particles). The experimental was designed using two platinum wires of 0.153 and 0.063 m long which were dipped in the sample containing nanofluids. These wires are then connected to wheatstone bridge which are responsible for measurement of highly sensitive deflection in the resistance, additionally a digital voltmeter was used convert the signal into thermal conductivity.

Using the hot-wire apparatus, the thermal conductivity of the fluid can be calculated as:

$$\Delta T(r_0, t) = \frac{q}{4\pi k} \ln\left(\frac{4at}{r_0^2 C}\right) \tag{1}$$

The above equation is ideal equation, where letters have the usual meaning such as ΔT is temperature difference, t is time, q is heat flux per unit length, k is thermal conductivity and r is radial variable. By plotting the slope of the line ΔT versus ln t, the thermal conductivity of the fluids is obtained.

Equation (1) is the fundamental equation which can be simplified as Eq. (2)

$$q = hA\Delta T \tag{2}$$

where q is the heat flow, h is the heat transfer coefficient, A is the heat transfer area, and ΔT is the temperature difference that results in heat flow. It can be stated from this equation that the increased heat transfer can be achieved by: (i) increasing ΔT;

(ii) increasing A; (iii) increasing h. A greater temperature difference ΔT can lead to increase the heat flow, but ΔT is often limited by process or materials constraints. For example, the maximum temperature in a nuclear reactor must be kept below a certain value to avoid runaway reactions and meltdown. Therefore, increased ΔT can only be achieved by decreasing the temperature of the coolant. However, this would reduce the rate of the nuclear reaction and decrease the efficiency of the process (Beck 2008).

Nanofluids can be used in the wide variety of industries, ranging from transportation to energy production, in electronics systems like microprocessors, Micro-Electro-Mechanical Systems (MEMS) and in the field of biotechnology. Recently, the number of companies that observe the potential of nanofluids technology and their focus for specific industrial applications are increasing rapidly. In the transportation industry, nanocars, GM and Ford, along with others are focusing on nanofluids research projects (Choi 1995; Das et al. 2003, 2007; Lee and Choi 1997; Wang and Mujumdar 2008a, b; Mokhatab et al. 2006). Nanofluids can also be used for increasing the dielectric strength and life of the transformer oil by dispersing nanodiamond particles (Sridhara et al. 2009; Sridhara and Satapathy 2011; Botha 2006). In the present scenario the nanofluids can be used in following specific areas as depicted in the Table 2.

5 Thermal Conductivity of Nanofluids

Thermal conductivity is an important parameter for enhancing heat transfer performance of a heat transfer fluid. Various researchers have reported experimental studies on the thermal conductivity of nanofluids. Numerous methods has been deployed to measure the thermal conductivity of nanofluids such as the transient hot wire (THW) method, temperature oscillation and the steady-state parallel plate method. Among all THW method is among widely used method (Xuan and Li 2000; Hong et al. 2005; Murshed et al. 2005). The thermal conductivity of solid nanoparticles is much higher than that of fluids, the suspended particles are expected to be able to increase thermal conductivity and heat transfer performance. The comparisons of thermal conductivity increment of some nanofluids reported in literature are depicted in the Fig. 3.

Various classical models for thermal conductivities have been proved that cannot predict the anomalously high enhancements accurately as described in the review of Murshed et al. (2008a). Several mechanisms have been proposed, to date, to explain these phenomena. Also previously Keblinsky et al. (2002) classified the four different mechanisms for heat transfer and from review of Murshed et al. (2008a) also justified by Nieto de Castro et al. (2010) that the effect of the particle surface chemistry and the structure of the interphase particle/fluid are the major mechanisms responsible for the unexpected enhancement in nanofluids. Eastman et al. (2001), found that thermal conductivity of 0.3% copper nanoparticles in ethylene glycol nanofluids can lead to the increment up to 40% as compared to base fluid. Authors

Table 2 Application of nanofluids in specific areas

Area of application	References
Cooling of electronics	Vasu et al. (2009), Sridhara et al. (2009), Botha (2006), Liu et al. (2008)
Nanofluids in cameras, microdevices, and displays	Top News (2009), Nguyen et al. (2007)
Application in chillers	Wu et al. (2009)
Application in domestic refrigerator	Bi et al. (2008), Jiang et al. (2009), Wang et al. (2006), Wang et al. (2003), Li et al. (2006), Park and Jung (2007), Peng et al. (2009), Naphon et al. (2009)
Engine cooling/vehicle thermal management	Choi et al. (2008), Ollivier et al. (2006), Tzeng et al. (2005), Ma and Liu (2007), Leong et al. (2010)
Detection of knock occurrence in a gas SI engine	Ollivier et al. (2006)
Application as a coolant in machining	Srikant et al. (2009)
Cooling of diesel electric generator	Kulkarni et al. (2008)
Diesel combustion	Kuo et al. (2003), Pivkina et al. (2004), Risha et al. (2003), De Luca et al. (2005), Galfetti et al. (2007)
Boiler flue gas temperature reduction	Saidur et al. (2010)
Solar water heating	Natarajan and Sathish (2009), Tyagi et al. (2009), Tyagi (2008)
Cooling and heating in buildings	Yu et al. (2007)
Application in transformer	Choi et al. (2008), Davidson (2009)
Heat exchanger	Yang (2006), Das (2006)
In oscillating heat pipes	Naphon et al. (2008)
Space, defense and ships	Wang and Mujumdar (2008), Wilson (2006)
Medical applications	(Lee and Choi 1997)
The antibacterial activities	Zhang et al. (2008), Hirota et al. (2010)
Application in nuclear reactor	Buongiorno and Hu (2009)
Application in grinding	Shen (2008)
New sensors for improving exploration	Ying and Sun (1997), Scott et al. (2003)
Application in fuel cell	Islam et al. (2015)
Other applications	Mokhatab et al. (2006)

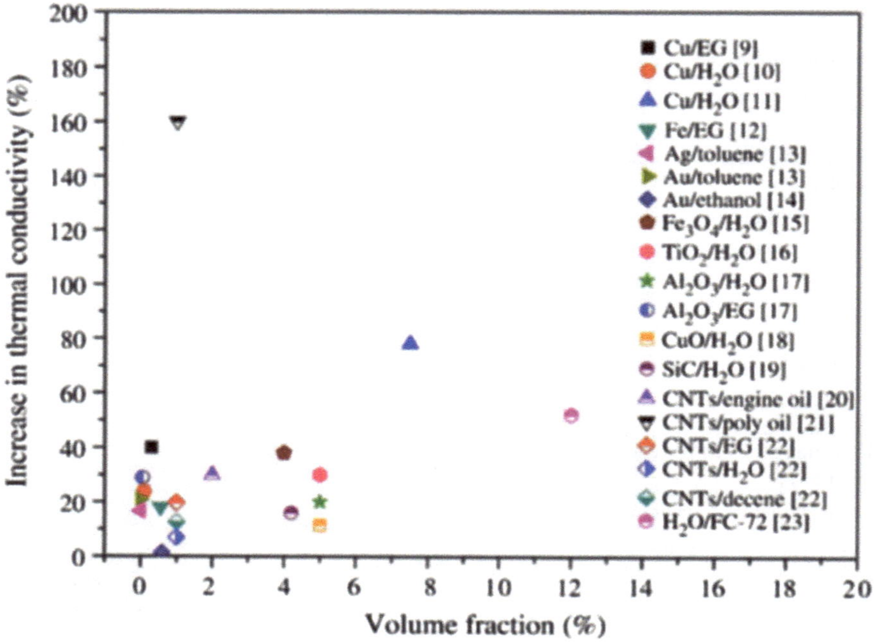

Fig. 3 Comparison of experimental data on thermal conductivity of nanofluids (Li et al. 2009)

stressed that, this property plays an important role in construction of energy efficient heat transfer equipment. Liu et al. (2006b), investigated the thermal conductivity of copper–water nanofluids produced by chemical reduction method. The results showed 23.8% improvement at 0.1% volume fraction of copper particles. Higher thermal conductivity and larger surface area of copper nanoparticles are attributed to this improvement. It was worthy to note that thermal conductivity increases with particles volume fraction but decreases with elapsed time. Hwang et al. (2006a), suggested that thermal conductivity enhancement of nanofluids was greatly influenced by thermal conductivity of nanoparticles and base fluid. For instance, thermal conductivity of water based nanofluids with multiwalled carbon nanotube (MWCNT) has noticeably higher thermal conductivity compared to SiO_2 nanoparticles in the same base fluid. However, Yoo et al. (2007), argued that surface to volume ratio of nanoparticles is a dominant factor that influences the nanofluids thermal conductivity rather than nanoparticles thermal conductivity. Surface to volume ratio is increased with smaller sizes of nanoparticles.

Choi et al. (Hong et al. 2005), reported a 150% thermal conductivity enhancement of poly(α-olefin) oil with the addition of multiwalled carbon nanotubes (MWCNT) at 1% volume fraction. Similarly, Yang et al. (Xie et al. 2003), reported a 200% thermal conductivity enhancement for poly(α-olefin) oil containing 0.35% (v/v) MWCNT. It is important to note that this thermal conductivity enhancement was accompanied by a three order of magnitude increase in viscosity. Eastman et al. (2001), found a

40% thermal conductivity enhancement for ethylene glycol with 0.3% (v/v) copper nanoparticles (10 nm diameter), although the authors added about 1% (v/v) thioglycolic acid to aid in the dispersion of the nanoparticles. The addition of this dispersant yielded a greater thermal conductivity than the same concentration of nanoparticles in the ethylene glycol without the dispersant. Jana et al. (2007) measured the thermal conductivity of a similar copper containing nanofluid, however the base fluid was water and laurate salt as a dispersant. Authors observed a 70% thermal conductivity enhancement for 0.3% (v/v) Cu nanoparticles in water. Kang et al. (2006) reported a 75% thermal conductivity enhancement for ethylene glycol with 1.2% (v/v) diamond nanoparticles of the diameters between 30 and 50 nm. Despite these remarkable results, some researchers have measured the thermal conductivity of nanofluids and have found no anomalous results. Also, those results can often be predicted by conventional thermal conductivity models (Timofeeva et al. 2007; Zhang et al. 2006a, 2007).

Lee et al. (2008) concluded that the thermal conductivity of nanofluids are affected by pH level and addition of surfactant during nanofluids its preparation. Better dispersion of nanoparticles is achieved with addition of surfactant such as sodium dodecylbenzenesulfonate. Optimum combination of pH and surfactant can leads to 10.7% thermal conductivity enhancement of 0.1% Cu/H_2O nanofluids. Thermal conductivity of ethylene glycol based ZnO nanofluids measured by transient short hot wire technique is found to be increased non-linearly with nanoparticles volume fraction (Yu et al. 2007, 2008). Mintsa et al. (2009) added thermal conductivity of nanofluids also depend on the nanoparticles size and temperature. Author Vajjha and Das (2009) also agreed that thermal conductivity is dependent not only on the nanoparticles concentration but also on the temperature. Authors concluded that, it will be more beneficial if nanofluids are used in high temperature applications.

It has been noticed that most authors agreed that nanofluids provide higher thermal conductivity compared to base fluids. Its value increases with particles concentration. Temperature, particles size, dispersion and stability do play important role in determining thermal conductivity of nanofluids (Murshed et al. 2009). Figure 3 shows the comparison of thermal conductivity of heat transfer fluids and nanofluids. And Fig. 4 shows comparison of the thermal conductivity of common liquids, polymers and solids. Table 3 shows the thermal conductivity ratio (i.e. thermal conductivity of solid to liquids) of nanofluids. These shows an indication that when solid particles are added in conventional liquids/coolants, thermal conductivity can be increased tremendously.

Research has shown that the thermal conductivity and the convection heat transfer coefficient of the fluid can be largely enhanced by suspended nanoparticles (Choi 1995; Xuan and Roetzel 2000; Choi et al. 2001b). Choi et al. (2001a) observed that the thermal conductivity of this nanofluid was 150% greater than that of the oil alone. Table 3 shows the thermal performances of different types (metallic, non-metallic, MWCNT) and concentrations of nanofluids.

Recently, tribology research shows that lubricating oils with nanoparticles additives (MoS_2, CuO, TiO_2, diamond, etc.) exhibit improved load-carrying capacity, anti-wear and friction-reduction properties (Xu et al. 1996; Verma et al. 2007).

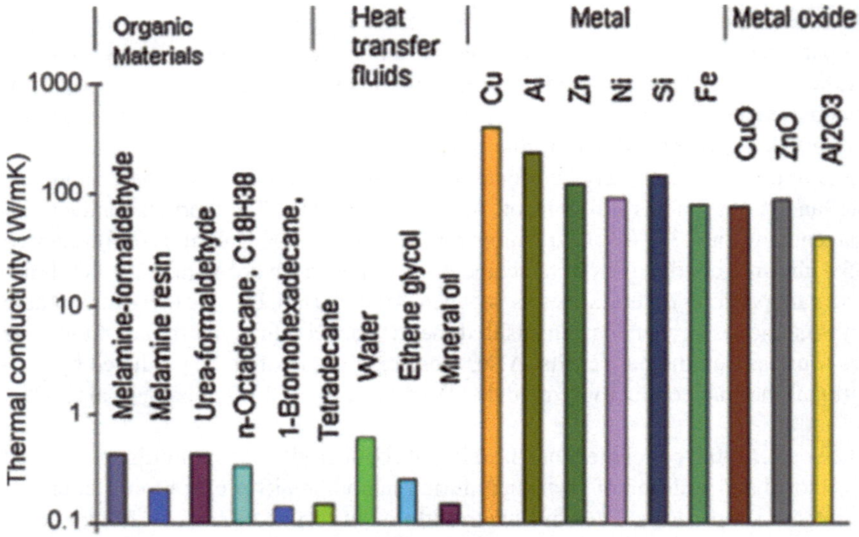

Fig. 4 Comparison of the thermal conductivity of common liquids, polymers and solids (Wen et al. 2009)

These features made nanofluids very attractive in some cooling and/or lubricating application in many industries including manufacturing, transportation, energy, and electronics, etc.

6 Thermal Conductivity of Ionanofluids

Ionic liquids (ILs) are salts composed of large organic cations and inorganic anions that cannot form an ordered crystal and thus remain liquid at or near room temperature. ILs have unique properties such as negligible vapor pressure at room temperature, low melting points, non-flammability, high thermal stability and their ability to dissolve organic and inorganic compounds, make them attractive for the scientific community and a variety of industrial applications. Chemically, ILs are composed by a bulky organic cation coupled with a disperse charge, and an asymmetric organic or inorganic anion (Wasserscheid and Welton 2002). Limitless combinations of cation and anion that form the ILs lead to different physical properties and phase behavior, thus allowing ionic liquids to be labeled as "designer" solvents. In the current, ILs has already been shown to be safe and sustainable alternatives for many applications in various chemical and industry manufacturing (Khan et al. 2014a, b, c, 2015; Passos et al. 2014; Khan et al. (Submitted); Carvalho et al. 2013; Khamooshi et al. 2013; Swatloski et al. 2002; Bin Dahbag et al. 2014). This characteristics factures of ILs make them possibly to be used as a new group of heat transfer fluids for heat exchange

Table 3 Thermal conductivities ratio of different types of nanofluids

Nanoparticles	Fluids	k_1/k_2	References
Al_2O_3	Water	66	Xie et al. (2002), Zhang et al. (2006a, (b, 2007), Das et al. (2003), Yoo et al. (2007), Timofeeva et al. (2007), Li and Peterson (2007), Kim et al. (2007), Wang et al. (1999b), Lee et al. (1999)
	Ethylene glycol	156	Xie et al. (2002a, b), Kim et al. (2007), Lee et al. (1999), Wang et al. 1999b, Trisaksri and Wongwises (2009)
	Glycerol	140	Xie et al. (2002)
	Oil	342	Xie et al. (2002, b), Wang et al. (1999a), Venerus et al. (2006)
CuO	Water	127	Zhang et al. (2006a, 2007), Das et al. (2003), Li and Peterson (2007), Wang et al. (1999a), Lee et al. (1999), Hwang et al. (2006b)
	Ethylene glycol	300	Xie et al. (2002a), Wang et al. (1999a), Lee et al. (1999), Hwang et al. (2006b)
TiO_2	Water	14	Zhang et al. (2006, 2007), Yoo et al. (2007), Kim et al. (2007)
	Ethylene glycol	33	Hong et al. (2005), Kim et al. (2007)
Fe_3O_4	Water	11.5	Zhu et al. (2006)
ZrO_2	Water		Zhang et al. (2006a)
WO_3	Ethylene glycol		Yoo et al. (2007)
ZnO	Water	48	Kim et al. (2007)
	Ethylene glycol	113	Kim et al. (2007)
SiO_2	Water	2.2	Hwang et al. (2006a), Kang et al. (2006)
	Ethylene glycol	5.2	Kang et al. (2006)
Cu	Water	655	Jana et al. (2007), Liu et al. (2006a)
	Ethylene glycol	1550	Eastman et al. (2001)
	Water+ Ethylene glycol		Jwo and Teng (2005)
Ag	Water	697	Kang et al. (2006)

(continued)

Table 3 (continued)

Nanoparticles	Fluids	k_1/k_2	References
Au	Water	518	Venerus et al. (2006)
	Ethanol	1830	Putnam et al. (2006)
	Toluene	2370	Zhang et al. (2007)
Fe	Water	132	Hong et al. (2005)
	Ethylene glycol	311	Hong et al. (2005), Yoo et al. (2007)
Alxcuy	Water		Manna et al. (2005), Chopkar et al. (2007)
	Ethylene glycol		Manna et al. (2005), Chopkar et al. (2007)
Agxcuy	Oil		Ceylan et al. (2006)
Agxal	Water		Chopkar et al. (2007)
	Ethylene glycol		Ceylan et al. (2006)
Carbon nanotubes	Water	3290	Jana et al. (2007), Hwang et al. (2006b), Assael et al. (2004), Wright et al. (2007), Wensel et al. (2008)
		7780	Assael et al. (2004), Velve Casquillas et al. (2007)
	Antifreeze	17,100	Yoo et al. (2007)
	Oil	14,300	Choi et al. (2001a), Assael et al. (2004), Shaikh et al. (2007)
Carbon nanofibers	Water	21	Zhang et al. (2006)
	Oil	111	Shaikh et al. (2007)
	Toluene	3	Putnam et al. (2006)
Graphite	Oil	1020	Velve Casquillas et al. (2007)
Diamond	Ethylene glycol	3500	Kang et al. (2006)

in chemical plants, absorption cooling cycle system (Khan et al. 2014c, 2015; Passos et al. 2014), and solar thermal power generation (Bai et al. 2010), where water and ethylene glycol may not be suitable for the application owing to the limitation of their thermophysical and chemical properties. Figure 5 shows the growth of number of publications on INFs which is searched by using the keyword "ionanofluids" in the web of science from the year 2010–2015.

In the recent year introduction of nanofluids based on ILs are growing significantly with the emphasis on various nanoparticles such as Au (Wang et al. 2011), CuO (Swadzba-Kwasny et al. 2012), Al_2O_3 (Bridges et al. 2011), and multi-walled carbon nanotubes (MWCNTs) (Murshed and Nieto de Castro 2014; Choi et al. 2001a; Nieto de Castro et al. 2010, 2012; Murshed et al. 2012; França et al. 2013; Ribeiro et al. 2013; Ferreira et al. 2013; Wang et al. 2010). Comprehensive review preparation and thermal performances of CNT nanofluids are available on the heat transfer

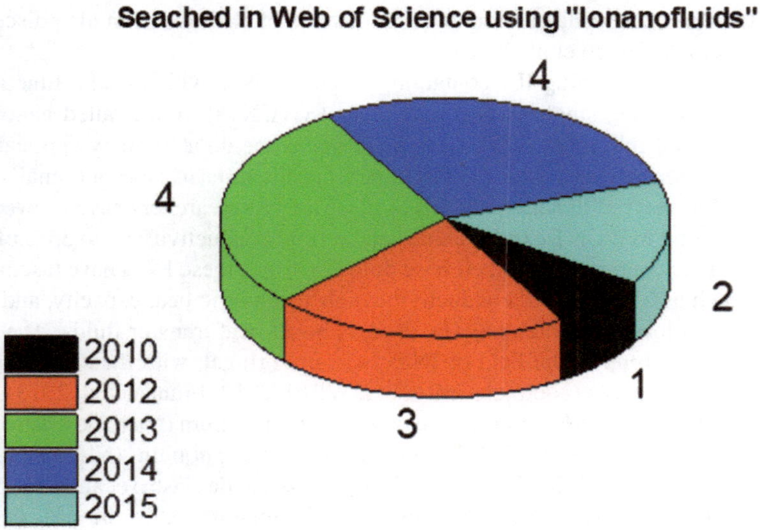

Fig. 5 Growth of publications on INFs from year 2010 to 2015 searched by keyword "ionanofluids" in web of science on October 1, 2015

characteristics of CNT nanofluids (Xie and Chen 2011). The current finding on the ILs based fluids i.e. Ionanofluids (INFs) depict that this INF exhibit enhanced thermal conductivity as compared with the pure ILs, which has again made this liquid versatile in the field of green chemistry with numerous application. Previous studies in nanofluids have used common liquids such as water, ethylene glycol, or oils. No innovation has been performed so far in the base fluids until the publication by Ribeiro et al. (Nieto de Castro et al. 2010).

The suspension of the nanomaterial in the green solvent called ILs is termed as INFs. The concept of INFs, which has been focused in the recent year by various research group (França et al. 2013; Nieto de Castro et al. 2009, 2012; Ferreira et al. 2013; Wang et al. 2012a; Rodríguez-Palmeiro et al. 2013; Shevelyova et al. 2013; Lee and Aida 2011). Firstly, this pioneer concept was developed by Fukushima and his co-workers (Fukushima et al. 2003) that carbon nanotubes (CNT) and room-temperature ionic liquids can be blended to form gels termed as "bucky gels of ILs". These blends were also found to exhibit high thermal conductivity, high heat capacity, non-volatile, designable, and green solvent characteristics. Thus this novel INFs opens the door in replacement to traditional nanofluids as advanced heat transfer fluids in numerous cooling technologies, heat exchangers, chemical engineering and green energy-based applications. Other attractive features of INF are that they are designable and fine-tunable through their base ILs for desired properties and tasks. These fluids can also be used for the development of new pigments for solar energy-based applications (Vieira et al. 2012). INFs have also showed application in the development of new pigments for paint coatings of solar collectors with their higher solar absorbance and thermal emissivity as compared to the base paint (Vieira et al.

2011). Comprehensive discussion and significance of INF has been also discussed earlier (Nieto de Castro et al. 2010).

The possibility of using ILs containing dispersed NPs with specific functionalization, for example, single-walled nanotubes (SWCNTs), multiwalled nanotubes (MWCNTs) and fullerenes (C60, C80, etc.), opens the door to many applications. The use of nanoparticles as heat transfer enhancers allows us to associate small quantities of different types of nanomaterials to ILs. Various researchers have showed that INFs containing MWCNT exhibit enhanced thermal conductivity up to 35% of specific heat capacity compared to their base fluid ILs. Since these INFs have fascinating features such as high thermal conductivity, high volumetric heat capacity, and nonvolatility therefore, can potentially be used as novel heat transfer fluids. The most active research group in the field of INFs is from Portugal, with their INFs can be found in the references (Nieto de Castro et al. 2010, 2012; França et al. 2013).

Various studied on INFs reported are based on imidazolium (Nieto de Castro et al. 2010, 2012; França et al. 2013; Mohammad 2012), phosphonium (Nieto de Castro et al. 2010; Ferreira et al. 2013) and pyrrolidinium (Nieto de Castro et al. 2010) based ILs with suspended MWCNTs as a function of temperature. In the past decade, the works of the CNTs nanofluids focused on the enhancement of their thermal conductivity (Xie et al. 2003; Choi et al. 2001a; Wensel et al. 2008; França et al. 2013; Yang et al. 2006; Han et al. 2007) (Nieto de Castro et al. 2010, 2012; Ferreira et al. 2013; Wang et al. 2010, 2012; Shevelyova et al. 2013; Neo and Ouyang 2012). Recently, several studies on covalent modification of MWCNTs with imidazolium cation-based ILs have been reported with improved dispersibility of MWCNTs in these corresponding ILs (Yu et al. 2006; Wang et al. 2008; Park et al. 2006). The IL-functionalized MWCNTs (F-MWCNTs) show good dispersion, switchable solubility in corresponding solvents including ILs (Yu et al. 2006; Park et al. 2006), high charge-transfer activity (Zhang et al. 2006c), and high electronic conductivity (Zhao et al. 2004). ILs suspensions containing F-MWCNTs can be considered as "nanofluids" because of their favorable homogeneity and stability, which is expected to combine the advantages of CNTs and ILs.

In the year 2010, França et al. (2013) used two different ILs ([C_4mim][$(CF_3SO_2)_2N$] and [C_2mim][$EtSO_4$]) with MWCNT to prepare INFs and measured its thermal conductivities. Although the thermal conductivity of these two MWCNT-INFs increases considerably over base ILs with MWCNT concentration but [C_4mim][$(CF_3SO_2)_2N$] based INFs exhibit larger enhancement of thermal conductivity. Studies shows that the effect of water on the purity of [C_4mim][$(CF_3SO_2)_2N$] at any given temperature give decrement to the thermal conductivity decreases with the increment of amount of water but the effect increases with slightly increment in temperature. Later in the year 2010, again Nieto de Castro et al. (2010) prepared five different INFs using 1-hexyl-3-methylimidazolium tetrafluoroborate ([C_6mim][BF_4]), 1-butyl-3-methylimidazolium hexafluorophosphate ([C_4mim][PF_6]), 1-hexyl-3-methylimidazolium hexafluorophosphate ([C_6mim][PF_6]), 1-butyl-3-methylimidazolium trifluoromethanesulfonate ([C_4mim][CF_3SO_3]), 1-butyl-1-methylpyrrolidinium bis(trifluoromethyl)sulfonylimide ([C_4mpyrr][$(CF_3SO_2)_2N$]) with 1 wt% MWCNT and found enhancements of 9% on thermal conductivity and observed a weak dependence of this phenomena on temperature. Nieto de Castro et al. (2010) also measured the heat capacity of INFs with 1-butyl-3-methylimidazolium hexafluorophosphate ([C_4mim][PF_6]) with (1 and 1.5) % baytubes with enhancement (8%).

In the same year 2010, Wang et al. (2010) studied the rheological behavior of nanofluids containing functionalized multi-walled carbon nanotubes (F-MWCNTs) with [C_4mim][PF_6] which shows shear thinning behavior at low concentrations of F-MWCNTs (up to 0.1 wt%). The phenomenon was explained by the high specific size ratio and flexibility of F-MWCNTs and the formation of transient network through CNT–CNT and CNT-liquid matrix interactions. The shear viscosity of these INFs was lower than that of pure IL especially for high shear rates, and the behavior was attributed to some self-lubrication of F-MWCNTs. In the past decade, the works of the CNTs nanofluids focused on the enhancement of their thermal conductivity (Xie et al. 2003; Choi et al. 2001a; Wensel et al. 2008; Yang et al. 2006; Han et al. 2007). In the year 2012, other research group from China reported (Wang et al. 2012b) stable graphene (GE) and MWCNTs-based INFs with IL ([C_6mim][BF_4]) as base fluids without any surfactant. The GE-dispersed INFs were more homogeneous and stable than those containing MWCNTs at the same nano-additive loading. The remarkable thermal conductivity enhancement ratio of GE based INFs were reported to be larger than those of the MWCNT-based INFs at the same nano-additive loading. Also Wang et al. reported (2012b) that the INFs exhibit lower viscosity than their base fluids and the specific heat of the GE and MWCNT dispersed INFs were found to be very close to that of the pure IL which is beneficial for their application as heat transfer fluids.

Another author Neo and Ouyang (2012) studied interaction behavior of 1-propyl-3-methylimidazolium iodide ([C_3mim][I]) with carboxylic group F-MWCNTs with viscosity measurement. The viscosity of the IL decreased from (1380 to 400) mPa s at $T = 298.15$ K after a load of 0.1 wt% of the F-MWCNTs which indicates that the blends of MWCNTs with the IL only decrease the viscosity of the IL with loadings up to 0.1%, and beyond this concentration the viscosity raised abruptly to approximately four times that of the pure IL. In the year 2013, Ferreira et al. (2013) selected phosphonium based ILs such as trihexyltetradecylphosphonium phosphinate ([(C_6)$_3$PC_{14})][Phosph]), tretrabutylmethylphosphonium methylsulfate ([(C_6)$_3$PC_{14}][NTf_2]) and trihexyltetradecylphosphonium trispentafluoroethyltrifluorophosphate [(C_6)3PC_{14}][FAP] as base fluids and MWCNTs as nano additives to fabricate INFs suspensions with small loadings of CNT, concluded that the thermal conductivity of INFs with mixture of [(C_6)$_3$PC_{14})][Phosph] and [(C_6)$_3$PC_{14}][NTf_2] with MWCNT increase slightly and decrease linearly with increasing MWCNT concentration and temperature, respectively. The viscosity of INFs reported significantly lower in comparison to the base ionic liquid. Also it has been reported that the enhancement of heat capacity of [(C_6)$_3$PC_{14})][Phosph]-based INFs which increases negligibly with temperature whereas in case of [(C_6)$_3$PC_{14}][NTf_2]-based INFs the same heat capacity (increment of 13%) was found almost constant over the entire studied temperature and end with the conclusion that like ILs the studied INFs are also thermally stable.

Table 4 Different properties of INFs studied by various authors

Ionanofluids	Properties studied	References
$[C_4mim][(CF_3SO_2)_2N]$/ MWCNT	Thermal conductivity	França et al. (2013)
$[C_2mim][EtSO_4]$/ MWCNT		França et al. (2013)
$[C_4mim][NTf_2]$/MWCNT	Thermal conductivity	Nieto de Castro et al. (2012)
$[C_4mim][CF_3SO_3]$/MWCNT	Thermal conductivity	Nieto de Castro et al. (2012)
$[C_6mim][NTf_2]$/MWCNT	Thermal conductivity	Nieto de Castro et al. (2012)
$[C_8mim][NTf_2]$/MWCNT	Thermal conductivity	Nieto de Castro et al. (2012)
$[C_4mim][PF_6]$	Specific heat capacity	Nieto de Castro et al. (2012)
$[C_6mim][BF_4]$ + MWCNT	Thermal conductivity,	Nieto de Castro et al. (2010)
$[C_4mim][CF_3SO_3]$ + MWCNT	Thermal conductivity	Nieto de Castro et al. (2010)
$[C_4mpyrr][(CF_3SO_2)_2N]$ + MWCNT	Thermal conductivity	Nieto de Castro et al. (2010)
$[C_4mim][PF_6]$ + MWCNT	Thermal conductivity	Nieto de Castro et al. (2010)
$[C_6mim][PF_6]$ + MWCNT	Thermal conductivity	Nieto de Castro et al. (2010)
$[C_4mim][PF_6]$ + Baytubes	Heat capacity	Nieto de Castro et al. (2010)
$[C_4mim][PF_6]$ + F-MWCNT	Rheological behaviors	Wang et al. (2010)
$[C_3mim][I]$ + F-MWCNT	Viscosity	Neo and Ouyang (2012)
$[(C_6)_3PC_{14})][Phosph]$ + MWCNT	Thermal conductivity, heat capacity	Ferreira et al. (2013)
$[(C_6)_3PC_{14}][NTf_2]$ + MWCNT	Thermal conductivity, heat capacity	Ferreira et al. (2013)
$[(C_6)3PC_{14}][FAP]$ + MWCNT	Thermal conductivity, heat capacity	Ferreira et al. (2013)
$[C_4mim][BF_4]$ + MWCNT	Heat capacity	Shevelyova et al. (2013)
$[C_4mim][PF_6]$ + MWCNT	Heat capacity	Shevelyova et al. (2013)
$[C_6mim][BF_4]$ + Graphene	Thermal conductivity, heat capacity and VISCOSITY	Wang et al. (2012b)
$[C_6mim][BF_4]$ + MWCNT	Thermal conductivity, heat capacity and viscosity	Wang et al. (2012b)
$[P_{6,6,6,14}]Cl$ + AgI	Electric conductivity, heat capacity refractive index, and viscosity	Rodríguez-Palmeiro et al. (2013)

Another author Rodríguez-Palmeiro et al. (2013) reported the stability of silver iodide-trihexyl(tetradecyl)phosphonium chloride (AgI–[$P_{6,6,6,14}$]Cl) INFs. The low electric conductivity of the IL was attributed to the large quantity of Ag Inanoparticles (highly conducting) dispersed with in the maximum. The result states that for 20% w/w Ionanofluids the electrical applications of the synthesized INFs are limited. Shevelyova et al. reported (Shevelyova et al. 2013) heart capacity of INFs with MWCNT and ILs ([C_4mim][BF_4] and [C_4mim][PF_6]) over the entire temperature range, found specific heat capacity to be an additive quantity of the components. It was also found that the INFs are liophobic and provide the opportunity to predict thermodynamic properties of similar INFs from the data on individual components. Different properties studied by authors are listed in Table 4.

The actual cost of ILs is higher than conventional heat transfer fluids like water, ethylene and silicon oils, the future production of higher quantities can make their use competitive, especially if a target price of 25US$/kg is achieved (Ribeiro et al. 2013; Murshed et al. 2008b). And may be the near future cost of ILs decease because of its growing industrial application in various sector leading to boom in INFs.

7 Conclusion

There are numerous challenges facing the nanofluids community ranging from formulation, practical application. It can be concluded that ionanofluids show great advancements over the traditional fluids which promises to be used as innovative heat transfer fluids and can serve as novel media for many green energy-based applications. The stated result shows that ionanofluids exhibit superior thermophysical properties compared to base ILs. This chapter concludes that ionanofluids are better heat transfer fluids for heat exchangers or other heat transfer devices than traditional ILs.

Acknowledgements The author's are thankful to their respective universities for providing internet facilities for collecting the research paper. The author I. Khan acknowledges to Department of Chemistry, Sultan Qaboos University for Internal grant, SQU (IG/SCI/CHEM/19/03).

References

Assael MJ, Chen CF, Metaxa I, Wakeham WA (2004) Thermal conductivity of suspensions of carbon nanotubes in water. Int J Thermophys 25(4):971–985. https://doi.org/10.1023/B:IJOT.0000038494.22494.04

Bai L, Zhu J, Chen B, Li C, Fei W (2010) Applications of ionic liquids in heat transfer and heat storage process. CIESC J 12:000

Beck MP (2008) Thermal conductivity of metal oxide nanofluids. ProQuest

Bi S-s, Shi L, Zhang L-l (2008) Application of nanoparticles in domestic refrigerators. Appl Therm Eng 28(14–15):1834–1843. https://doi.org/10.1016/j.applthermaleng.2007.11.018

Bin Dahbag M, AlQuraishi A, Benzagouta M (2014) Efficiency of ionic liquids for chemical enhanced oil recovery. J Petrol Explor Prod Technol 1–9. https://doi.org/10.1007/s13202-014-0147-5

Botha SS (2006) Synthesis and characterization of nanofluids for cooling applications. University of the Western Cape

Bridges NJ, Visser AE, Fox EB (2011) Potential of nanoparticle-enhanced ionic liquids (NEILs) as advanced heat-transfer fluids. Energ Fuels 25(10):4862–4864. https://doi.org/10.1021/ef2012084

Buongiorno J, Hu L-W (2009) 8. Innovative technologies: two-phase heat transfer in water-based nanofluids for nuclear applications final report. Massachusetts Institute of Technology Cambridge, MA 02139–4307

Carvalho PJ, Khan I, Morais A, Granjo JFO, Oliveira NMC, Santos LMNBF, Coutinho JAP (2013) A new microebulliometer for the measurement of the vapor-liquid equilibrium of ionic liquid systems. Fluid Phase Equilib 354:156–165. https://doi.org/10.1016/j.fluid.2013.06.015

Ceylan A, Jastrzembski K, Shah SI (2006) Enhanced solubility Ag-Cu nanoparticles and their thermal transport properties. Metall Mater Trans A 37(7):2033–2038. https://doi.org/10.1007/BF02586123

Choi SUS (1995) Enhancing thermal conductivity of fluids with nanoparticles. American Society of Mechanical Engineers, Fluids Engineering Division (Publication) FED, pp 99–105

Choi SUS (2009) Nanofluids: from vision to reality through research. J Heat Transfer 131(3):033106–033106. https://doi.org/10.1115/1.3056479

Choi SUS, Zhang ZG, Yu W, Lockwood FE, Grulke EA (2001a) Anomalous thermal conductivity enhancement in nanotube suspensions. Appl Phys Lett 79(14):2252–2254. https://doi.org/10.1063/1.1408272

Choi S, Yu W, Hull JR, Zhang Z, Lockwood F (2001b) Nanofluids for vehicle thermal management. SAE Technical Paper

Choi C, Yoo HS, Oh JM (2008) Preparation and heat transfer properties of nanoparticle-in-transformer oil dispersions as advanced energy-efficient coolants. Curr Appl Phys 8(6):710–712. https://doi.org/10.1016/j.cap.2007.04.060

Chol S (1995) Enhancing thermal conductivity of fluids with nanoparticles. ASME Publ Fed 231:99–106

Chopkar M, Kumar S, Bhandari DR, Das PK, Manna I (2007) Development and characterization of Al_2Cu and Ag_2Al nanoparticle dispersed water and ethylene glycol based nanofluid. Mater Sci Eng B Solid-State Mater Adv Technol 139(2–3):141–148. https://doi.org/10.1016/j.mseb.2007.01.048

Das SK (2006) Nanofluids—the cooling medium of the future. Heat Transf Eng 27(10):1–2. https://doi.org/10.1080/01457630600904585

Das SK, Putra N, Thiesen P, Roetzel W (2003) Temperature dependence of thermal conductivity enhancement for nanofluids. J Heat Transfer 125(4):567–574. https://doi.org/10.1115/1.1571080

Das SK, Choi SU, Yu W, Pradeep T (2007) Nanofluids: science and technology. Wiley, London

Davidson J (2009) Nanofluid for cooling enhancement of electrical power equipment. Vanderbilt University-Department of Electrical Engineering, Nashville, TN, www vanderbilt edu/technology transfer Bulletin

De Luca LT, Galfetti L, Severini F, Meda L, Marra G, Vorozhtsov AB, Sedoi VS, Babuk VA (2005) Burning of nano-aluminized composite rocket propellants. Combust Explos Shock Waves 41(6):680–692. https://doi.org/10.1007/s10573-005-0080-5

Eastman J, Choi U, Li S, Thompson L, Lee S (1996) Enhanced thermal conductivity through the development of nanofluids. In: MRS proceedings, Cambridge University Press, p 3

Eastman JA, Choi SUS, Li S, Yu W, Thompson LJ (2001) Anomalously increased effective thermal conductivities of ethylene glycol-based nanofluids containing copper nanoparticles. Appl Phys Lett 78(6):718–720. https://doi.org/10.1063/1.1341218

Ferreira AGM, Simões PN, Ferreira AF, Fonseca MA, Oliveira MSA, Trino ASM (2013) Transport and thermal properties of quaternary phosphonium ionic liquids and IoNanofluids. J Chem Thermodyn 64:80–92. https://doi.org/10.1016/j.jct.2013.04.013

França JMP, Vieira SIC, Lourenço MJV, Murshed SMS, Nieto de Castro CA (2013) Thermal conductivity of [C4mim][(CF3SO2)2 N] and [C2mim][EtSO4] and their IoNanofluids with carbon nanotubes: experiment and theory. J Chem Eng Data 58(2):467–476. https://doi.org/10.1021/je301183r
Fukushima T, Kosaka A, Ishimura Y, Yamamoto T, Takigawa T, Ishii N, Aida T (2003) Molecular ordering of organic molten salts triggered by single-walled carbon nanotubes. Science 300(5628):2072–2074. https://doi.org/10.1126/science.1082289
Galfetti L, DeLuca LT, Severini F, Colombo G, Meda L, Marra G (2007) Pre and post-burning analysis of nano-aluminized solid rocket propellants. Aerosp Sci Technol 11(1):26–32. https://doi.org/10.1016/j.ast.2006.08.005
Ghosh MM, Roy S, Pabi SK, Ghosh S (2011) A molecular dynamics-stochastic model for thermal conductivity of nanofluids and its experimental validation. J Nanosci Nanotechnol 11(3):2196–2207
Godson L, Lal DM, Wongwises S (2010) Measurement of thermo physical properties of metallic nanofluids for high temperature applications. Nanoscale Microscale Thermophys Eng 14(3):152–173. https://doi.org/10.1080/15567265.2010.500319
Haddad Z, Abid C, Oztop HF, Mataoui A (2014) A review on how the researchers prepare their nanofluids. Int J Therm Sci 76:168–189. https://doi.org/10.1016/j.ijthermalsci.2013.08.010
Han ZH, Yang B, Kim SH, Zachariah MR (2007) Application of hybrid sphere/carbon nanotube particles in nanofluids. Nanotechnology 18(10):105701
Hirota K, Sugimoto M, Kato M, Tsukagoshi K, Tanigawa T, Sugimoto H (2010) Preparation of zinc oxide ceramics with a sustainable antibacterial activity under dark conditions. Ceram Int 36(2):497–506. https://doi.org/10.1016/j.ceramint.2009.09.026
Hong T-K, Yang H-S, Choi CJ (2005) Study of the enhanced thermal conductivity of Fe nanofluids. J Appl Phys 97(6):064311. https://doi.org/10.1063/1.1861145
Huaqing X, Jinchang W, Tonggeng X, Yan L (2001) Study on the thermal conductivity of SiC nanofluids. J Chin Ceram Soc 29(4):361–364
Hwang YJ, Ahn YC, Shin HS, Lee CG, Kim GT, Park HS, Lee JK (2006a) Investigation on characteristics of thermal conductivity enhancement of nanofluids. Curr Appl Phys 6(6):1068–1071. https://doi.org/10.1016/j.cap.2005.07.021
Hwang Y, Park HS, Lee JK, Jung WH (2006b) Thermal conductivity and lubrication characteristics of nanofluids. Curr Appl Phys 6(Suppl. 1):e67–e71. https://doi.org/10.1016/j.cap.2006.01.014
Islam MR, Shabani B, Rosengarten G, Andrews J (2015) The potential of using nanofluids in PEM fuel cell cooling systems: a review. Renew Sustain Energy Rev 48:523–539. https://doi.org/10.1016/j.rser.2015.04.018
Jana S, Salehi-Khojin A, Zhong W-H (2007) Enhancement of fluid thermal conductivity by the addition of single and hybrid nano-additives. Thermochim Acta 462(1–2):45–55. https://doi.org/10.1016/j.tca.2007.06.009
Jiang W, Ding G, Peng H (2009) Measurement and model on thermal conductivities of carbon nanotube nanorefrigerants. Int J Therm Sci 48(6):1108–1115. https://doi.org/10.1016/j.ijthermalsci.2008.11.012
Jwo CS, Teng TP (2005) Experimental study on thermal properties of brines containing nanoparticles. Rev Adv Mater Sci 10(1):79–83
Kang HU, Kim SH, Oh JM (2006) Estimation of thermal conductivity of nanofluid using experimental effective particle volume. Exp Heat Transf 19(3):181–191. https://doi.org/10.1080/08916150600619281
Keblinski P, Phillpot SR, Choi SUS, Eastman JA (2002) Mechanisms of heat flow in suspensions of nano-sized particles (nanofluids). Int J Heat Mass Transf 45(4):855–863. https://doi.org/10.1016/S0017-9310(01)00175-2
Keblinski P, Eastman JA, Cahill DG (2005) Nanofluids for thermal transport. Mater Today 8(6):36–44. https://doi.org/10.1016/S1369-7021(05)70936-6
Keblinski P, Prasher R, Eapen J (2008) Thermal conductance of nanofluids: is the controversy over? J Nanopart Res 10(7):1089–1097

Khamooshi M, Parham K, Atikol U (2013) Overview of ionic liquids used as working fluids in absorption cycles. Adv Mech Eng 2013:1–7. https://doi.org/10.1155/2013/620592

Khan I, Taha M, Ribeiro-Claro P, Pinho SP, Coutinho JAP (2014a) Effect of the cation on the interactions between alkyl methyl imidazolium chloride ionic liquids and water. J Phys Chem B 118(35):10503–10514. https://doi.org/10.1021/jp5057495

Khan I, Kurnia KA, Sintra TE, Saraiva JA, Pinho SP, Coutinho JAP (2014b) Assessing the activity coefficients of water in cholinium-based ionic liquids: experimental measurements and COSMO-RS modeling. Fluid Phase Equilib 361:16–22. https://doi.org/10.1016/j.fluid.2013.10.032

Khan I, Kurnia KA, Mutelet F, Pinho SP, Coutinho JA (2014c) Probing the interactions between ionic liquids and water: experimental and quantum chemical approach. J Phys Chem B 118(7):1848–1860. https://doi.org/10.1021/jp4113552

Khan I, Batista ML, Carvalho PJ, Santos LM, Gomes JR, Coutinho JA (2015) Vapor-liquid equilibria of imidazolium ionic liquids with cyano containing anions with water and ethanol. J Phys Chem B 119(32):10287–10303. https://doi.org/10.1021/acs.jpcb.5b03324

Kim J, Kang YT, Choi CK (2007) Soret and Dufour effects on convective instabilities in binary nanofluids for absorption application. Int J Refrig 30(2):323–328. https://doi.org/10.1016/j.ijrefrig.2006.04.005

Kulkarni DP, Vajjha RS, Das DK, Oliva D (2008) Application of aluminum oxide nanofluids in diesel electric generator as jacket water coolant. Appl Therm Eng 28(14–15):1774–1781. https://doi.org/10.1016/j.applthermaleng.2007.11.017

Kuo KK, Risha GA, Evans BJ, Boyer E (2003) Potential usage of energetic nano-sized powders for combustion and rocket propulsion. In: MRS Proceedings. Cambridge University Press, p AA1.1

Lee J, Aida T (2011) "Bucky gels" for tailoring electroactive materials and devices: the composites of carbon materials with ionic liquids. Chem Commun 47(24):6757–6762

Lee S, Choi S (1997) Application of metallic nanoparticle suspensions. ANL

Lee S, Choi SUS, Li S, Eastman JA (1999) Measuring thermal conductivity of fluids containing oxide nanoparticles. J Heat Transf Trans Asme 121(2):280–289. https://doi.org/10.1115/1.2825978

Lee J-H, Hwang KS, Jang SP, Lee BH, Kim JH, Choi SUS, Choi CJ (2008) Effective viscosities and thermal conductivities of aqueous nanofluids containing of Al_2O_3 low volume concentrations nanoparticles. Int J Heat Mass Transf 51(11–12):2651–2656. https://doi.org/10.1016/j.ijheatmasstransfer.2007.10.026

Leong KY, Saidur R, Kazi SN, Mamun AH (2010) Performance investigation of an automotive car radiator operated with nanofluid-based coolants (nanofluid as a coolant in a radiator). Appl Therm Eng 30(17–18):2685–2692. https://doi.org/10.1016/j.applthermaleng.2010.07.019

Li CH, Peterson GP (2007) The effect of particle size on the effective thermal conductivity of Al_2O_3-water nanofluids. J Appl Phys 101(4):044312. https://doi.org/10.1063/1.2436472

Li P, Wu X-M, Li H (2006) Pool boiling heat transfer experiments of refrigerants with nanoparticle TiO_2. In: Proceedings of the 12th symposium on engineering thermophysics, Chinese Institute of Engineering Thermophysics, Chinese Academy of Sciences, Beijing, China, pp 325–328

Li L, Zhang Y, Ma H, Yang M (2008) An investigation of molecular layering at the liquid-solid interface in nanofluids by molecular dynamics simulation. Phys Lett A 372(25):4541–4544

Li Y, Je Zhou, Tung S, Schneider E, Xi S (2009) A review on development of nanofluid preparation and characterization. Powder Technol 196(2):89–101. https://doi.org/10.1016/j.powtec.2009.07.025

Liu M-S, Ching-Cheng Lin M, Huang IT, Wang C-C (2005) Enhancement of thermal conductivity with carbon nanotube for nanofluids. Int Commun Heat Mass Transfer 32(9):1202–1210. https://doi.org/10.1016/j.icheatmasstransfer.2005.05.005

Liu M-S, Lin MC-C, Tsai CY, Wang C-C (2006a) Enhancement of thermal conductivity with Cu for nanofluids using chemical reduction method. Int J Heat Mass Transf 49(17–18):3028–3033. https://doi.org/10.1016/j.ijheatmasstransfer.2006.02.012

Liu MS, Lin MCC, Huang IT, Wang CC (2006b) Enhancement of thermal conductivity with CuO for nanofluids. Chem Eng Technol 29(1):72–77. https://doi.org/10.1002/ceat.200500184

Liu H, Wang X, Zhang M, Zhang W, Liu J (2008) Heat transfer simulation of nanofluids in micro channel cooler. In: International Conference on electronic packaging technology & high density packaging, ICEPT-HDP 2008, 28–31 July 2008, pp 1–5. https://doi.org/10.1109/icept.2008.4606947

Ma K-Q, Liu J (2007) Nano liquid-metal fluid as ultimate coolant. Phys Lett A 361(3):252–256. https://doi.org/10.1016/j.physleta.2006.09.041

Manna I, Chopkar M, Das PK (2005) Nanofluid- a new concept in heat transfer and thermal management. Trans Indian Inst Met 58(6):1045–1055

Mintsa HA, Roy G, Nguyen CT, Doucet D (2009) New temperature dependent thermal conductivity data for water-based nanofluids. Int J Therm Sci 48(2):363–371. https://doi.org/10.1016/j.ijthermalsci.2008.03.009

Mohammad A (2012) Green solvents I: properties and applications in chemistry, vol 1. Springer Science & Business Media

Mokhatab S, Fresky MA, Islam MR (2006) Applications of nanotechnology in oil and gas E&P. J Petrol Technol 58(04):48–51

Murshed SMS, Nieto de Castro CA (2014) Superior thermal features of carbon nanotubes-based nanofluids—a review. Renew Sustain Energy Rev 37:155–167. https://doi.org/10.1016/j.rser.2014.05.017

Murshed SMS, Leong KC, Yang C (2005) Enhanced thermal conductivity of TiO_2—water based nanofluids. Int J Therm Sci 44(4):367–373. https://doi.org/10.1016/j.ijthermalsci.2004.12.005

Murshed SMS, Leong KC, Yang C (2008a) Thermophysical and electrokinetic properties of nanofluids —A critical review. Appl Therm Eng 28(17–18):2109–2125. https://doi.org/10.1016/j.applthermaleng.2008.01.005

Murshed SMS, Leong KC, Yang C (2008b) Investigations of thermal conductivity and viscosity of nanofluids. Int J Therm Sci 47(5):560–568. https://doi.org/10.1016/j.ijthermalsci.2007.05.004

Murshed SMS, Leong KC, Yang C (2009) A combined model for the effective thermal conductivity of nanofluids. Appl Therm Eng 29(11–12):2477–2483. https://doi.org/10.1016/j.applthermaleng.2008.12.018

Murshed SMS, Castro CANd, Lourenço MJV, Lopes MLM, Santos FJV (2012) Current research and future applications of nano- and ionano-fluids. J Phys: Conf Ser 395(1):012117

Naphon P, Assadamongkol P, Borirak T (2008) Experimental investigation of titanium nanofluids on the heat pipe thermal efficiency. Int Commun Heat Mass Transfer 35(10):1316–1319. https://doi.org/10.1016/j.icheatmasstransfer.2008.07.010

Naphon P, Thongkum D, Assadamongkol P (2009) Heat pipe efficiency enhancement with refrigerant–nanoparticles mixtures. Energy Convers Manag 50(3):772–776. https://doi.org/10.1016/j.enconman.2008.09.045

Natarajan E, Sathish R (2009) Role of nanofluids in solar water heater. Int J Adv Manuf Technol 1–5. https://doi.org/10.1007/s00170-008-1876-8

Neo CY, Ouyang J (2012) Functionalized carbon nanotube-induced viscosity reduction of an ionic liquid and performance improvement of dye-sensitized solar cells. Electrochim Acta 85:1–8. https://doi.org/10.1016/j.electacta.2012.08.041

Nguyen CT, Roy G, Gauthier C, Galanis N (2007) Heat transfer enhancement using Al_2O_3-water nanofluid for an electronic liquid cooling system. Appl Therm Eng 27(8–9):1501–1506. https://doi.org/10.1016/j.applthermaleng.2006.09.028

Nieto de Castro CA, Lourenço MJV, Ribeiro APC, Langa E, Vieira SIC, Goodrich P, Hardacre C (2009) Thermal properties of ionic liquids and IoNanofluids of imidazolium and pyrrolidinium liquids†. J Chem Eng Data 55(2):653–661. https://doi.org/10.1021/je900648p

Nieto de Castro CA, Lourenço MJV, Ribeiro APC, Langa E, Vieira SIC, Goodrich P, Hardacre C (2010) Thermal properties of ionic liquids and ionanofluids of imidazolium and pyrrolidinium liquids. J Chem Eng Data 55(2):653–661. https://doi.org/10.1021/je900648p

Nieto de Castro CA, Murshed SMS, Lourenço MJV, Santos FJV, Lopes MLM, França JMP (2012) Enhanced thermal conductivity and specific heat capacity of carbon nanotubes ionanofluids. Int J Therm Sci 62:34–39. https://doi.org/10.1016/j.ijthermalsci.2012.03.010

Ollivier E, Bellettre J, Tazerout M, Roy GC (2006) Detection of knock occurrence in a gas SI engine from a heat transfer analysis. Energy Convers Manag 47(7–8):879–893. https://doi.org/10.1016/j.enconman.2005.06.019

Park K-J, Jung D (2007) Boiling heat transfer enhancement with carbon nanotubes for refrigerants used in building air-conditioning. Energ Build 39(9):1061–1064. https://doi.org/10.1016/j.enbuild.2006.12.001

Park MJ, Lee JK, Lee BS, Lee Y-W, Choi IS, Lee S-g (2006) covalent modification of multiwalled carbon nanotubes with imidazolium-based ionic liquids: effect of anions on solubility. Chem Mater 18(6):1546–1551. https://doi.org/10.1021/cm0511421

Passos H, Khan I, Mutelet F, Oliveira MB, Carvalho PJ, Santos LMNBF, Held C, Sadowski G, Freire MG, Coutinho JAP (2014) Vapor-liquid equilibria of water plus alkylimidazolium-based ionic liquids: measurements and perturbed-chain statistical associating fluid theory modeling. Ind Eng Chem Res 53(9):3737–3748. https://doi.org/10.1021/Ie4041093

Patel HE, Das SK, Sundararajan T, Sreekumaran Nair A, George B, Pradeep T (2003) Thermal conductivities of naked and monolayer protected metal nanoparticle based nanofluids: Manifestation of anomalous enhancement and chemical effects. Appl Phys Lett 83(14):2931–2933. https://doi.org/10.1063/1.1602578

Peng H, Ding G, Jiang W, Hu H, Gao Y (2009) Heat transfer characteristics of refrigerant-based nanofluid flow boiling inside a horizontal smooth tube. Int J Refrig 32(6):1259–1270. https://doi.org/10.1016/j.ijrefrig.2009.01.025

Pivkina A, Ulyanova P, Frolov Y, Zavyalov S, Schoonman J (2004) Nanomaterials for heterogeneous combustion. Propellants, Explos, Pyrotech 29(1):39–48. https://doi.org/10.1002/prep.200400025

Putnam SA, Cahill DG, Braun PV, Ge Z, Shimmin RG (2006) Thermal conductivity of nanoparticle suspensions. J Appl Phys 99(8):084308. https://doi.org/10.1063/1.2189933

Ribeiro APC, Vieira SIC, Goodrich P, Hardacre C, Lourenço MJV, de Castro CAN (2013) Thermal conductivity of [$C_n$mim][$(CF_3SO_2)_2N$] and [C_4mim][BF_4] IoNanofluids with carbon nanotubes-measurement, theory and structural characterization. J Nanofluids 2(1):55–62. https://doi.org/10.1166/jon.2013.1035

Risha GA, Boyer E, Evans B, Kuo KK, Malek R (2003) Characterization of nano-sized particles for propulsion applications. In: MRS proceedings, Cambridge Univ Press, p AA6.6

Rodríguez-Palmeiro I, Rodríguez-Cabo B, Rodil E, Arce A, Saiz-Jabardo JM, Soto A (2013) Synthesis and characterization of highly concentrated AgI–[P6,6,6,14]Cl ionanofluids. J Nanopart Res 15(9):1–8. https://doi.org/10.1007/s11051-013-1881-1

Saidur R, Ahamed JU, Masjuki HH (2010) Energy, exergy and economic analysis of industrial boilers. Energ Policy 38(5):2188–2197. https://doi.org/10.1016/j.enpol.2009.11.087

Saidur R, Leong KY, Mohammad HA (2011) A review on applications and challenges of nanofluids. Renew Sustain Energy Rev 15(3):1646–1668. https://doi.org/10.1016/j.rser.2010.11.035

Sankar N, Mathew N, Sobhan C (2008) Molecular dynamics modeling of thermal conductivity enhancement in metal nanoparticle suspensions. Int Commun Heat Mass Transfer 35(7):867–872

Sarkar S, Selvam RP (2007) Molecular dynamics simulation of effective thermal conductivity and study of enhanced thermal transport mechanism in nanofluids. J Appl Phys 102(7):074302

Scott S, Crudden C, Jones C (2003) Nanostructured catalysts. Nanostructure science and technology series. Springer, New York

Shaikh S, Lafdi K, Ponnappan R (2007) Thermal conductivity improvement in carbon nanoparticle doped PAO oil: an experimental study. J Appl Phys 101(6):064302. https://doi.org/10.1063/1.2710337

Shen B (2008) Minimum quantity lubrication grinding using nanofluids. ProQuest

Shevelyova MP, Paulechka YU, Kabo GJ, Blokhin AV, Kabo AG, Gubarevich TM (2013) Physicochemical properties of imidazolium-based ionic nanofluids: density, heat capacity, and enthalpy of formation. J Phys Chem C 117(9):4782–4790. https://doi.org/10.1021/jp3059432

Sridhara V, Satapathy LN (2011) Al2O3-based nanofluids: a review. Nanoscale Res Lett 6(1):1–16

Sridhara V, Gowrishankar BS, Snehalatha Satapathy LN (2009) NanoFluids—A new promising fluid for cooling. Trans Indian Ceram Soc 68(1):1–17

Srikant RR, Rao DN, Subrahmanyam MS, Krishna VP (2009) Applicability of cutting fluids with nanoparticle inclusion as coolants in machining. Proc Inst Mech Eng Part J J Eng Tribol 223(2):221–225. https://doi.org/10.1243/13506501jet463

Swadzba-Kwasny M, Chancelier L, Ng S, Manyar HG, Hardacre C, Nockemann P (2012) Facile in situ synthesis of nanofluids based on ionic liquids and copper oxide clusters and nanoparticles. Dalton Trans 41(1):219–227. https://doi.org/10.1039/C1DT11578B

Swatloski RP, Spear SK, Holbrey JD, Rogers RD (2002) Dissolution of cellose with ionic liquids. J Am Chem Soc 124(18):4974–4975. https://doi.org/10.1021/ja025790m

Timofeeva EV, Gavrilov AN, McCloskey JM, Tolmachev YV, Sprunt S, Lopatina LM, Selinger JV (2007) Thermal conductivity and particle agglomeration in alumina nanofluids: experiment and theory. Phys Rev E 76(6):061203. https://doi.org/10.1103/physreve.76.061203

Top News (2009) http://www.topnews.in/nanofluids-be-used-make-new-types-cameras-microdevices-and-displays-221378

Trisaksri V, Wongwises S (2009) Nucleate pool boiling heat transfer of TiO_2-R141b nanofluids. Int J Heat Mass Transf 52(5–6):1582–1588. https://doi.org/10.1016/j.ijheatmasstransfer.2008.07.041

Tyagi H (2008) Radiative and combustion properties of nanoparticle-laden liquids. ProQuest

Tyagi H, Phelan P, Prasher R (2009) Predicted efficiency of a low-temperature nanofluid-based direct absorption solar collector. J Sol Energy Eng 131(4):041004–041004. https://doi.org/10.1115/1.3197562

Tzeng SC, Lin CW, Huang KD (2005) Heat transfer enhancement of nanofluids in rotary blade coupling of four-wheel-drive vehicles. Acta Mech 179(1–2):11–23. https://doi.org/10.1007/s00707-005-0248-9

Vajjha RS, Das DK (2009) Experimental determination of thermal conductivity of three nanofluids and development of new correlations. Int J Heat Mass Transf 52(21–22):4675–4682. https://doi.org/10.1016/j.ijheatmasstransfer.2009.06.027

Vasu V, Krishna KR, Kumar ACS (2009) Heat transfer with nanofluids for electronic cooling. Int J Mater Prod Technol 34(1–2):158–171. https://doi.org/10.1504/IJMPT.2009.02241

Velve Casquillas G, Le Berre M, Peroz C, Chen Y, Greffet JJ (2007) Microlitre hot strip devices for thermal characterization of nanofluids. Microelectron Eng 84(5–8):1194–1197. https://doi.org/10.1016/j.mee.2007.01.187

Venerus DC, Kabadi MS, Lee S, Perez-Luna V (2006) Study of thermal transport in nanoparticle suspensions using forced Rayleigh scattering. J Appl Phys 100(9):094310. https://doi.org/10.1063/1.2360378

Verma P, Chaturvedi P, Rawat JSBS, Kumar M, Pal S, Bal M, Rawal DS, Harsh Vyas H, Ghosal P, Bhatnagar PK (2007) Elimination of current non-uniformity in carbon nanotube field emitters. J Mater Sci: Mater Electron 18(6):677–680. https://doi.org/10.1007/s10854-006-9079-2

Vieira S, Ribeiro A, Lourenço M, Nieto de Castro C (2011) Paints with ionanofluids as pigments for improvement of heat transfer on architectural and heat exchangers surfaces. In: Proceedings of the 25th European symposium on applied thermodynamics, Saint Petersburg, Russia

Vieira SI, Lourenço MJV, Alves JM, de Castro CAN (2012) Using ionic liquids and MWCNT's (Ionanofluids) in pigment development. J Nanofluids 1(2):148–154. https://doi.org/10.1166/jon.2012.1017

Wang L, Fan J (2010) Nanofluids research: key issues. Nanoscale Res Lett 5(8):1241–1252. https://doi.org/10.1007/s11671-010-9638-6

Wang X-Q, Mujumdar AS (2008a) A review on nanofluids—Part I: theoretical and numerical investigations. Braz J Chem Eng 25(4):613–630

Wang X-Q, Mujumdar AS (2008b) A review on nanofluids—Part II: experiments and applications. Braz J Chem Eng 25(4):631–648. https://doi.org/10.1590/s0104-66322008000400002

Wang X, Xu X, Choi SUS (1999a) Thermal conductivity of nanoparticle—fluid mixture. J Thermophys Heat Transf 13(4):474–480. https://doi.org/10.2514/2.6486

Wang XW, Xu XF, Choi SUS (1999b) Thermal conductivity of nanoparticle-fluid mixture. J Thermophys Heat Transf 13(4):474–480. https://doi.org/10.2514/2.6486

Wang R, Hao B, Xie G, Li H (2003) A refrigerating system using HFC134a and mineral lubricant appended with n-TiO_2 (R) as working fluids. In: Proceedings of the 4th international symposium on HAVC, Tsinghua University Press, Beijing, China, pp 888–892

Wang K-J, Ding G-L, Jiang W (2006) Nano-scale thermal transporting and its use in engineering. In: Proceedings of the 4th symposium on refrigeration and air conditioning, Southeast University Press, Nanjing, China, pp 66–75

Wang Z, Zhang Q, Kuehner D, Xu X, Ivaska A, Niu L (2008) The synthesis of ionic-liquid-functionalized multiwalled carbon nanotubes decorated with highly dispersed Au nanoparticles and their use in oxygen reduction by electrocatalysis. Carbon 46(13):1687–1692. https://doi.org/10.1016/j.carbon.2008.07.020

Wang B, Wang X, Lou W, Hao J (2010) Rheological and tribological properties of ionic liquid-based nanofluids containing functionalized multi-walled carbon nanotubes. J Phys Chem C 114(19):8749–8754. https://doi.org/10.1021/jp1005346

Wang B, Wang X, Lou W, Hao J (2011) Gold-ionic liquid nanofluids with preferably tribological properties and thermal conductivity. Nanoscale Res Lett 6(1):1–10

Wang F, Han L, Zhang Z, Fang X, Shi J, Ma W (2012a) Surfactant-free ionic liquid-based nanofluids with remarkable thermal conductivity enhancement at very low loading of graphene. Nanoscale Res Lett 7(1):314. https://doi.org/10.1186/1556-276x-7-314

Wang FX, Han LJ, Zhang ZG, Fang XM, Shi JJ, Ma WS (2012b) Surfactant-free ionic liquid-based nanofluids with remarkable thermal conductivity enhancement at very low loading of graphene. Nanoscale Res Lett 7(1):1–7. https://doi.org/10.1186/1556-276X-7-314

Wasserscheid P, Welton T (2009) Ionic liquids in synthesis, vol. 1, 2nd Edn. WILEY-VCH Verlag GmbH & Co. KGaA: Darmstadt, Federal Republic of Germany

Wen D, Lin G, Vafaei S, Zhang K (2009) Review of nanofluids for heat transfer applications. Particuology 7(2):141–150. https://doi.org/10.1016/j.partic.2009.01.007

Wensel J, Wright B, Thomas D, Douglas W, Mannhalter B, Cross W, Hong H, Kellar J, Smith P, Roy W (2008) Enhanced thermal conductivity by aggregation in heat transfer nanofluids containing metal oxide nanoparticles and carbon nanotubes. Appl Phys Lett 92(2):023110. https://doi.org/10.1063/1.2834370

Wilson CA (2006) Experimental investigation of nanofluid oscillating heat pipes. University of Missouri–Columbia

Wright B, Thomas D, Hong H, Groven L, Puszynski J, Duke E, Ye X, Jin S (2007) Magnetic field enhanced thermal conductivity in heat transfer nanofluids containing Ni coated single wall carbon nanotubes. Appl Phys Lett 91(17):173116. https://doi.org/10.1063/1.2801507

Wu S, Zhu D, Li X, Li H, Lei J (2009) Thermal energy storage behavior of Al_2O_3-H_2O nanofluids. Thermochim Acta 483(1–2):73–77. https://doi.org/10.1016/j.tca.2008.11.006

Xie H, Chen L (2011) Review on the preparation and thermal performances of carbon nanotube contained nanofluids. J Chem Eng Data 56(4):1030–1041. https://doi.org/10.1021/je101026j

Xie HQ, Wang JC, Xi TG, Liu Y, Ai F, Wu QR (2002a) Thermal conductivity enhancement of suspensions containing nanosized alumina particles. J Appl Phys 91(7):4568–4572. https://doi.org/10.1063/1.1454184

Xie H, Wang J, Xi T, Liu Y (2002b) Thermal conductivity of suspensions containing nanosized SiC particles. Int J Thermophys 23(2):571–580. https://doi.org/10.1023/A:1015121805842

Xie H, Lee H, Youn W, Choi M (2003) Nanofluids containing multiwalled carbon nanotubes and their enhanced thermal conductivities. J Appl Phys 94(8):4967–4971. https://doi.org/10.1063/1.1613374

Xu J, Zhang J, Du Y, Zhang X, Li Y (1996) Ultrasonic velocity and attenuation in nano-structured Zn materials. Mater Lett 29(1–3):131–134. https://doi.org/10.1016/S0167-577X(96)00130-9

Xuan YM, Li Q (2000) Heat transfer enhancement of nanofluids. Int J Heat Fluid Flow 21(1):58–64. https://doi.org/10.1016/s0142-727x(99)00067-3

Xuan Y, Roetzel W (2000) Conceptions for heat transfer correlation of nanofluids. Int J Heat Mass Transf 43(19):3701–3707. https://doi.org/10.1016/S0017-9310(99)00369-5

Khan I, Batista MLS, Carvalho PJ, Santos LMNBF, Gomes JRB, Coutinho JAP (Submitted) Vapor-liquid equilibria of imidazolium ionic liquids with cyano containing anions with water and ethanol. J Phys Chem B
Yang Y (2006) Carbon nanofluids for lubricant applications, vol 68(04)
Yang B, Han ZH (2006) Thermal conductivity enhancement in water-in-FC72 nanoemulsion fluids. Appl Phys Lett 88(26):261914. https://doi.org/10.1063/1.2218325
Yang Y, Grulke EA, Zhang ZG, Wu G (2006) Thermal and rheological properties of carbon nanotube-in-oil dispersions. J Appl Phys 99(11):114307. https://doi.org/10.1063/1.2193161
Ying J, Sun T (1997) research needs assessment on nanostructured catalysts. J Electroceram 1(3):219–238. https://doi.org/10.1023/A:1009931726749
Yoo D-H, Hong KS, Yang H-S (2007) Study of thermal conductivity of nanofluids for the application of heat transfer fluids. Thermochim Acta 455(1–2):66–69. https://doi.org/10.1016/j.tca.2006.12.006
Yu W, Xie H (2012) A review on nanofluids: preparation, stability mechanisms, and applications. J Nanomater 2012:17. https://doi.org/10.1155/2012/435873
Yu B, Zhou F, Liu G, Liang Y, Huck WTS, Liu W (2006) The electrolyte switchable solubility of multi-walled carbon nanotube/ionic liquid (MWCNT/IL) hybrids. Chem Commun 22:2356–2358. https://doi.org/10.1039/B603878F
Yu W, France DM, Choi SU, Routbort JL (2007) Review and assessment of nanofluid technology for transportation and other applications. Argonne National Laboratory (ANL)
Yu W, France DM, Routbort JL, Choi SUS (2008) Review and comparison of nanofluid thermal conductivity and heat transfer enhancements. Heat Transf Eng 29(5):432–460. https://doi.org/10.1080/01457630701850851
Zhang X, Gu H, Fujii M (2006a) Experimental study on the effective thermal conductivity and thermal diffusivity of nanofluids. Int J Thermophys 27(2):569–580. https://doi.org/10.1007/s10765-006-0054-1
Zhang X, Gu H, Fujii M (2006b) Effective thermal conductivity and thermal diffusivity of nanofluids containing spherical and cylindrical nanoparticles. J Appl Phys 100(4):044325. https://doi.org/10.1063/1.2259789
Zhang C, Liu R, Zhang J, Chen Z, Zhou X (2006c) Reactivity of lanthanocene hydroxides toward ketene, isocyanate, lanthanocene alkyl, and triscyclopentadienyllanthanide complexes. Inorg Chem 45(15):5867–5877. https://doi.org/10.1021/ic0602998
Zhang X, Gu H, Fujii M (2007) Effective thermal conductivity and thermal diffusivity of nanofluids containing spherical and cylindrical nanoparticles. Exp Therm Fluid Sci 31(6):593–599. https://doi.org/10.1016/j.expthermflusci.2006.06.009
Zhang L, Ding Y, Povey M, York D (2008) ZnO nanofluids—a potential antibacterial agent. Prog Nat Sci 18(8):939–944. https://doi.org/10.1016/j.pnsc.2008.01.026
Zhao F, Wu X, Wang M, Liu Y, Gao L, Dong S (2004) Electrochemical and bioelectrochemistry properties of room-temperature ionic liquids and carbon composite materials. Anal Chem 76(17):4960–4967. https://doi.org/10.1021/ac0494026
Zhu H, Zhang C, Liu S, Tang Y, Yin Y (2006) Effects of nanoparticle clustering and alignment on thermal conductivities of Fe3O4 aqueous nanofluids. Appl Phys Lett 89(2):023123. https://doi.org/10.1063/1.2221905

MIX
Papier | Fördert gute Waldnutzung
FSC® C083411

Kolibri 360 GmbH
Ferdinand-Jühlke-Straße 7
99095 Erfurt, Deutschland
produktsicherheit@kolibri360.de